The Contamination of the Earth

History for a Sustainable Future

Michael Egan, series editor

The Contamination of the Earth

A History of Pollutions in the Industrial Age

François Jarrige and Thomas Le Roux
translated by Janice Egan and Michael Egan

The MIT Press
Cambridge, Massachusetts
London, England

This book was set in Sabon by Westchester Publishing Services. Printed and bound in the United States of America.

Library of Congress Cataloging-in-Publication Data

Names: Jarrige, François, 1978- author. | Le Roux, Thomas, author.
Title: The contamination of the earth : a history of pollutions in the industrial age /
   François Jarrige and Thomas Le Roux ; translated by Janice Egan and Michael Egan.
Other titles: Contamination du monde. English
Description: Cambridge, Massachusetts : The MIT Press, [2020] | Series: History for a
   sustainable future | Translation of: La contamination du monde : une histoire des
   pollutions à l'âge industriel by Éditions du Seuil in Paris, 2017. | Includes
   bibliographical references and index.
Identifiers: LCCN 2019028395 | ISBN 9780262043830 (hardcover)
Subjects: LCSH: Pollution—History. | Industries—History. | Environmental engineering.
Classification: LCC TD174 .J37713 2020 | DDC 363.7309—dc23
LC record available at https://lccn.loc.gov/2019028395

10 9 8 7 6 5 4 3 2 1

Contents

Series Foreword

Michael Egan

The ravages of pollution are ubiquitous across the contemporary landscape. Even as we become increasingly concerned about parts per million and parts per billion of particulate matter in our air, soil, and water—contaminants that we cannot see—modern life is accompanied by the implicit understanding that we have committed grave violences against our home. The legacy of this violence is older and goes back further than many might assume. And, indeed, so does our awareness of it. For all the popular preoccupation with their starry nights, leisurely garden scenes, and water lilies, the French Impressionist painters of the second half of the nineteenth century were fascinated with industrialization. As a remarkable exhibit at the Art Gallery of Ontario demonstrated in early 2019, factories, chimney stacks, and smoke abounded throughout the paintings of such luminaries as Claude Monet, Vincent Van Gogh, Camille Pissarro, Edgar Dégas, Gustave Caillebotte, Maximilien Luce, Paul Signac, and several others. In many cases, as in Pissarro's famous 1896 paintings from Rouen, the bustle of the industrial world became the centerpiece of their fine paintings. Even earlier, in 1883, Pissarro demonstrated a fascination with the factories and their plumes of smoke at Pontoise. In other instances, however, the shadow of industrialization is more subtle. Monet's 1872 depictions of Argenteuil include quiet chimneys towering over the church's steeple; chimney stacks line the horizon of an otherwise idyllic pastoral river scene in Caillebotte's *The Seine at Argenteuil* (c. 1892).[1]

Explicit or implicit, however, the Impressionists captured a common feature of the new landscape: pollution. The impressionists' interest in smoke came from a deeper interest in modernity. They were consumed by movement, light, and sensation.[2] In 1830, the entirety of France's railroad lines consisted of 22 kilometers of track. By 1876, that number had swollen to 22,673 kilometers. As the French art critic Champfleury argued in 1857, "Industry mixed with nature has its poetic side: this point is to see it and be inspired." Steam trains, engines, and factories pocked the land-scape and transformed not just industry but also labor.

There is an element of pastoral elegy in some depictions of peasants watching steam engines cut across the landscape (Pascal Adolphe Jean Dagnan-Bouveret's *Peasants in the Fields Watching a Steam Train* (1879) comes to mind), but there is no such wistfulness in the darker renderings of industrial labor, like Luce's Charleroi factory paintings from 1895, or Monet's *Coalmen* (c. 1875).

The nature of change in art and industry during the second half of the nineteenth century was the main thrust of the AGO exhibit. But it is difficult for an environmental historian who has long been interested in pollutants and their histories to witness such an exhibit without con-centrating upon the smoke and the emergence of a contaminated earth. Where Van Gogh's smoke plumes at Clichy appear almost whimsical, my twenty-first-century sensibilities are troubled by this early chapter of a story that has come to be an all-consuming feature of contemporary life.

For François Jarrige and Thomas Le Roux, this is a history steeped in the histories of science and technology, but they argue that it is also an economic history and a legal one. Their analysis is global, but they iden-tify important consistencies in the social and legal structures of Europe, North America, and Asia that sanctioned and empowered industrial growth and widespread environmental contamination. Indeed, pollution problems were a critical driver in shaping the modern world. In the inter-est of public health, factories were pushed outside of city limits: a sen-sible notion, but it resulted in the establishment of gargantuan industrial complexes whose appetite for excess transformed pollution from a local problem into something much grander. "Out of sight and out of mind" accelerated the scales of resource extraction by orders of magnitude so gargantuan that many extractive terrorscapes have taken on almost oth-erworldly qualities.

A quick word or two on the work's translation. For me, this title has offered the very personal pleasure of collaborating with French special-ist and technical translator, Janice Egan, my wife (or, rather, I am her husband). One of the telling evidences of a longstanding partnership is a shared idiolect, collected and cultivated over years of mutual experi-ence. Shared idiolect is built from inside jokes, popular culture catch-phrases that enjoy lasting and intimate resonance for whatever reason, and children's precious misspoken vocabularies that become entrenched family lexicon. That special vocabulary is a frequently overlooked and underappreciated facet of everyday life; this translation project—with its regular interactions over and reflections on language—provided ample opportunity to recognize the value and virtue of that shared, private vernacular. Translator and historian approached this work from different backgrounds and with different priorities: one with a commitment to literal accuracy, the other intent on ensuring historical clarity. As the work developed, we came to appreciate our shared passion for language, meanings, and fidelities with the text, with the past, and with its topics and themes. Sometimes this was contentious, and sometimes one or the other may have conceded a point a little too easily.

Though charmed—and frequently tempted—by Jorge Luis Borges's impish suggestion that an original text is sometimes unfaithful to its translation, I have learned that academic translation permits (requires) greater adherence to the original's literal intent and meaning. At the same time, the notion that the "the dictionary is based on the hypothesis—obviously an unproven one—that languages are made up of equivalent synonyms," which also comes from Borges, rings true.[3] Borges took this as license to impose his will on the myriad texts he translated. Many of them benefited from his vision and his grasp of their true soul. But Borges was also Borges—and at one end of a larger translation spectrum. The literary critic James Woods classified the opposing perspectives of this continuum. Originalists, he wrote, "honor the original text's quiddities, and strive to reproduce them as accurately as possible in the translated language." In contrast, activists "are less concerned with literal accuracy than with the transposed musical appeal of the new work." Naturally, such a spectrum is not an either-or kind of proposition. "Any decent translator must be a bit of both," Woods insists.[4] Scholarly translation likely blanches at the more activist predispositions of literary translation, however. That is not to say that there is less spirit (or literary merit) in academic prose, but there is a different kind of obligation in honestly conveying the empirical findings of the original authors. We delighted in the intellectual challenges associated with Borges's truism: that the interchange between languages is filled with nuance; deftness, we learned, is a key attribute in selecting the correct words, phrases, and sentence structures to be as literally accurate as possible—especially when French tends toward the passive voice more than is acceptable in English writing. Nevertheless, "translations are a partial and precious documentation of the changes the text suffers."[5] In this, rather than inventing new meanings or taking liberties with the original text, we have elected to adapt the language of translation to better suit the original work's needs. We have made an important concession to the French that warrants comment here, because we feel it might help to enlighten and enliven English-speaking environmental studies. Throughout the French text, the authors refer to "pollutions" in the plural. Typical English practice is to singularize the multitude of contaminants emitted into air, soil, water, and living bodies; Anglo scholarship refers to "pollution," *tout court*. But the modern world is afflicted by *pollutions*. A big part of the challenge in writing so ambitious a survey as this surrounds reconciling the different forms, sources, and discoveries of pollutions across time, space, and chemical content. Theme and chronology overlap in such complicated ways that it is a tricky business trying to organize the flows of heavy metals, chlorine-based contaminants, and radioactive elements into a coherent narrative. Further, English-speaking historians should be more forthright in acknowledging to their readers that the contaminated earth is a product of myriad forms of contamination, each with their own production and consumption patterns, their own technological and economic origins, and their own ecological consequences. *Pollution* burns off the nuance in this important history. Following the good example set by the authors, we have embraced *pollutions*, and used it through the text. We hope *pollutions* becomes acceptable and accepted practice in English works.

For the purposes of this series, Jarrige and Le Roux's history of pollutions is most welcome. To properly interrogate history in order to achieve a more sustainable future, it is important to recognize the full extent of the historical trajectories under examination. Too frequently, pollutions are regarded as a recent problem or one that began only with the rise of industrialization. Jarrige and Le Roux acknowledge the introduction of new contaminants during these more modern periods—as well as the rapid increase in scale of many pollutions—but they also provide a valuable *longue durée* accounting of the tide of organic and synthetic detritus that has created the contamination of the world. And of human bodies. Studies suggest that air pollutions in Europe have contributed to 800,000 deaths annually, while the annual global number is around 8.8 million.[6] While I have no interest in distracting attention from climate change and its hazards, pollutions are their own slow disaster of the greatest significance. Jarrige and Le Roux put this reality in stunning perspective. This is not a history of unanticipated consequences or environmental disasters born from accidents. Rather, this is in large part a story of deliberate planning and intellectual histories of ideals surrounding cleanliness and progress. The fault—and this is the important lesson—lies not with the structure of a particular chemical compound and its relationship with the physical environment, but with the economic structure that invited its creation, the scientific structure that ignored its prospective harms, and the legal and political structures that exported those harms to the disenfranchised masses. Hubris, yes, but there is something too simplistic in that explanation, one that absolves the callous slow violence perpetrated by the individuals and institutions. Just as Jarrige and Le Roux inspire discussion of pollutions in the plural, they remind us to reflect on the causes and drivers of these environmental crises (also in the plural), and not just the consequences. This is a lesson and a structure that remain incontestably relevant in the twenty-first century.

Acknowledgments

The result of long-term work, this book is drawn from several years of courses and seminars at the École des Hautes Études en Sciences Sociales (EHESS), at the University of Burgundy in Dijon and at Oxford University. The exchanges we had with the students—their insatiable curiosity and their interest in unfortunately recurring news—convinced us to try this historical synthesis of pollution on a world scale in the industrial age. We will not hide the fact that we want to take this issue out of the academic circles to which it is all too often confined. Beyond the discussions of seminar rooms, we wish to offer to a wider public the historical narrative of a phenomenon which is now more and more clearly akin to a flight forward and a race toward the abyss. If pollutions have become ubiquitous, to the point that they sometimes arouse resigned lassitude and stupefaction, the history of the phenomenon remains poorly known. Yet it is one of the major stakes in the relationship that humanity weaves with its environment, which is both ecological and public health.

This manuscript has been the subject of numerous exchanges with colleagues, friends, and students, without whose inputs and insights we could not have succeeded. Our warm thanks go especially to those who have lent themselves to the constructive game of readings, re-readings, and informal discussions: Renaud Bécot, Christophe Bonneuil, Thomas Bouchet, Patrick Bret, Jean-Luc Demizieux, Quentin Deluermoz, Patrick Fournier, Stéphane Frioux, Moritz Hunsmann, Anne Le Huérou, Geneviève Massard-Guilbaud, Charles-François Mathis, Hervé Mazurel, Raphael Morera, Adrien Normand, Xavier Vigna, Julien Vincent, Denis Wornonoff, and Alexis Zimmer.

Our thanks also go to David Valageas of the Georges-Chevrier Center at the University of Bourgogne for his help in drawing up the graphs. And of course we are grateful for the support and encouragement received from Éditions du Seuil where this book was first published in France, and in particular from Caroline Pichon and Séverine Nikel for their repeated confidence and patience.

Last, but not least, we warmly thank our two Canadian translators, Janice and Michael Egan, for their accurate and clever work and their efforts to make this book accessible to the English-speaking world.

Introduction

Pollutions have become one of the principal concerns of our time. According to the World Health Organization (WHO), millions of people die prematurely every year as a result of air pollutions. Never have so many chemical products—the safety of which generate widespread uncertainty—been in circulation. Chemical contamination is a feature of the entire planet. The use of fertilizers and pesticides has impoverished agricultural lands. In 1997, the "seventh continent of plastic" was discovered; nondegradable microwastes continue to accumulate in the earth's oceans and seas. However, in response to these serious problems, international mobilization remains limited.

Is today's world more polluted than it was in the nineteenth century? Is diesel more toxic than gasoline? Can we continue to safely bathe in rivers? Are fish still fit for consumption? The promise of increasing wealth and life expectancy provides reassurance that learning to live with contaminations means that these and so many other similar questions often go unasked and unanswered. We are often not aware of pollution's presence or effects; its most dangerous forms cannot always be seen or smelled. Even those who are lucky enough or rich enough not to live near an oil refinery or downstream from a metallurgical factory or near a battery recycling plant will still, over the course of a day, come into contact with several sources of pollution. Whether they breathe in fine particles from diesel, perhaps work in an office insulated with asbestos, handle plastics, eat tuna containing mercury or an apple dosed with pesticides, wear deodorant or makeup, work in close proximity to Wi-Fi, or make several calls on their cell phones, human bodies have become unwitting consumers of contaminations. Private spaces are no less polluted. Indoor air quality—saturated with the polyphenol compounds found in furniture—is typically even poorer than outdoor air quality. And more stuff: the generation of new goods has witnessed a concomitant (and

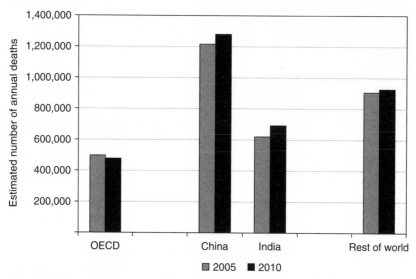

Figure 0.1
The human cost of ambient air pollutions (2005 and 2010). *Source:* OECD, "The Cost of Air Pollution: Health Impacts of Road Transport," 48. According to the WHO, air pollution is the cause of several million premature deaths each year, making it the main environmental risk to people's health. It mainly affects low-income countries; two thirds of these deaths occur in Southeast Asia. In France, fine particle pollution is said to be responsible for 48,000 premature deaths each year.

cataclysmic) rise in waste. Recyclable and nonrecyclable wastes all need to be industrially treated. Amid ubiquitous contaminants produced by waste, modern society further suffers an almost debilitating apathy or indifference or lacks an individual sense that one might have done one's part by not printing those notes from the 11 a.m. meeting, or separating recyclables from waste, or maybe even eating organic food.

The fragmentation of knowledge conducive to denial, the deep functioning inequality of the global economic system—as well as the inadequacies of regulatory policies—explains the paradox that connects the sense of a clean environment in our Western societies with the permanent and ubiquitous existence of modern pollution. To cope with this, public policies—local, national, and international—are modeled on a managerial logic whose administrative rationale is grounded in neoclassical analysis; it believes that only the market can guarantee optimal use of resources and best manage nuisances through fair price fixing. In this context, all pollutions are assimilated into a "negative externality," which is to be minimized and rendered acceptable by rational calculation costs that should lead to an "optimal pollution" level. Confidence in technical progress is the corollary to this way of understanding the problem, and it continues to shape how we see pollution: economic growth and innovations remain the only apparent remedies for the threats that weigh so heavily on the biosphere.[1]

For a long time, pollutions were considered a minor problem—almost a taboo issue, in fact—but they have become a critical subject, receiving extensive attention from numerous fields of expertise throughout the physical and social sciences. Apart from pollution experts, the phenomenon has been taken up and studied by chemists, ecologists, and toxicologists; and more recently by increasing numbers of historians, sociologists, philosophers, and anthropologists. But this expansion of expertise is double-edged: while it provides a significant amount of data—to the point where we cannot fail to find information about the environmental state of the world—it is also making it more and more difficult to see the whole picture, particularly from a historical point of view. There is some consensus that the nature and extent of pollutions remained relatively limited until the advent of industrial civilization. Industrialization and its mechanisms created the contemporary pollution crisis; as pollutions continued to expand and evolve along complex pathways, they have saturated the world and our perception of it.

Nuisances and Pollutions: Language Mutations

Pollution is a fluid and elusive subject: its meaning has changed over the course of time and effective categorization has been slow to emerge. Today, legal experts still stumble to determine and establish its role or place in numerous controversies. Indeed, there remains no universal or fixed understanding of pollutions or their limits.[2] Of course, pollutions might be generally defined as the degradation of an environment following the introduction of unwanted substances that cause adverse change—of varying degrees—to the ecosystem. But even this definition did not exist before the end of the nineteenth century; older societies favored notions of "nuisance," "corruption," or "insalubrity," which can take on quite different meanings. Indeed, prior to the nineteenth century, pollution was typically treated in a religious or moral context. Etymologically, the term derives from the Latin *pollutio* and *polluere*, which figuratively denote "desecration" and "to defile." In the religious sense, they express elements of profanity.[3] Further, seventeenth- and eighteenth-century dictionaries define pollution as the "desecration of a temple," or as "the soil left on one's body after shameless touching."[4]

Although pollution's religious connotations declined in the seventeenth century, its contemporary meaning—as the alteration of natural environments by human action—took two centuries to evolve and gain acceptance. After a long period of limited initial use, the term pollution reached its peak in the 1970s, which coincided with the widespread feeling that the planet was suffering from an environmental crisis. This modern usage of the word first spread throughout the United Kingdom in the early 1800s. In 1804, for example, "the pollution of the stream" from tanneries was explicitly mentioned in a trial in Scotland. It would appear that "pollution" was first used in court in the United States in 1832, to describe the condition of waterways. Between 1850 and 1860, river contamination could no longer be ignored, and a Royal Commission on River Pollution was created in Great Britain to study the problem's magnitude. The term spread as a result of myriad similar inquests throughout the European continent and the United States. In North America, "river pollution" was used more widely in the last quarter of the nineteenth century. As a method of describing air contaminants, "the smoke problem" was not uncommon.[5] The contemporary meaning of the French word *pollution* and the Spanish word *polución*, appeared later, around 1860, in translations of British scientific works about the pollution of London's waterways by sewers.[6] Later, references increased in number in scientific and industrial periodicals.[7] In 1874, the *Journal Officiel* used the term in relation to waters sullied by waste. In 1883, an article in *The Lancet* was translated under the title "Pollution de la Tamise par la vidange," and in 1889 the International Congress of Hygiene and Demography published a report on the "pollution des résidus industriels."[8] The French language dictionary, *Littré,* confirmed the existence of the word in 1880 and defined pollution as "the action of defiling through garbage." In the 1890s, the Pollution and Water Association was well known in sanitation engineering circles. But it was especially after 1900 that the term was also adopted to describe air contaminations.[9] In addition, depending on the language in question, more emphasis was put either on pollution's risks to health—as with the German term *Verschmutzung*—or on the aesthetic dimension of the phenomenon.[10]

In the first half of the twentieth century, pollution was a term used principally in scientific publications. After 1945 it began to be used in a wide variety of contexts to designate phenomena that were ever broader and more diverse. Similarly, the French verb *polluer,* which only foreshadowed its modern meaning at the end of the nineteenth century (1895), was used to translate the English term *to pollute* around the middle of the twentieth century, before it was more abundantly employed in the 1970s. In France, the Association pour la Prévention de la Pollution Atmosphérique (APPA) was established in 1958; it published a journal called *Pollution atmosphérique, climat, santé, société* [Air Pollution, Climate, Health, Society], intended for professionals in the sector—doctors, urban planners, and environmentalists. "Pollution" became so commonplace that Guy Debord was able to assert in the opening lines of his book, *A Sick Planet* (1971), that "'Pollution' is in fashion today, exactly in the same way as revolution: it dominates the whole life of society, and it is represented in illusory form in the spectacle. It is the subject of mind-numbing chatter in a plethora of erroneous and mystifying writings and speech, yet it really does have everyone by the throat."[11]

If pollution can be defined as the degradation of an environment by the introduction of substances—the fruit of some technical act—it also entails the study of larger historical processes and tendencies, such as production systems and their transformation. To understand the emergence of the contemporary idea of pollution is to observe that it progressively distinguished itself from the legal category of "nuisance" over the course of the nineteenth century. Two developments mark this separation. First, this was a period during which industrialization sank its claws into society; second, the physical and biochemical sciences made it possible to measure pollutants. While it is sometimes difficult to distinguish nuisance from pollution, especially given the uncertainty surrounding pollution's reach and impact, nuisance can be characterized by law and social processes, while pollution depends more on scientific judgment, which can often be controversial. In continental Roman law and Anglo-Saxon common law, European nuisances were registered in a litigation system between private individuals and well-identified communities. Biochemical criteria, which tend to muddy liability, made nuisance law inoperative and allowed technology to dominate. Moving between the two terms also reflected the change in scale that has occurred over the last three hundred years.

Writing the History of Pollutions

There are many ways to think about pollutions: as a benchmark of technological deficiencies requiring improvement, a fundamental byproduct of all productive activity, an essentially chemical issue requiring further understanding of the molecular makeup of the world; or, more profoundly, as a critical intersection at the crossroads of political, economic, social, and scientific issues, locally and globally. Pollutions insidiously impact everyday life. Their invisibility and dilution render them controversial and subject to denial, fueled by "merchants of doubt," who interfere with and manipulate scientific data and public discourse.[12] Pollutions are legion and various: widespread or accidental, preexisting or unforeseen, they are rarely adequately understood. Their place in modern history is amplified or underestimated by scandal, pressure groups, expert discussions, ecologists' actions, and ambitious politicians and corporations. Their magnitude and complexity, much like their cause and effect, are difficult and controversial topics to encompass. That is undoubtedly why no single work has attempted to present a historic panorama of pollutions in order to provide a greater understanding of their workings. Such a study would need to discern what is at stake and follow the global spread of pollutions in time and space, and all the necessary preconditions for responding and reacting to them.[13] Between paralyzing catastrophism and reassuring explanations, it is difficult to know how to consider pollutions, when each expert and specialist imposes her or his own vision of the subject.

Over the last few decades, environmental history has become a well-established subject, initially in the United States, then gradually on a global scale. As a field, it has helped give a new and heretofore overlooked perspective on pollutions. Histories of contamination have been central in the emergence and development of the field since its inception in the 1970s. By and large, these histories of industrial trespasses into the environment have constituted a "tragic" environmental history, effectively an accounting of the damage caused to natural environments by human activities.[14] Subsequently, this history gave way to other approaches that focused more on exchanges and interactions between humans and nature, relegating pollutions to the background as subsidiary to larger systems of organization. In France, very few works have explored the history of pollutions—and those that have do not examine the dynamics of industrial pollution in any depth, at best considering it a partial aspect of greater environmental transformations.[15] Over the last decade, however, many new monographs have markedly expanded the historical understanding of pollutions. They have permitted a more focused exploration of the subject's depths, complexities, and chronologies—and topics requiring still further consideration. But this literature is generally not very accessible, and tends to concentrate on specific industries, regions, or countries, thus offering something of a scattered or incomplete overview. While this book relies heavily on these studies, it attempts a global synthesis by linking the traditions of social and economic history, the history of law and the state, and the histories of science and technology, with new directions in environmental history.

While it is evidently not possible to cover everything—an encyclopedia would be needed—this book endeavors to bring together often segmented approaches, illuminating key dynamics at work while pointing out new challenges and issues that need to be explored. By proposing a social and political history of pollutions since the beginning of the industrial age, the aim is to present a global, historic vision of what has polluted different environments—air, soil, and water—over a period of three centuries. Special attention is paid to relations of social power, which shaped successive pollution cycles, their appearance and disappearance. Too, it takes especial care to map the contours of cultural changes, such as various regulatory strategies designed to deal with pollutions.

Temporalities and Scales of Pollution in the Industrial Age

This survey of global contaminations focuses on the period between the rise of industrial capitalism in the eighteenth century and the early 1970s. While pollutions are an historical constant in all productive activity—human society has always engaged in altering the earth's air, soil, and water—their reach, scale, and severity reached unprecedented levels with the advent of the industrial age. Indeed, it is important to recall that pollutions predate modern industrialization. Throughout the Middle Ages, blast furnaces spread pollutions across Asia and Europe; Europe experienced a semblance of industrialization as early as the thirteenth century. But the industrial capitalism that evolved from the eighteenth century onward was new, and it profoundly altered the nature of pollution, its scale, and how it was broached and understood. As a result, the twentieth century suffered from pollution on an even grander scale thanks to industrialization's growing intensity and its array of new contaminants, despite strong efforts to reduce and displace nuisances. Concluding this work in the early 1970s rather than bringing it up to the present is a deliberate choice: with globalization and the development of political ecology, the transformation of the monetary system, the end of colonial wars, the swing in environmental sanctions, the emergence of the "risk society," and—above all—the geographic and neoliberal redistribution of the production system, the period after 1970 deserves a volume of its own to do it justice. Moreover, the contemporary period has introduced unprecedented configurations from which historical analysis might need greater distance in order to sound its depths effectively, beyond sketching broad outlines for inquiry and improvement. The epilogue of this book is devoted precisely to making those first, tentative steps into some analysis of pollutions in the contemporary world.

Thus, this book's chronology follows industrialization from its earliest beginnings through to the 1970s. Its narrative, which spans three hundred years, is neither homogeneous nor strictly linear. The history of pollutions cannot be reduced to a simple account of the continual, concomitant increase in the number of nuisances that took place in step with industrialization. Nor is it the simple story of a problem that could be controlled by continued technical progress and better regulatory processes. Far from being a monolithic issue, pollutions form part of the successive "cycles" that accompany changes in modes of production and goods being produced. They are also intrinsically linked to social actions and reactions.[16] While a more holistic reading of pollutions accepts that they exist on a global level, local examples have always been in flux, on occasion disappearing entirely but often reappearing in another form. This idea of novelty is therefore important: whatever the period, the distinction must be made between places into which industry was introduced—thus provoking strong reactions—and environments that were constructed by and for industry, where a sense of acceptance and resignation tended to rule.

The relocation and containment of pollutions effectively follows strategies for the social distribution of risk. One of the few incontrovertible historical laws is that pollutions will always do the most harm to the poorest people, the poorest neighborhoods, working-class cities, and countries in the Southern hemisphere. Throughout the history of pollution, it is the logic of domination and exclusion, of hierarchy and inequality, that needs to be thoroughly interrogated. In this regard it would seem unusual then to focus exclusively only on what occurs inside an industrial plant, or outside it. Rather, it is the interaction between activities on both sides of those walls that is most deserving of attention. If medical, environmental, and legal literature tended to distinguish the toxic effects of industrial production inside and outside of the factory, this distinction obscured the reality of the problem. It artificially separated those impacted by contaminations to their air, soil, and water from those responsible for those nuisances. Further, that separation established combative strategies to confine pollutions to the private sphere: within the factory and among its workers, who contracted with their employer for work. Again, if there is a recurring theme, it is that of pollutions affecting the workers above all else, poisoned before their neighbors by air saturated with dust and fumes.

As with most concerns in environmental history, pollutions lend themselves poorly to a national framework or lens. The earth's water and carbon cycles are, in essence, global phenomena; polluting substances circulate across borders despite the efforts of states to regulate the phenomenon. This book's approach means to stress this historical reality, even if Europe and North America—home to the world's most severe production of pollutants prior to the 1970s—occupy the lion's share of its analysis. The history of pollutions since the eighteenth century is one of successive changes in scale: broadly speaking, this book traces the passage from dispersed and small-scale industry, situated on a single site, to the large industrial complexes and massive proliferation of pollution sources that typify the contemporary pollutions regime. The globalization of the economy, which prompted a radical reorganization of the production system, plays a decisive role in the study of pollution. An examination of legal developments, for example—one indicator amongst many of this larger evolution—shows that local regulations dominated until the beginning of the nineteenth century, when national legislation combined with local regulation became more common. Transnational approaches to controlling pollutions became preponderant after the Second World War. However, the globalization of contamination, which acknowledges the processes through which a wide-scale increase in toxic substances spreads pollutions to every corner of the planet, should not mask pollution's local dimensions, which do not manifest themselves in the same way everywhere or disperse evenly across the landscape. Pollutions are invariably most intense near their sources of emission. To wit, before being diluted in larger spaces, automobile pollution mainly affects highly trafficked conurbations; the front line of factory emissions is the neighborhoods or river basins immediately downstream.

In the Anthropocene, or rather of the "Capitalocene," the physical impacts of which are gradually being felt across the entire globe, industrial pollutions now affect every part of the planet.[17] More than simply regrettable, indirect effects of certain forms of production and consumption, pollutions have become a decisive element in the functioning of the capitalist world-system.[18] To follow these pollutions' historical trajectory is therefore also to think about the conflicts and organization of powers in the industrial age—social power relations—while reconstructing the dynamics that shaped modernity. The preindustrial history of pollutions begins with small, widely dispersed, and localized forms of pollution, which intensified with increased production and the liberalization of environmental regulation at the end of the eighteenth century. After 1830, industry became more and more the condition of progress, a new fetish that helped to naturalize industrial nuisances as beneficial and inevitable phenomena, despite repeated complaints and protests. The twentieth century, which began with the emergence of total war, inaugurated a period of particularly disproportionate means and methods of pollution. The recently concluded century witnessed increased and redistributed contaminations according to a logic still prevalent today. The mechanism of the "polluter's paradise," championed by the era of market globalization, ensures that the most polluting products tend to leave countries with strong environmental regulations for poorer countries or countries with more lax controls (which typically amount to the same thing).[19] The aim of this synthesis study is to examine whether there are social regimes of industrial pollution, to understand their political sources and dimensions, their economic and social dynamics, and to identify their constants while putting them in context in order to best articulate the main points of inflection and rupture. The premise of this work is not only a question of offering a synthesis or bird's-eye view of the planet's ecological state and the historical development of pollutions, but also of proposing a political interpretation for the future of the industrial world through the study of the emanations and residues that encumber it in ever greater measure.

I

The Industrialization and Liberalization of Environments (1700–1830)

From the Renaissance to the end of the Enlightenment, the human population grew from 500 million to 900 million people, a demographic growth that was accompanied by large-scale commercial growth driven by European explorations in America and Asia. This fiscal transformation marked the origin of a new "world economy."[1] The effects of this first globalization, driven by European monarchies in search of wealth and imperial conquest, were manifold: if the American continent was subjugated, Asia, on the other hand, was only affected by reduced contact.[2] In this connected world, increase in trade went hand in hand with the intensification of artisanal and industrial production. Much of this work remained rural, but similarities existed on a global scale. In the eighteenth century, the great divergence between Europe and the rest of the world was just beginning, but the same types of pollution were found everywhere in the world.[3] Transformation accelerated especially around the 1800s: trade and industry put mounting pressure on the environment; their interests contributed to the liberalization of old regulations.

In the middle of the eighteenth century, more than 80 percent of the global human population were peasants, even though urban numbers increased in some regions, particularly in northwestern Europe and some coastal areas of China and Japan. The great agrarian empires, such as Qing China, Mughal India, Tokugawa Japan, and the Russian, Ottoman, and Hapsburg empires, fostered concentrations in population. Whether they were devoted to rice-growing or producing cereals or livestock on an increasing scale, peasant populations shared common experiences despite the profound heterogeneity of their social and ecological structures. But the early modern period also witnessed the acceleration of the great domestication of nature: under the effect of the rationalization of certain agrarian exploitations, territories were remodeled in favor of a production more and more controlled by the nation-state and the scientific and administrative knowledge they wielded. After 1700, from the Indonesian islands to Scandinavia, from Siberia to the vast plains of North America, nomadic and traditional populations competed with peasant farmers and sedentary pioneers. In many parts of the world, grasslands and steppes were tilled, forests cleared, marshes drained. These environmental changes profoundly altered the relationships people had with their environment.[4] Throughout the early modern period, this rural world engaged in a series of essential, proto-industrial craft activities that polluted their immediate surroundings.

Historians of the early modern period generally favor a vision of these agrarian societies as generally being incapable of overcoming the Malthusian limits imposed by the biological constraints of the ecosystems in which they lived. The subsequent emergence of the "Industrial Revolution" completely rejected natural limits to human, economic, and productive growth.[5] Industrialization is regarded as the Rubicon of human development. In other words, historians of the preindustrial period try to understand what possibilities natural environments offered its inhabitants and how that symbiotic relationship evolved, from the point of view of "progress" of material civilization and of valuing environments rather than insisting on their transformation. But by ignoring the environmental impact of even these more modest, proto-industrial production sites, such a narrative prevents us from grasping how the most traditional artisans and many new sectors changed the metabolism of societies, and how the relationship of societies to their environment began their transformation. Two things: pollutions were common on the landscape prior to industrialization and the logic of industrial production was a necessary condition for its realization. Indeed, an "industrious revolution" gradually influenced and changed the preindustrial world. The exchange and transportation of materials increased; a growing intensification of work occurred on all four continents.[6] Urbanization and consumption, the growth of trade and finance, and the increasing intervention of the state all anticipated a world in the process of industrializing—in varied rhythms—where several parallel worlds coexisted on a global scale.[7] All these changes were accompanied by an increase in emissions of pollutants and a deep rethinking of what it meant to be clean or dirty, healthy or unhealthy.

Before 1830 and the industrial epic that was consolidated by the railway, societies were already very concerned about industry's deleterious effects. They set up coercive devices against the discharges and fumes they considered toxic: the fight against dirty and unhealthy introductions into their environments was a constant preoccupation that outweighed economic and industrial growth in a manner that has still not received careful historical consideration (chapter 1). The dynamics of capitalism, however, carved out new spaces and practices for production and demanded a new appreciation of growth and progress. In so doing, it insisted on dismissing questions of environmental change and risks to health. In the early phases of industrial capitalism, these pockets of pollution were still limited or punctuated across the landscape: around mines or certain cities mainly. But the spread of contamination on larger scales was already under way (chapter 2). Around 1800, a new pollution regime was born in Europe, particularly in the United Kingdom and France, thanks to the rise of liberalism and increased confidence in the emancipatory virtues of industry. Pollution subsequently became the inevitable side effect of industrialization, which was universally perceived as beneficial (chapter 3).

1

Sketches: An *Ancien Régime* of Pollution

What would a preindustrial history of pollution look like? In the predominantly rural and agrarian worlds of the early modern period, how would that narrative unfold—and how might we understand the relationship between pollution and the governing powers that fostered and tried to control it? The tanning industry seems as good a case study as any for sketching the contours of a response to those questions. The production of leather was present on all continents, everywhere animals were slaughtered for meat, and it was universally disparaged for creating noxious odors and "corrupting" waters. Leather served many purposes: as clothing, in military equipment, for tools and harnesses, as furniture, and so forth. It was a key product of the preindustrial world, and tanners enjoyed honorable status in the hierarchy of *métiers*. In rural spaces, tanning was a side venture among more affluent farmers and landowners. These tanneries were small-scale ventures that used just one or two cisterns for the tanning process. In comparison, urban tanneries employed multiple artisans and collaborated on grander leatherworks, and their access to consumers often extended well beyond the city walls. More expensive and rarer leathers reached more diverse, international markets. In both cases, tanning depended upon the supply and demand of local economies.

In order to transmogrify animal skin into rot-resistant leather, skins were steeped in tanks filled with water and tanbark (boiled and ground) for more than a year. Even with processes involving minerals instead of bark, tanning required animals, water, and forests: the use and treatment of organic materials were central to early industry. Despite diverse and evolving methods, tanning remained an art that changed little before the technical transformations and chemical advances of the 1820s. Whatever the product's quality, the transition from animal hide to leather created forms of pollution that raised questions about the panoply of technical

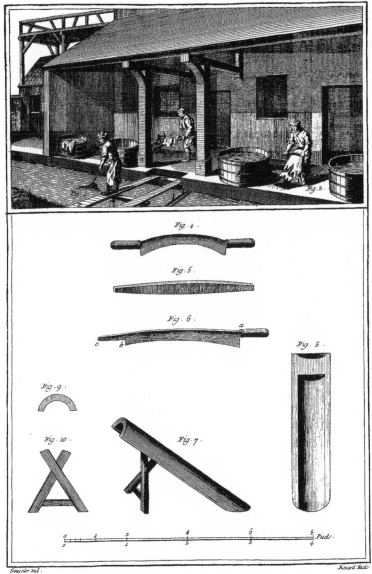

Figure 1.1

The tanners from *L'Encyclopédie* (1751–1772). The representations in *L'Encyclopédie* were inspired by Parisian installations. The image here was probably inspired by a tannery found on the Bièvre River. In order to make their leather, tanners situated their factories at the edge of small rivers, where they could wash their skins and capture the water for their maceration basins. Due to water pollution, they were located downstream of residential areas.

processes, their coexistence with other activities, changes in the physical environment, and new understandings of health and medicine. Pollution encountered public responses designed to preserve public health and clean water; the reactions served as the foundation for regulatory control. As local communities became concerned about the health hazards associated with tanning effluent in their environments, tanneries were pushed to more remote settings.[1]

Leather production demonstrates the ubiquity, continuity, and persistence of environmental nuisance dating back to the Middle Ages, even if the scale of pollutions resulting from multiple artisanal and industrial processes remained modest.[2] An examination of such chronic pollution suggests a preindustrial world that was both diverse and dynamic. Before studying the vectors of change that triggered a new relationship with pollution beginning at the end of the eighteenth century, an examination of what historian Fernand Braudel called "the structures of everyday life" exposes the density and diversity of the problems posed by the production regimes of the early modern period. A few key elements become clear. First, a large proportion of the early pollutants originated in rural environments, which hosted the bulk of preindustrial production. As centers of population and consumption, cities attracted artisans while at the same time imposing stricter regulations on nuisances derived from their crafts. The challenges associated with separating domestic and artisanal spaces and the endless search for sinks into which byproducts of craft could be dumped instigated new medical ideas about insalubrity as it related to air, soil, and water. Moreover, fears of dirty and unwholesome living spaces inspired a series of strict and structured nuisance laws, which indicate a longstanding tradition of environmental governance and control before the French Revolution.

Everyday Impurities

In the early modern period, the concept of pollution was rooted in religious cosmology and its ideas about purity and impurity. This dichotomy had a longstanding influence on how premodern societies imposed order by stigmatizing dirt vectors.[3] In Asia, Buddhism, Taoism, Confucianism, Shintoism, and Hinduism were wary of contaminants. Everything that degraded the environment was considered impure. Humans were to live in harmony with nature and preserve nature's ecological integrity.[4] Similarly, in Islam, pollution was a moral aberration and a sin. In the Western world, sin and pollution merged around carnal activities deemed

immoral: sexual fluids such as menstrual blood and nonprocreating ejac-
ulate were included in both concepts. All over the world, pollution had a
moral value before it had an ecological one. In pollution was the profane.
It is hardly surprising, then, that inferior classes and social outcasts were
tasked with cleaning and waste collection. In Japanese cities, the *hinin*
(which translates literally as nonhuman, but implied banished or exiled)
were the waste collectors. In China, a class of subaltern sweepers was
responsible for cleaning neighborhoods. And in India, castes were bound
to their proximity to dirt, waste, and pollution: the "untouchables," symbolically associated with death, were the cleaners of society's detritus.[5]
In Europe, ragpickers fulfilled a similar task and inhabited a comparable
social status: they were dangerous, ostracized, shunned.

The fear of pollution was not only a device for articulating social hierarchies. It was also rooted in the biological survival mechanisms of populations. Potential carriers of disease—corpses, the sick, the poor—were
considered dangerous and inspired an instinctive repulsion. The lone
antidote to pollution and the fears it provoked was pure and clean water,
which explains the continual mistrust around stagnant waters (in particular, wells) as being "dirty," "corrupted," and "unhealthy." Confronted
daily by bad odors, many urban populations found access to clean water
difficult; the idea of contamination gradually doubled as a moral connotation of impurity. Around 1750 it was still difficult to distinguish
between domestic and industrial or artisanal contaminants. On the one
hand, this lack of distinction stemmed from the low intensity of pollutants coming from both sources. On the other hand, in the early modern
period, domestic and industrial spheres collided in ways that made it
difficult to separate the one from the other. Only a more recent understanding of technical systems and production procedures has allowed for
the pinpointing of primary causes and sources of production-oriented
pollution.[6]

In these essentially agrarian worlds, rural running waters bore the
brunt of these pollutants. With hydraulic energy as the basis for developments linking technique, production, and environment, the watermill
appeared all over the countryside from the time of the hydraulic revolution in the Middle Ages. In addition to grinding wheat in the flourmill,
waterpower was harnessed for crushing ores, and for grinding tanning
barks, paper, and oil. It also fueled trip hammers, textile looms, mill-
stones for sharpening blades, bellows for forges, and other mechanical saws. The mill's power output became a sort of standard of energy
measurement in the preindustrial world. In Europe, there were roughly

500,000 mills by the eighteenth century. Asia was equally dependent on hydraulic energy production. Most rivers were converted, diverted, channeled, or constrained.[7] Inasmuch as hydraulic energy was renewable and nonpolluting, waterpower enabled new operations and modes of production that generated wastes downstream.

The textile industry was a particular catalyst of sullied water systems. Amongst common textiles in the early modern period—and typically in rural environments—degreasing wool required considerable quantities of ash, lye, and other plant byproducts. Their burned wastes were typically dumped into the environment after use. One less environmentally damaging but very common practice consisted of using fermented urine in boilers. Oil for greasing and soap (or urine) for cleaning constituted the gamut of materials removed from manufacturing spaces, which then became pollutants in the environment. Such contaminants posed minor hazards, to be sure, but their quantity warrants attention. The geographically diverse nature of wool manufacturing—all before the advent of the dyeing industry, which consolidated industry near and within the walls of emerging cities—disseminated small-scale pollutions across rural Europe.[8] In certain textile centers, such as Reims or Abbeville, urine was systematically collected and used in treating wool fibers.[9]

Other less common textiles were no less environmentally intrusive. For example, silk—in which China possessed a quasi-monopoly before the thirteenth century, and which became a worldwide expression of luxury fashion in the early modern period—was produced on rural silkworm farms. Silk manufacturing also emerged in India, Japan, and in the Mediterranean basin—especially in Italy and the south of France. Silkworm farms require access to mulberry leaves, the primary food source for silkworms. The high concentration of silkworms emitted unpleasant odors, notably from the fermentation of their excrement. By the end of the eighteenth century, silkworm farms in rural regions had become a focus of public health concerns.[10]

Separating flax and hemp fibers was an even more common source of water contamination in rural settings. Water-retting demanded underwater maceration over the course of several days, which resulted in the eutrophication of rivers and ponds. The fermentation process released putrid odors, while the impoverished oxygen content of the water systems affected surrounding flora and fauna. Locally, such fiber making tended to be a modest endeavor, but it was omnipresent across Europe: in Russia (the largest global exporter in the eighteenth century), Ireland, France, and the Baltic countries. Flax and hemp were also popular materials in the Americas and in Asia. Along with wool, before cotton's ascent to dominant textile, they were the most common textile fibers in the world. Throughout flax- and hemp-producing regions, most farms had their water-retting pits.[11] Their environmental impact was similarly ubiquitous. The amount of pollution flax and hemp manufacturing produced intensified during the Enlightenment, especially as hemp production increased in relation to its naval applications in sails and ropes. Hemp's important contribution to the rapidly increasing naval production during this period marked its apogee as a critically important resource.[12] Other plant fibers were recruited for ropes and military garments—notably broom, which was commonly retted in Languedoc and southern Italy. If it is quantitatively difficult to compare broom's environmental impact with hemp's, it might suffice to note that any such retting polluted water systems and provoked numerous environmental conflicts.[13]

Maceration was also a feature of the papermaking industry. Inasmuch as its raw materials constituted a form of recycling for used cellulose textiles, papermaking was another downstream polluter whose softening and pressing practices produced considerable waste. Paper was a Chinese innovation. It spread rapidly during the Renaissance after the invention of the printing press and the concomitant popularity and spread of books. In every European city, ragmen collected spent textiles to supply pulp as raw material for rural paper manufactories. Rotting, degreasing, and cleaning—fermenting organic materials in methods similar to other pre-industrial maceration practices—gave paper its desired texture and produced a pulp slurry that was disposed of in local waterways. At the height of this process, from the middle of the fifteenth century to the beginning of the eighteenth century, the slurry waste had an especially deleterious impact on local rivers. This was exacerbated by the added waste coming from gluemaking workshops, which also employed preparation methods derived from using—and then dumping—stagnant water.[14]

Many other industries relied on hydraulic equipment. For example, gunpowder factories dotted the length of river sides, using mill power to grind their three base materials: saltpeter, sulfur, and charcoal. Another Chinese invention, explosive gunpowder, spread to Europe in the fifteenth century and remained a pivotal military technology throughout the early modern period. The hazards of explosion meant that production was kept away from population centers. The fifteen factories in France, for example, were all situated more than twenty kilometers from big cities. Because storing gunpowder was potentially calamitous, its production waxed and waned in concert with wars, which subsequently shifted the concentration of gunpowder production around Europe. In the sixteenth century, Venice was the biggest producer, but it was soon matched and exceeded at different times by English, French, Russian, or Swedish production. Taking advantage of its extensive saltpeter resources in the Ganges Valley and in Bengal, India, under the control of European imperial powers became a primary producer in the eighteenth century. By 1790, the gunpowder factory at Ichapur in Bengal was one of the biggest manufactories in the world, its 2,000 laborers living in daily fear of detonation.[15] In England, the royal gunpowder factory at Waltham Abbey in north London was one of the biggest industrial complexes in Europe in the eighteenth century.[16]

If the war industry required water for energy and for cleaning and separating ores, it was equally dependent on biomass to burn. This demand constituted the first major source of atmospheric carbon emission from human industry. And where the need for sources of combustible or water energy explained the initial dissemination of metallurgical factories all across rural Europe, industrial concentrations started to emerge. Two hundred arms factories with 50,000 ironworkers were congregated in Brescia under Venetian authority at the end of the fifteenth century. Similarly, Styria, surrounding Graz, became another military-industrial center. Forges working in conjunction with blast furnaces produced cast iron through a process known in Europe since the end of the Middle Ages, and which spread during the early modern period—first to Styria, and then to the southeast of France and the region surrounding Liège. By the seventeenth century, bronze cannons, made of a copper and tin alloy, were being produced in blast furnaces in Sweden and in increasing quantities. Between 1500 and 1700, the production capacity of blast furnaces doubled. This necessitated more stringent resource management. Forest ordinances prescribed strict regulations on tree cutting, defining maximum annual cuts surrounding metalworking factories, which demanded ever more wood as fuel. Restrictions were made to control resource extraction; no controls were imposed on the carbon gases entering the atmosphere.[17]

In this new rural-woodland interface, combustion-dependent factories became increasingly concentrated—metallurgical plants, foundries, breweries, refineries, glassworks, pottery works, tile factories, brickworks, salt manufactures—and put greater demands on biomass that could be burned and transformed into charcoal. Charcoal pits decomposed the biomass without oxygen, removing its impurities through a process called pyrolysis, but it released acetic acid into the air, along with methane, tarring vapors, and various other gaseous microparticles. The perpetual appetite for charcoal left marks across the landscape. At the beginning of the seventeenth century, Russian salt manufacturing in the Kama River basin plundered forests to a 300-kilometer radius from its center. Complaints and arbitrations between industries and local users inspired new regulations. After 1724, the famous salt mines at Wieliczka in Poland were barred from burning their waste in light of the devastation to the surrounding forests. While some regions saw woodland clearing stabilize over time, others witnessed precipitous declines in forest resources. Forests covered 50 percent of Poland in the fifteenth century. By 1790, that number had plummeted to 25 percent. In France, a standard blast furnace powering two forges would consume 100 hectares (almost 250 acres) of wood. It needed to have 2,000 hectares of wood at its disposal for consistent production, which strained relations with surrounding communities in competition for depleted resources. Drawing on wood from more than 30 kilometers from the factory was prohibitively expensive. In Carinthia, metalworks consumed 220,000 cubic meters of wood in the sixteenth century. By 1768, 300 blast furnaces had devoured 700,000 cubic meters of wood. Growing demand for the materials of war—firelocks, muskets, rifles, cannons, swords, and other blades—mobilized a rapid increase in iron production. In 1525, 100,000 metric tons of iron were produced across Europe; 180,000 metric tons in 1700; and just over a million metric tons in 1800.[18]

Urban Centers

With the exception of water retting, which was generally not permitted in cities, no manufacturing industry was exclusively rural because most drew their labor force and their markets from urban centers. Cities grew rapidly between 1500 and 1800. London and Paris both grew from 300,000 inhabitants in the sixteenth century to 800,000 and 600,000, respectively, by the end of the eighteenth century. Though urban development in the East was generally slower than in Europe, Istanbul had 700,000 inhabitants in the seventeenth century, and Beijing's population was at least 3 million by the end of the eighteenth century. In Japan, Osaka saw the most significant growth in population, from 300,000 to 500,000, surpassing Kyoto. But Edo (Tokyo), with more than a million inhabitants, had the highest population of any Japanese city by 1800. Given the growing urban concentration, combined with slow and unreliable modes of transportation, it was inevitable that more and more production became situated in urban centers.

All cities had to deal with pollution problems. Controlling waste disposal—human and animal excrement, residue from artisanal work-shops—to assure water quality was a ubiquitous challenge.[19] In this respect, there is little doubt that cities in Asia were cleaner and healthier than those in Europe. During the Tokugawa period (1600–1868) in Japan, urban water collection systems and sanitation techniques far exceeded anything comparable in Europe at the time. The collection of waste was more systematic; it was recycled as fertilizer for agriculture and adapted for reuse in myriad urban projects. Better waste management at the source meant that Japanese rivers were hardly contaminated, despite a techni-cally less sophisticated system than in Europe. Consequently, Edo, though much more densely populated than either Paris or London, was far less polluted. This efficiency postponed the need for water closets and sew-ers until the beginning of the twentieth century.[20] Similarly, in Delhi the drainage system for water and waste meant that waterways were not too polluted.[21] Nevertheless, cities in Europe had their ragpickers to collect detritus, or managed depots along certain major public thoroughfares for the collection of reusable waste. In France and the Flemish and Walloon provinces, some cities developed policies for the salvaging and efficient cleaning of waste: a form of recycling before the formal invention of the term.[22]

Environmental quality in European cities seemed to deteriorate during the early modern period up until the 1750s. Growing populations and artillery advancement transformed the space in and around them, at least in northern Europe, where medieval city walls defined urban and rural spaces, leading to overflowing neighborhoods. Defense fortifications became more horizontal in nature. A vast no-man's-land surrounded cit-ies. Stagnant water pits, leveled earth, and general wastelands abutted the areas where the putrefying industries such as tanneries, dyeing, and saltpeter factories, and textile bleaching were located.[23] These industries manipulated organic matter and shaped the urban economy, creating an environment where putrefaction was frequent, with its odors and waste notably located along urban rivers. To save the Seine, Parisians actively decided to relocate polluting industries to the smaller Bièvre River. In Brussels, the Senne fell victim to butchers, tanners, tallow foundries, tripe merchants, brewers and dyeing factories before 1800. Similar professions in London sullied the water of the small Fleet River, which emptied into the Thames.[24]

Whether derived from plants or animals, pollution was a constant feature of the organic economy in the majority of cities.[25] Because the typical European diet called for fresh meat, abattoirs and butchers peppered every community in the typical city. Other parts of these butchered animals (tallow, gelatin, horns, hoofs, entrails, bone, hide, hair) were used to produce necessary items such as candles, fuel oil, sausage skin, musical instrument strings, leather, and many more products besides. Such was the organic economy's network that multiple trades developed symbiotic relationships. Located next to the abattoirs, candlemakers exploited the rejected inedible tallow. Their labors emitted intolerable stenches and greasy smoke into the air and posed a perpetual fire risk that threatened the entire city. While many such works were small in size, their number constituted an environmental hazard. The center of Paris, for example, counted fifty tallow foundries within the confines of the city limits in the eighteenth century. Despite efforts by cities like Paris and London to relocate manufacturing facilities outside the city, nuisances from animal-based manufactories remained omnipresent.[26] Plant-based industry was similarly widespread in its negative effects. Beer and flour starch were both produced through fermentation of grains in vats; its waste was dumped straight into the rivers. If artisanal techniques changed little during the early modern period, the quantities of production increased markedly, which in turn profoundly impacted and influenced major changes in working landscapes. For example, before the end of the seventeenth century, starch was imported from the Netherlands into France. After 1700, however, Paris was producing its own starch in some thirty factories.[27]

While drapers, silk weavers, and paper makers were staples of the rural landscape, they were drawn to cities for their consumer markets and emerging bourgeois classes. Nîmes, capital of French silk-making in the seventeenth and eighteenth centuries, suffered from its own success, and had to deal with the polluting byproducts of its spinning mills as the industry grew.[28] Tanneries and dyeing factories, the worst offenders for dirty byproducts, became entrenched in cities the world over in the early modern period. Along with other leather workers—taw and alum tanners, chamois manufacturers, curriers, furriers, and leather shopkeepers—tanners employed a variety of different practices. Plant-based tanning was the most common form in Europe, while mineral-based tanning was more popular in dry regions like the Middle East, or for use with finer skins that were treated with alum. A grease-based tanning technique was dominant in more northerly regions, or where chamois leather was produced. Indigenous populations in North America combined a water treatment with the smoking of skins.[29] Despite this great diversity in tanning methods, one feature common to all was odors and the waste released into nearby waterways, which could also include arsenic, quicklime, verdigris, or iron sulfate. As a result, tanners around the world were frequently clustered outside and downstream of cities. Tanneries were also omnipresent in rural areas. In France at the end of the eighteenth century, there were 5,000 tanneries in the countryside, with some operating more than one macerating pit. Whether found in Pennsylvania in 1650, Montreal in 1770, or Bordeaux in 1800, their stench was universally condemned.[30]

Dyeing factories were often found in close proximity to tanneries. This final step in textile production required a certain expertise and an urban consumer market. Dyeing was equally ubiquitous, though it depended somewhat on fibers that were grown regionally. In France, therefore, silk dyers clustered in and around Lyon, while woolen drapers were found in the north and in Languedoc. Dyeing required copious amounts of water and was an extremely polluting activity, which emitted its effluent into local rivers. To prepare fabrics for dyeing, water was combined with various substances such as diethyl ether, modified turpentine, soap, chalk, lime, and clay. Dyes were subsequently added but often not without the aid of further chemical products, such as mordants, which helped to fix the colors in the dyeing process. Alum was the most commonly used mordant, but during the eighteenth century more toxic substances, such as metallic oxides and ammonia salts, were also introduced.

The production of colorants for dyeing, whether derived from plants or minerals, was a source of pollution in and of itself. The famous Prussian blue tint, produced in Berlin during the eighteenth century, combined blood, alum, and sulfuric acid in one of the most feared urban crafts of the Enlightenment period. Its red or carmine counterpart was produced in baths of nitric acid and tin combined with ground-up cochineal insects. Another famous shade of scarlet red was created in Paris, at the Gobelins dyeworks on the edge of the Bièvre. But the most spectacular tint, Adrianople or Turkey red, was also the most toxic to produce. It was first made in Greece (Smyrna and Salonica) and Turkey (Edirne), and was composed of lead chromate, lead oxide, and eosin. Not only was it extremely toxic but it was also putrid, with its ingredients comprising rancid fats, oil, urine, lime, alum, excrement, and blood. Advances in chemistry by the end of the eighteenth century meant that the art of dyeing was becoming ever more refined.[31]

Some cities specialized in particular products. Marseille, for example, was famous for its quality soaps and dominated the market in Europe until the end of the eighteenth century. Mixing fats and lye in boiling vats to make soap emitted strong, odorous smoke along with voluminous residues of carbonate and sulfur from lime, and other byproducts. Sulfur mixed with air produced a sulfurous gas, which could cause the soap waste to spontaneously combust. The soap factories were initially confined to a district on the south bank of the port, relatively distant from Marseille's center, but they were gradually engulfed by an expanding web of urban expansion. On the eve of the French Revolution, forty-eight factories were producing 20,000 metric tons of soap per year in Marseille. The residues of its manufacture were collected and stored in an outlying district of the city, along the coast, which rapidly prompted warnings from fishermen and naturalists that the industrial discharge was harming maritime wildlife.[32]

The preservation of state power fostered its own manufacturing landscapes. Naval building sites, metallurgical factories, and munitions factories appeared in urban and coastal regions, and had significant pollution and resource impacts. For example, 3,000 one-hundred-year-old oak trees were required to build a single ship. In 1669, Jean-Baptiste Colbert's French forest code introduced a series of resource laws that facilitated the sustained extraction of lumber to benefit state commerce, military needs, and imperial expansion. The expansion of high-sea navigation gave considerable impetus to the development of the metallurgic trades. By the end of the seventeenth century there were more than 5,600 cannons on board French ships and 8,000 aboard English vessels. Several factories, often situated beside iron and copper mines, discharged smoke into the urban atmosphere. This was most obvious in central Europe, near Cologne, Regensburg, Nördlingen, Nuremberg, and Suhl. The dockyards, which combined construction of maritime and military equipment, became major industrial complexes. With 3,000 laborers, the dockyards at Venice were for a long time the archetype: its foundries and factories dominated and controlled the surrounding hinterland. On occasion, dockyard construction became the impetus for building a city. Beginning in 1703, Peter the Great commissioned and willed into existence the city of St. Petersburg. In spite of its unfavorable siting, built upon scattered islands at the mouth of the Neva River, St. Petersburg was originally designed as a military port. It counted 75,000 inhabitants in 1750, and more than 200,000 by 1789: rapid growth that suggested a radical and unprecedented mobilization of people and transformation of the physical environment. In the subsequent century, the garrison city was a perpetual construction site, defined and buttressed by its seaport on the west and its cannon foundry and munitions factories on the east. Given St. Petersburg's concentration and specialization as a manufacturer of military industries, its concomitant smoke and waste were a prominent feature of the city's character and identity.[33]

Enlightenment Imaginaries and Sensibilities

This brief survey of the principal preindustrial polluting activities suggests that environmental contamination is not strictly a recent phenomenon. While in most cases the scale and extent of the pollution remained limited and localized compared with that of later periods—and while many types of polluting activities happened intermittently, whether due to seasonal work or the practices themselves only occurred over short periods of time—industrial expansion during the early modern period forces us to rethink the manner in which contamination affected more and more people. Religious connotations of the sacred and the profane—purity and impurity, the clean and the unclean—shifted moral values to encompass environmental conditions and emerging medical understandings of health. Godliness and healthfulness: cleanliness was next to both in equal portion. Frontiers of healthfulness and unhealthfulness, of clean and dirty, were undergoing epistemological revolution. The civilizing process described by the sociologist Norbert Elias manifested itself in a kind of popular obligation toward a collective attention to a physical cleanliness of bodies and places over the course of the eighteenth century. A new health-specific lexicon—"insalubrity" or "insalubrious," for example—emerged, which marked an evolution with respect to environmental sensibilities.[34] To escape unhealthy air and dirty cities, the English aristocracy acquired a passion for the beauty, simplicity, and safety of the countryside. They cultivated an aesthetic of a wild nature, unsullied by human activities.[35]

It is hardly surprising that this evolution was most visible in Europe. Whereas most religions, East and West, shared a mentality that stigmatized the impure as immoral, Christianity was alone in insisting that humans were masters or stewards over nature and were free to rationalize its exploitation.[36] At the same time, however, Europeans linked human health to the environment, where struggle was waged with an adversarial physical environment—the quintessential other—to improve living conditions for its populace. Greek thought was recruited to transmit this tension.[37] Western thought retained Aristotle's assertions about the balances and mobilities of physical elements. Hippocrates drew on these balances to infer that environment was one of the key indicators of health.[38] In the Italian Renaissance, physicians such as the Venetian Andrea Marini condemned tainted air as a health threat. Similar anxieties permeated most of northern Italy.[39] Beginning at the end of the seventeenth century, such health concerns spread more widely. In Modena and Padua, Bernardino Ramazzini worked to identify "poisonous spirits" that manifested in wheat, comparing resulting symptoms in the plants after spraying various acid and alkaline solutions on them.[40] In his famous 1700 treatise, *De Morbis Artificum Diatriba* (Diseases of Workers), Ramazzini adopted the same methods to study associations between workers and health hazards in various workplaces. This epidemiological data collection systematically revealed incontrovertible correlations between specific diseases and exposures to specific fumes, molds, chemicals, dusts, metals, and other agents in specific working environments. In effect, many artisans were being poisoned by the substances with which they came in contact in their trades. Among the fifty-three professions Ramazzini investigated, he denounced, for example, the awful stenches associated with retting hemp and tanning hides, comparing them with the far more visibly mortal hazards suffered by miners and metallurgical laborers. Though his theories on occupational health were translated and read across Europe, Ramazzini's body of work was hardly questioned before the 1830s.[41] In Britain, the natural philosopher John Evelyn, an active founding member of the Royal Society, was also intent on establishing connections between health and air quality. His famous 1661 book, *Fumifugium* (or *The Inconvenience of the Aer and Smoak of London Dissipated*), was the first to specifically engage with London's air pollution. Evelyn was critical of the extensive coal burning in and around London, which he argued was responsible for the problem. He recommended that all furnaces be removed from the capital. Evelyn's recommendations resonated with the newly restored English monarchy. King Charles II embraced such environmental reforms as an opportunity to renew the state and reaffirm his authority after a decade of republicanism.[42] Charles and his emissaries called on scientific and medical expertise to tackle the array of epidemic and chronic diseases that afflicted the country and to get at the root of their origins.[43] As a result, the eighteenth century reinforced the significance of medical investigations into environmental causes. And as Montesquieu linked social hierarchy to air quality in France, a sort of neo-Hippocratism emerged.[44] In 1726, in his *Elementa Physica* (*Elements of Natural Philosophy*), the Dutch physician Petrus Van Musschenbroek revisited Aristotle's theories of balance:

The atmosphere is composed of air, vapours, and emissions. The latter two are composed of smaller volatile parts, separated from terrestrial bodies.... Neither these vapours, nor these emissions are safe for respiration: there are no poisons so dangerous nor so pernicious as those that come from the vapours and emissions derived from burning organic substances. Doesn't burnt sulfur instantly kill all kinds of animals? ... Trapped in a sealed glass, heated copper emissions quickly suffocate whatever animal we place in the receptacle with it. The emissions from wood and peat charcoal are equally deadly.... Many brewers have been found dead in their cellars, suffocated by the fermentation process.... And don't we learn all the time of coal and metal miners who are killed on the job after they release such horrid emissions from the depths?[45]

Cited in Diderot's *Encyclopédie*, Van Musschenbroek's work expanded European curiosity about the environment. In 1751, the British doctor John Arbuthnot, another member of the Royal Society, established further evidence of a direct link between air quality and human health in his *Essay Concerning the Effects of Air on Human Bodies*. Thirty years later, the Frenchman Jean-Jacques Menuret de Chambaud's *Essai sur l'action de l'air dans les maladies contagieuses* won a prize from the French Royal Society of Medicine for drawing connections between artisanal waste and contagious illnesses. The 1776 creation of this new Royal Society of Medicine revived erstwhile interest in Hippocratic theories of environment and health. From its very inception, the Society opened inquiries that succeeded in advancing scientific notions of how the environment played a central role in the propagation of epidemics.[46] This medical knowledge was widely shared and accepted throughout Europe, especially in Great Britain and in the Germanic states, where the links between illness and environment, air, and pollutants became accepted as incontrovertible fact.[47] The key indicator of deleterious emanations remained bad smells, which prompted the scientific vocabulary around "miasmas." A growing aversion to the stenches coming from swamps, marshes, stagnant waters, and animal and manufacturing wastes spanned the second half of the eighteenth century. Miasma was understood as a vector for disease, contaminating air and soil, and harming animal and human bodies. Pollution from hemp retting was considered responsible for widespread epidemics, and by the beginning of the 1780s the Royal Society of Medicine launched a major inquiry into its dangers. In the same manner, blood and its vile, putrefied odors provoked increasing public horrors that helped to explain a general indignation toward animal slaughterhouses.[48] Other studies on English and Italian cities traced in similar fashion popular fears of miasmas.[49]

As a result, urban inhabitants complained regularly against what they considered to be an affront to their health and comfort. That was the case in sixteenth-century Venice, where the writer Tomaso Garzoni castigated the city's tanners, members of a "rank and stinky" trade, for their "excessive filth."[50] Complaints of this nature multiplied in the eighteenth century. In Normandy, fishermen and coastal peasants alerted the Agricultural Society of Rouen to the "pestilential" smoke emanating from the caustic soda furnaces that lined the beaches. Soda production, which burned kelp, was essential for glass and soap manufacturers. According to those living nearby, the smoke had deleterious effects on surrounding trees, plants, and wheat, not to mention aquatic sea life.[51] In Paris, frequent complaints from city-dwellers led to the relocation of inner-city workshops that spat out bothersome and unhealthy odors and wastes to beyond the city limits. But this spurred new upsets in the outlying districts that inherited the polluting industries. In 1725, seven plant oil factories, one pottery works, and one plaster oven were destroyed in Passy and Auteuil, communes just outside Paris, "because the ovens emitted so much smoke as to irritate passers-by." By the end of the 1720s, inhabitants of Passy managed to stop the installation of a new glassworks intended for their village after it had been shut down in Chaillot.[52]

In the early modern period, protests from riverside populations against polluting factories raised concerns over the potential loss of vital resources. Defense of communal rights to forest, heath, and moor, and to bogs, marshes, and peatlands figured prominently in popular insubordination.[53] Even if such protests were rarely very large, local inhabitants knew how to defend their rights against monopolies and polluters. Such complaints divided social protests into struggles for justice against the unjust, challenging monopolies against individual rights, and developing a kind of moral economy of the crowd, as described by the historian E. P. Thompson.[54] In the Germanic principalities, the very first industrial factories provoked vociferous disputes.[55] Numerous court grievances from 1789 recorded hostile contempt for industries that were perceived as being wasteful of resources, but also because of their malodorous emissions and threats to public health. In Lorraine, in the book of grievances from Daspich and Ebange, inhabitants complained about the terrible pollution from discharges from forge residues, mills, and copper works, in the Fenche. The river, they claimed, was "infected by a mixture of murky water, yellow and full of iron rust to the point that even gardeners cannot water their gardens with it."[56] Popular representations supplemented medical knowledge to develop new values and understandings of clean and dirty during the latter half of the Enlightenment. The alteration of elements was understood to be a "corruption," and already, in Great Britain, the word *pollution* had acquired its contemporary meaning. In legal circles, it appears to have been adopted in 1804 for the first time in a Scottish case charging tanners with having altered a river with their waste. "Pollution of the stream" and "pollution of water" featured in the records.[57] Linked to these social and medical perceptions—and in mirror reflection of social interests—nuisance laws became a particularly important regulation tool.

Nuisance Laws

The standard historical narrative states that preindustrial societies were passive in response to their environment suffering from dirt and stench, but the record demonstrates that this was not the case. Rather, the regulatory measures they took against pollution were shocking in their rigor and impact, especially when compared with similar laws adopted after 1800 in Europe.[58] Public health authorities were local: whether municipal or judiciary, they enforced these laws under the juridical concept of "nuisances," which encompassed any harm to others. This included what we could today call pollution, but also noise, fires, and other risks and inconveniences.[59] The struggle against sanitation and environmental threats was grounded in part in a longstanding heritage of medieval law, but also in the conceptual evolution of pollution and its health risks. Limiting pollution through improved technology was never a mode of action against environmental nuisances. Health authorities preferred banning polluting activities or exiling them to remote areas. In efforts to preserve good health, preindustrial health authorities engaged in a range of prescriptive and harsh measures while demonstrating a capacity for negotiation in a world where neighborhood and interdependence were essential components of community survival. The threat of pestilence and epidemic were common enough features in provoking careful vigilance. Public hygiene was taken very seriously; quarantines were commonplace, instituted through a series of codes introduced in cities in northern Italy at the end of the Middle Ages. Lazarettos—quarantine stations to isolate the sick and contagious—were implemented to create a kind of sanitary customs control before travelers and goods could debark in Mediterranean port cities and followed the same kind of separation imposed on infection vectors.[60]

On the municipal level—and designed to check the worst pollution—methods of regulation were predominantly judiciary. In approaching nuisances as a legal phenomenon, the courts were typically left to resolve conflicts. In common-law countries, nuisance laws were based on the maxim "*Sic utere tuo ut alienum non laedas*," which encouraged a freedom to benefit from one's property so long as it did not bring harm to others or their property. This maxim was at the heart of nuisance law dating back to the "Assize of Nuisance" from the fourteenth century, and it was frequently invoked in nuisance cases. Nuisance cases could be public, punishable by injunctions—fines, closures, or prison—or private matters settled by compensation for damages.[61] In London, "someone ... who wishes to make sulfuric acid or nitric acid will corrupt the air and affect the health of his neighbours should be brought to justice and suffer a penalty proportional to the damage he has committed." For the magistrate Nicolas Des Essarts, who compiled the French penal code before the French Revolution, such guidelines were necessary. "The English legislation contains a valuable principle for maintaining order," he wrote. "Somebody who makes something that causes harm—or neglects to do something that the general good requires as being necessary—is guilty of an offence, which is called a nuisance."[62] Because the principle of common law didn't exist in France, respect for residential property was a more intangible concept. Instead, a hybrid legal system, which integrated Roman law and local customs, interpreted nuisance laws as a function of neighborhood codes and responsibilities, which made up servitude law and followed the medieval principle "*Si servitus vendicetur*": "Inasmuch as a property owner may do as he wishes in his own backyard, he may not engage in works that inhibit his neighbour from enjoying the same freedoms, or that might cause his neighbour harm."[63] As a result, smoke emissions in private yards were forbidden where servitude law did not explicitly state otherwise. Reflecting on the relationships between society, pollution, and the conventions that bind them, the law operated in a system where local oversight played an important role in emerging cities with expanding industries. Individuals were empowered to take action through grievances, petitions, or the local courts' defense of the right of possession.

In his impressive *Traité de la Police* from the beginning of the eighteenth century, the police commissioner Nicolas Delamare outlined the tenets of police jurisprudence. The work was widely admired across Europe and showed that local police worried about the potential harm emanating from some artisanal and industrial production methods. The general interest, the work showed, supported the importance of preserving public health.[64] In 1784, the jurist Joseph Guyot explained the French legislative system's position with respect to civil codes of conduct. It involved arbitrating between the individual interest and the public interest: protecting the former but making sure it yielded to the latter. As much as possible, when an industry's outputs were considered locally undesirable, it was banned by public interest measures because its neighbors deserved "reciprocal respect .... It should not, therefore, be permitted to engage in activities which could render uninhabitable its neighbors' homes."[65] Public interest as a method of engaging with pollution was equally a feature of English law. While some professions were deemed necessary— candlemakers, tanners, and brewers, for example—their necessary products could be manufactured in places removed from densely populated areas.[66] The common law's standing on water and rivers was comparable to their treatment in nuisance and river property laws: any activity that threatened to pollute a neighbor's water quality was carefully supervised and severely reprimanded.[67]

Distancing or removing causes of infection was the general rule, and it seemed that in prerevolutionary societies the perceived link between industrial miasma and public health frequently succeeded in banning undesirable workshops and factories. As early as the sixteenth century, strict responses to pollution were already common in France. In 1510, for example, following a lumber shortage, forges around Rouen turned to coal to power their work. The coal fires triggered a local lawsuit; the court ordered the industries removed from the city just like other harmful artisans (dyers, leather curriers, etc.). Similar fears emerged in England following lumber shortages there, too.[68] After 1700 in Metz, local authorities consulted with neighboring communities before establishing a new metalworks. Meanwhile in Dijon, areas on the outskirts of town were designated for vinegar makers and police commissioners in Lyon assigned similar zoning locations outside the city for certain professions, and did not hesitate to sanction artisans who contravened the new regulations.[69] In Paris, the police maintained a careful eye on potential polluters. Over the course of the eighteenth century, they intensified their extradition of polluting industries to dedicated sites outside the city. Delamare devoted a whole chapter in his *Traité* to "Removing Professions That Can Contaminate the Air from the Middle of Cities." Beginning in 1738, inquiries into *commodo* and *incommodo* were imposed on tanners, dyers, butchers (animal guts were deemed particularly problematic), and other manufacturers. The inquiries, which turned to local residents and police expertise for input, increased police powers to prevent contaminations. As one commissioner recalled in 1773, there were several "police regulations that order professions which employ foodstuffs and pharmaceuticals that can contaminate the air and harm public health to withdraw to districts removed from more populated areas in order to continue their work."[70] By the French Revolution, Antoine-François Prost de Royer, who served as lieutenant-general of the Lyon police, could write without fear of being contradicted that:

> Every time that a warehouse, a trade, or any other operation might taint the air in a manner that threatens public health, the magistrate must ban or expel that practice.... The commodity, the profits derived from it, count for nothing: all that matters, and everything upon which justice hangs, is clean air and public health. *Salus populi, suprema lex esto.*[71]

Similar processes were in effect elsewhere in Europe. In Venice, the archetypal industrial city of the Renaissance, health officials banished many industries from the city proper. For example, those glassmakers and tinters who produced the famous Turkey red were among the first to be expelled, in 1255 and 1413 respectively. When pestilence threatened, tanners were automatically forbidden to work at all. Throughout the sixteenth century, butchers, slaughterhouse workers, leather curriers, and even the city's mint, originally located in the Piazza San Marco, were moved outside the city limits into neighborhoods farther afield. Only the arsenal, which enjoyed distinguished status in the Venetian Republic, was exempted from such banishment.[72]

Similarly, London manufacturers of quicklime, glass, and alum—pioneers in coal use—were removed from the inner city because of the sulfurous odors resulting from the burning of coal.[73] Certain banished trades reestablished themselves in particular districts around the city. For example, tanners and leather artisans settled in and around Bermondsey, milliners could be found in Bridwell, tin-glazing potters at Billings-Gate and Bishop-Gate, while Southwark became the center for various artisanal workshops (dyers, brewers, lime burners, glass and paint makers).[74] Cities—such as Bath, Oxford, York, and Edinburgh—witnessed similar expulsions and the creation of areas dedicated to industry outside of town. Brickwork construction in the Netherlands was subject to city authorities. More often than not, brickworks were expelled to more isolated regions because of their coal smoke. Anvers, in particular, had an impressive collection of city ordinances, trials, and fines that attest to active pollution regulation.[75]

The European example was mirrored all over the world. In North America, English law governed manufactories, most notably those powered by hydraulic energy. Factories and workshops were strictly regulated and in general needed a permit from local authorities to be allowed to operate. This was the case in Massachusetts for potters working with kilns, as well as bakers and chocolate makers. Similar regulations applied to ironworks in Philadelphia. In New York, both rum distillation and quicklime manufacture (from the boiling of oyster shells) were banned outright from the beginning of the eighteenth century, because of fear that their emissions caused illness.[76] Although specific studies do not exist, we know that similar trends were followed in cities in the Muslim world, whereby craftsmen were shunted to the outskirts of the city and more undesirable industries (curriers, blacksmiths, potters, dyers) were pushed beyond city limits.[77] And in Asia it appears that the most polluting factories were pushed as far as physically possible from urban centers. In Japan, for example, in 1712, central authorities (*Shogun*) ordered that ironworks in Osaka be built two kilometers from the city center because of their implicit risks and nuisances.[78] In colonial India, the concept of nuisance was also embraced and applied until late in the nineteenth century. Here it was defined as any "noise or smell considered a violation, danger, inconvenience, affront to the senses" which threatened "life, health, or property." Preindustrial European law clearly influenced authorities worldwide and gave them, in theory at least, the power to relegate industry and craftsmen whose wastes were released into the air, soil, and water.[79]

In these legal systems, therefore, the courts and authorities' governance predominated. Strict controls against manufactories that emitted unsavory odors indicated that public health was privileged over economic development. In one court case in Paris in 1768 against a nitric acid manufacturer, the commissioner Lemaire, the right-hand man of the lieutenant-general, recalled that:

> These vapours ... are capable of making people seriously ill; ... an establishment that poses such risk is not welcome in any residential area, especially in Paris, when that work can take place elsewhere; ... and since the protection of its citizens is the police's primary objective—and since there are a sizeable number of regulations that aim to ban from the capital all sorts of polluting industries that damage air quality—such infractions should fall under police jurisdiction.

This was evidently a trend. Numerous records point to artisans being called into police stations, frequently after neighbors complained about their emissions or waste. And because such infractions took on an increasingly legal character, appeals to sovereign courts were possible. In Paris, the Parlement of Paris was the nexus of power in such cases. It was not uncommon for the council to order the destruction of sanctioned furnaces listed in the lieutenant-general's rulings.[80] Such cases were equally common in London. In 1754, a Londoner complained that one of his neighbors had converted his house into a small glassworks, which burned a substantial amount of coal. The smoke constituted a nuisance and produced "noysome and filthy vapors" which destroyed the plants and one hundred trees in the plaintiff's orchard. The complaint involved a demand for £500 in damages; the jury awarded him £40, less than 10 percent of his claim, but fairly stiff compensation nevertheless.[81] Inasmuch as nuisances were regarded as public issues, actions in criminal court were fairly frequent. In Paris, in 1750, a nitric acid manufacturer was accused of being responsible for the illnesses in his neighborhood as a result of the acid's vapors emanating from his workshop. Twenty-four witnesses testified before the court on the matter. The lieutenant-general, accompanied by the crown prosecutor, was present to confirm the existence of this threat to public order and public health.[82] In Great Britain, numerous nuisance cases were entrusted to the King's Bench, the sovereign court. Criminal proceedings often followed suit.[83] In North America, the penal system was frequently involved in industrial complaints.[84]

* * * *

In the early modern period, pollutions were typically restricted to areas in close proximity to their source. They permeated both rural and urban landscapes but were essentially limited to transformations of organic matter that could be absorbed within the environment. Nevertheless, odors, fumes, and contaminated waters provoked fear that explained why citizens and authorities were reticent to live in industrial areas. With few exceptions (munitions factories, for example), nuisance laws responded to public concerns by adopting precautionary measures that privileged public health over economic development. By exiling and disseminating polluting practices to the rural landscape, early modern populations were generally protected from the pollutions emanating from their industrial world. However, new dynamics—stimulated and imposed by merchant capitalism and the acceleration of industrialization—undermined these processes. In some localized cases or for strategic products, new forms of pollution appeared and began reconfiguring the political and legal relations established between societies and their environment.

2

New Polluting Alchemies

The moral economy and environmental regulations established in the preindustrial era began to unravel first in Europe. The spirit of the Renaissance created an intellectual movement that encouraged rather than prohibited environmental exploitation. Natural philosophy and science united under the premise of controlling nature. The "mechanical arts," championed by Francis Bacon in England, René Descartes in France, and Gottfried Leibniz in the Germanic principalities validated a new mode of approach to interacting with nature and anticipated the Enlightenment. The naturalists of the eighteenth century believed that nature existed to serve humankind and that humans could exploit the natural world well beyond the limits of mere necessity.[1] Such rationale spurred the rise of industry and the material gains to be drawn from it after 1750.

New polluting alchemies emerged from the expansion of global capitalism. Mining was one of the first. Already far removed from populous areas and situated at the outer limits of common law, mines were the laboratories where the rules governing pollution were gradually bent or mutated in the interest of profit. Protecting the environment seemed much less of a concern in this new industrial era. After the discovery of the New World, the extraction of precious metals in Latin America contributed to the birth of a "world economy." New trading routes were forged, maritime transport was on the rise, and the first capitalist institutions emerged. New frontiers reconfigured the distribution of production around the world. In 1500, Central Europe produced 85 percent of the world's silver, but within fifty years the New World was producing more precious metals than the rest of the world combined. Mining really took off during the second half of the eighteenth century; Peru, Bolivia, and Mexico dominated the world's silver market, while Brazil alone produced 80 percent of the world's gold extraction.[2]

Figure 2.1
The mine and refining factories at Potosí (1585). *Source:* The Hispanic Society of America, *Atlas of Sea Charts*, K3. This watercolor on parchment is one of the oldest representations of the Potosí silver mine. In the background, the Cerro Rico, surmounted by a cross, dominates the city from which the miners and llamas begin their labors. The refining facility is showcased with Indian workers, grinding machinery, mercurial ponds, and stacks of ore and silver. The site in the southern highlands of Bolivia is now a UNESCO World Heritage Site.

Mineral extraction, however, came at a high, polluting cost. Separating silver from ore required a great quantity of lead, while mercury was used to extract silver and gold. Lead and mercury mines were thus linked to gold, silver, and copper mining. The residual waste from these mining processes comprised extremely toxic heavy metals threatening long-lasting contamination. While these pollutions were at first confined to specific pockets of industry, they soon became more widespread as more industrial sectors took advantage of these material sources and the chemical and metallurgical operations that accompanied them. Coal as an energy source and other chemical products necessary to expand industrial operations accelerated widespread environmental alterations. By the beginning of the nineteenth century the "great divergence" between Europe's attitude toward the environment and that of Asia grew ever wider.

Mines and Mining around the World

Metallurgic operations were no modern world invention. Metallurgies in copper, bronze, and iron had been in existence since 3000 BCE and contributed to the rise of ancient civilizations. The Chinese mastered iron metallurgy between 200 BCE and 200 CE. The city of Athens thrived thanks to the silver mines at Laurion, while the Roman Empire benefited from the silver and copper mines in Spain and the tin mines of Cornwall. In the eleventh century, the Song dynasty in China enjoyed a considerable imperial expansion due to its access to an abundance of metal resources. Justifying the extraction of these minerals was in direct conflict with Confucian principles, however, which condemned the overexploitation of resources. Mining was subsequently halted because of the Mongolian invasion.[3]

The first large-scale pollution from mining took place in the Roman Empire. In his book *Geography,* Strabo reports that silver foundries in Spain were built with very high chimneys to minimize exposure to their "lethal" fumes. Lead, arsenic, and mercury were also mined and used by various craftsmen. Isotopic tracing of lead in river deposits and glacier ice caps show that vapors from these Roman foundries traveled as far as the polar regions. These metals not only poisoned the environment, and those who worked with them, but they could also have contributed to the decline of the Roman Empire.[4]

Central Europe (Saxony, Bohemia, Silesia, and Tyrol) was the mining capital of the world when the Americas were discovered. The smelting and refining of copper, silver, and gold took place through liquation and cupellation, with both processes requiring a lot of lead. The lead mine at Tarnowitz, near Krakow, was at a crossroads on the metal thorough-fares. Its towering smelting factories filled the air with so much smoke and stank to such an extent that it was compared to Mount Etna.[5] In 1600, around 50 kilograms of lead were needed to produce a single kilo-gram of purified silver. The lead was never recovered. Between 1500 and 1800, the mines in Bohemia and Saxony produced 10 metric tons of silver annually; 150,000 metric tons of lead were released into the environment. To this day there are abnormally high levels of lead found in the bog lands of this region.[6]

Mining was big business and received investment from the great financiers of the time, such as Jacques Coeur, the Medicis, and especially the Fuggers.[7] Rules were bent to accommodate the industry, and the work-force was forcibly managed. In effect, the new world that put metallurgy at its epicenter transformed people's habits, beliefs, and the administrative structures that organized them. Social and environmental interests were made subsidiary to financial gain: the mines' financiers created the context for wide-scale environmental exploitation at the expense of other activities, leaving local peasants lamenting the despoliation of resources and landscapes. Mining's appetite was insatiable. The perpetual need for fuel meant that the mines' factories regularly had to relocate after they had depleted local resources through deforestation.[8]

Georgius Agricola's work, *De re metallica* (1556), was in part a response to the contested views of mining at the time. Local and financial interests clashed over what might be properly regarded as an early ancestor of the environmental impact statement. A mine could not be opened without an inquiry into its practices and property, the potential pollution of groundwater and other waterways, the destruction of flora and fauna, or the impact of altering the landscape by new throughways to access it. Other points of contention stemmed from deforestation, the building of annexes to house more industry, and the danger of the collapse of the mines. Agricola knew his topic.[9] He was at the time a physician and burgomaster in Chemnitz, the center of Saxony's mining industry. His seminal work on the techniques of early modern mining began with an earnest justification for mining, legitimizing even its most polluting activities. Even though he used terms such as "pestilential" and "noxious" to describe lead, Agricola stressed the social benefits derived from its application in metallurgy and how the metallurgical arts were critical to advancing chemical innovation.[10] Indeed, around this time the alchemist Paracelsus, based in Carinthia, discovered the potential for metals to be used in therapeutic treatments. He, too, had ties to mining, and experimented with such toxic substances as arsenic, sulfur, copper, and mercury in the name of medicine.[11]

Some of the environmental problems associated with metal mining in central Europe dissipated with the discovery of incomprehensible mineral wealth in the New World. Rich veins of gold and silver across the vast American continent enjoyed the added benefit (for financiers) of a region devoid of social and environmental regulations that inhibited exploitation. New World production rapidly surpassed production anywhere else in the world, with particular expansion throughout the eighteenth century.

At Potosí, in Peru (modern-day Bolivia), at an altitude of more than 4,000 meters, the first major silver mine opened in 1545 and provoked a rush on precious metals. After only fifty years, this previously untouched

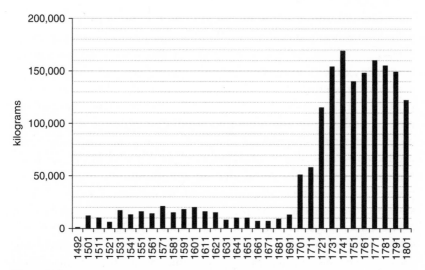

Figure 2.2a
Gold production (in kg) in the Americas, 1492–1801.

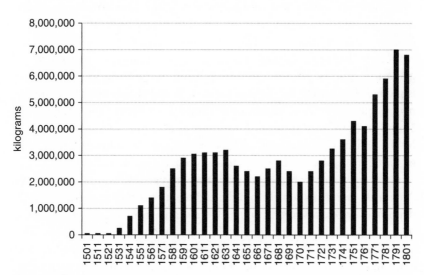

Figure 2.2b
Silver production (in kg) in the Americas, 1501–1801. *Source:* John J. TePaske, *A New World of Gold and Silver*, edited by Kendall W. Brown (Leiden: Brill, 2010): 188–191.

and uninhabited area had evolved into a booming metropolis of 100,000 people—the most populated in the Americas—whose sole purpose was to extract silver from the mountainside. By the end of the sixteenth century, silver was extracted using a mercury amalgam. Known as *patio,* this process mixed the crushed ore with mud, salt, mercury, and iron or copper pyrites in great 2,500-liter reservoirs. The silver amalgamated with the mercury and after several other treatments the mercury was evaporated off to leave a pure silver. It was an efficient process that greatly increased production. Production benefits involved the outsourcing of its costs. Mercury-based amalgamation involved the outrageous exploitation of people and environments. Appalling working conditions were only possible because indigenous peoples were forced to work for the Spanish crown under a system known as *mita.* Some 15,000 miners were conscripted into permanent, forced labor at Potosí. After a brief decline in the middle of the seventeenth century, mining production at Potosí grew to new heights during the eighteenth century. Between 1720 and 1820, 80 metric tons of refined silver were produced annually.[12]

As was the case at Potosí, mining in New Spain (Mexico) also depended on various practices to separate the valuable metals from other ores. Here, though, lead was also used. The extracted minerals were fused with lead monoxide (litharge), through a process called liquation, before the amalgamation process. Between 1600 and 1800, more than 300 mines were exploited. Production involved significant financial investment, a flawless management of mercury supplies, and forced labor. Between 1700 and 1750, New Spain produced twice as much silver as Peru, and three times as much by the end of the eighteenth century. A total of 40,000 metric tons were exported from America by the Spanish during the eighteenth century.[13]

Gold was also found and extracted in various regions across the Americas, mainly in Brazil from the eighteenth century onward. In 1692, deposits were discovered at Minas Gerais, and from 1720 until independence in 1822, 10 metric tons of fine gold were mined each year. This accounted for almost all gold production worldwide, even if illicit and contraband activities make such numbers underestimates. As with silver, amalgamation was necessary to separate the metal from its mineral, and mercury was the amalgam of choice.[14]

Inasmuch as gold and silver were the most coveted prize, mercury's importance to the extraction of the more valuable minerals should not be overlooked. In a quirk of historical chance, Spain already had access to ample mercury supplies at Almadén ("the mine" in Arabic), which contained the largest reserve of cinnabar (mercury ore) in the world, with a particularly high metal content of as much as 20 percent. To stress the hazards associated with working in the mercury mines, Almadén's workforce consisted of convicts and those condemned to death. Roughly 65,000 metric tons of mercury were exported to the New World between 1558 and 1816 (47,000 metric tons in the eighteenth century alone). In 1564, another cinnabar mine was discovered at Huancavelica, in Peru. It was manned with 3,000 miners under the *mita* system, and produced 62,000 metric tons between 1571 and 1812. These two mines, under the strict governance of the crown, supplied almost all of the world's mercury.[15]

After 1800, revolutionary wars, followed by Bolivarian wars of independence, put a stop to the mercury supply and led to a downturn in mining. Up to this point, however, mercury use constituted one of the most critical sources of pollution worldwide. First, cinnabar's extraction affected the miners, who were chronically poisoned and suffered from disorders of the nervous system. Observers at Huancavelica and Potosí likened these mines to "human abattoirs," and added that working there was essentially a death sentence. Hence the need for forced labor. Two thirds of those mining mercury died as a direct result of their work.[16] As for environmental damage, estimates suggest that 17,000 metric tons of highly toxic mercury vapors were emitted into the atmosphere at Huancavelica between 1565 and 1810. An additional 39,000 metric tons were discharged at Potosí during the same period. A total of 196,000 metric tons were released in Latin America between 1500 and 1800; 25 percent of all mercury vapors emanated from these two sites.[17] For every 1 kilogram of silver produced, between 1 and 1.5 kilograms of mercury were effectively lost through evaporation in the *patio* method. Another significant amount was released into waterways during the cleaning and mixing stages. It would then convert to methylmercury and work its way up the food chain. Ecotoxicology studies show that the remains of this pollution extend for several hundred kilometers around these mining sites, both downstream and in the air.[18]

The entire mining system in the New World depended on exploiting new frontiers, capitalist investment, state involvement, forced labor, and exorbitant levels of pollution. Because mining knowledge, techniques, and capital all circulated, it should come as no surprise that other mining centers followed a similar trajectory. In Sweden, Stora Kopparberg, the "mountain of copper," located in the boggy, wooded middle of the country, accounted for two thirds of Europe's copper supplies in the seventeenth century. The metal was extracted from its ore through a process called fusion. The ore contained a lot of sulfur, resulting in the emission of sulfurous smoke during fusion. After the mine's copper production peaked in 1650, when it produced 3,000 metric tons of copper—or 20 percent of the world's copper that year—its yield declined during the eighteenth century.[19] In Japan, mining at Ashio and Besshi, situated in uninhabited, mountainous regions, produced more than 1,500 metric tons of copper annually in the seventeenth century, during which they dominated copper mining in Asia. As in Sweden, though, the Japanese mines experienced a decline in the eighteenth century. The ores contained sulfurous pyrites, which polluted air and water with sulfur dioxide to such an extent that authorities exempted local victims from taxes in 1702.[20]

These declines were in part due to the growth of mines in Russia, China, and Great Britain. Peter the Great's embrace of relations with Western Europe ensured a vast market for Russian iron, copper, and silver after 1697. In order to operate the large Nerchinsk mining center in the Altai Mountains in Siberia, massive quantities of refined lead were transported from the Ukraine, 5,000 kilometers away. In China, a similar logic of pioneering frontiers enabled copper and silver mines at the periphery to prosper. Thus, while silver mines at the center of the Qing Empire were closed as the result of an imperial decree in 1435, a systematic exploitation of ores began in 1740 using the liquation and amalgam method, which linked it to the vast mercury mines in the neighboring province of Guizhou. Under the control of the military, production—difficult to quantify—had a similar environmental impact as those mines in the New World.[21]

Great Britain's story is somewhat different. After 1750, Great Britain became a great mining power, but its wealth did not derive from extracting precious metals. From the sixteenth century, under Elizabeth I, the state played a fundamental role in encouraging the introduction of German mining techniques.[22] In 1569, a royal decree relating to the mines and smelters in Keswick (Lake District, in the north of England) granted the private operator of a mine the income from his work. The subsequent boom in British mining could be explained in part by this combination of liberal exploitation, encouragement, and state management.[23] Whatever the reason, Great Britain dominated the tin mining and refining industry, carried out predominantly in Cornwall and Devon in the eighteenth century. Thanks to its expertise with explosives and mine shaft water pumps, it also led the way in worldwide copper production, the majority of which was refined at Swansea, in Wales. After 1778, copper became the most durable option for constructing the fairing of ships. Britain's was the most powerful fleet in the world on the high seas. Demand for copper continued to soar to ensure its supremacy. Production increased from 1,000 metric tons per year in 1730, to 2,500 metric tons during the 1760s, and again to 4,000 metric tons by the end of the century. Of course this rise in production meant that complaints about air pollution also increased, especially around urban smelting plants. Several common law trials ordered these factories to relocate outside of city limits. This occurred in Liverpool in 1770 and in Wales in 1796 and 1809. Britain's impressive power in the metal industry during the nineteenth century was already forged under a cloud of smoke.[24]

In general, every mine—even the smallest—in some way damaged the environment around it. This was the case at the lead and silver mines at Huelgoat and Poullaouen in Brittany, where the waterways had been polluted (probably with arsenic and sulfur). Harvests in the surrounding areas had been destroyed and after a lengthy trial between 1773 and 1776, the judge ruled that the farmers be compensated for their lost crops.[25] In Norway, a traveler in 1799 described a similar dilapidated landscape around the copper mine at Roros:

> The furnaces are near the city. One can see the mounds of ore which are to be burned and which, upon combustion, produces a quantity of very fine sulfur, but nobody bothers to collect it. It covers all the heaps that smoulder; the vapors frequently descend towards the city and fill the streets, which affects the respiration of its inhabitants. On passing by these piles on horseback, we found that these emissions were no less overwhelming than those released from Vesuvius' crater after an eruption. A doctor from the city confirmed that these gases produced the most pernicious effects. The mine's manager explained that the sulfur is not collected because it would be too costly to do so.[26]

"Dark Satanic Mills" and the British Enlightenment

The mine operators of the early modern period flung themselves wholeheartedly behind the new spirit of capitalism that swept Europe. The exploitation of nature driven by market demands, the intensification of human labor, the concentration of capital, the role of the state, imperial domination, and the ability to turn a blind eye to the deleterious effects of production all took place underground and in the centers that financed mineral extraction. From the 1750s, industrialization became an integral part of this new world that championed economic autonomy over other social phenomena. What Karl Polanyi called the "great transformation" destroyed a number of forms of social organization and justified—for many sectors—economic progress, even though it benefited only a limited few.[27] This new spirit took hold primarily in Europe, whose population grew from 140 million to 190 million between 1750 and 1800.

An extensive amount of literature has been written about the first Industrial Revolution. Historians continually discussed the intensification of work, how demand influenced production, technology, economic institutions, and ecological restraints. They also considered how ideology and culture influenced the practical aim of the British "Industrial Enlightenment," which groups like the Quakers embodied perfectly. The transition from proto-industrialization, where agrarian workers were hired to work as part of the production line, to urban-based manufacturing was put into perspective. On reading this research, it is clear that debates on the exact reasons that led to industrialization and the rate at which it developed have never been definitively settled. However, even if many theories exist surrounding the causes for industrial development, there is a general consensus that the birth of the industrial world took place in Great Britain between 1750 and 1830.[28] It is also worth noting that no histories of the road to industrialization put pollutions at the center of their research or narrative. In fact, the majority of them completely ignore the issue. It is as though industrialization was normalized along with its sources of pollution.[29]

The two great industrial sectors were textiles and metallurgy, and each produced its fair share of pollutants. The cotton industry was dominated by Great Britain and its empire, and production rose tenfold between 1770 and 1790, and again between 1790 and 1802.[30] Such dazzling progress ran concomitantly with significant technical innovations in spinning and weaving, which resulted in intense capital investment, larger factories driven by hydraulic mills, and soon, steam engines: all this ushered in the factory system. At the end of the eighteenth century, Great Britain boasted 700 spinning mills. While many employed fewer than fifty laborers, others operated on a much grander scale. One example: Robert Owen's textile mill at New Lanark, near Glasgow, employed 1,700 workers in 1816. By 1821 in Manchester, eleven spinning mills were seven stories high, most of them run on steam.[31] The mechanization of the textile industry developed alongside other industrial sectors, and created a demand for metal, notably iron and copper. Throughout the eighteenth century, industrial iron production depended on the quality of the available combustible fuel. Ores had to be melted in blast furnaces along with the burning fuel, which eliminated coal as a combustion option because it contained too much sulfur. In 1710, however, Abraham Darby

Sr. succeeded in removing the sulfur content from coal to produce coke. He did this at his forge in Coalbrookdale, near Birmingham, and overnight enhanced the potential of all blast furnaces. His son and grandson, Abraham Darby II and III, further refined the technique of the coke-making process, which generated a marked improvement in the quality of British iron production. As a result, from 1760 onward wooden rails in mines were replaced with steel rails, and in 1779 the first metal bridge in the world was built across the Severn, fittingly near the Coalbrookdale factory. Consequently, in 1784, Henry Cort invented a process called puddling, in which the cast iron is decarburized (carbon is removed) in a reverberatory furnace. From that point on, iron was a viable commodity, frequently replacing copper as the material of choice. The amount of iron produced in Great Britain doubled between 1786 and 1796. By 1820 British iron production had reached 400,000 metric tons annually.

Whatever the progress, advances in the metallurgical and textile industries provoked a sense of shock, because of the excess waste they produced. In 1776 the well-known agronomist, Arthur Young, visited Coalbrookdale, where he described the surrounding countryside as too beautiful

> to be much in unison with that variety of horrors art has spread at the bottom: the noise of the forges, mills etc. with all their vast machinery, the flames bursting from the furnaces with the burning of the coal and the smoak of the lime kilns, are altogether sublime and would unite well with craggy and bare rocks, like St. Vincent's at Bristol.[32]

Two years later the painter George Robertson visited Coalbrookdale and depicted the same contrast between the sublime and the horrific in six paintings. John Wilkinson's cannon factory—abutting Abraham Darby's, his former associate's ironworks—served as horrific archetype, spewing its infernal smoke and masking the sky. One of Robertson's paintings was a rendering of the factory's interior: the whole atmosphere was lugubrious, giving an apocalyptic vision of the industry. Philippe-Jacques de Loutherbourg's 1801 painting of the coking plant, forge, and steelworks at Coalbrookdale offered an even bleaker portrayal of the industrial landscape. Once again, the air appeared filled with warm colors and smoke fumes: simultaneously sublime and infernal.[33]

Such depictions gave life to the shock of industrial depredation, the terror of which had been discussed by the first generation of romantic writers. In her 1785 poem "Colebrook Dale," Anna Seward denounced the

tribes fuliginous [that] invade
The soft, romantic, consecrated scenes;
... Through thy coy dales; while red the countless fires,
With umber'd flames, bicker on all thy hills,
Dark'ning the Summer's sun with columns large
Of thick, sulfureous smoke, which spread, like palls,
That screen the dead, upon the sylvan robe
Of thy aspiring rocks;[34]

The next part of the poem offers an interesting semantic play on "pollute," where Seward gives the term a double meaning. On the one hand, she invokes sexual connotation, with the rape of the countryside. On the other, she observes environmental change in "pollute thy gales, and stain thy glassy waters." It warrants noting that Seward's poem likely marks the first express link between pollution and the new excretions of the industrial world.[35] William Wordsworth denounced the new industrial system in much the same vein in his collection of poems *The World Is Too Much with Us* (1807).[36] He lamented the loss of the traditional English countryside, consumed by industry's voracious appetite. William Blake's "dark satanic mills," from his poem *Jerusalem* (1804), resonated deeply. It was inspired by the first steam-powered grain mills, which appeared near his home in London in 1785.[37]

Coal-Fired Power: Stresses on the Environment

Coal and its black plumes of smoke over England became an entrenched, iconic symbol of the Industrial Revolution. This sedimentary rock, resulting from the degradation over geological time of the organic matter of plants, is known by many names depending on its carbon content. Peat contains less than 50 percent carbon, lignite around 75 percent, and industrial coal as much as 90 percent. The combustion of these various fuels produced substances that are more or less toxic. Carbon dioxide, sulfur oxide, and nitrogen oxide—all of which acidify the air—soot, elemental particulates such as cadmium, arsenic, mercury, bitumen, and other volatile organic compounds spread far and wide across the physical environment. Historically speaking, coal was exploited most in Asia, most notably in China, where its use was mastered in metallurgy in the eleventh century. However, in 1078 an imperial order banned all mining activities because mining in and of itself conflicted with Confucianist ethical beliefs. This ban combined, with the Mongolian invasions, put a halt to further advancements in the mining industry. Although coal was still mined in the early modern era, its quantities were limited.[38] In Japan, the demand for coal only developed in the eighteenth century, and even then its consequences instilled fear.[39]

For a long time, coal use remained relatively rare in Europe as well, confined to specific locations like Great Britain and the Liège Basin in Belgium. In Liège and in the vicinity of its zinc and copper mines, the alloy of which produced brass, outcropping coal deposits helped the region gain pre-eminence in military production and in a number of other metallurgical sectors during the Renaissance. Coal was also used in Holland starting in the seventeenth century, where it was burned with peat, which also emitted highly polluting particulates.[40] But coal was most widely used in England, where Newcastle-upon-Tyne was the principal producer until the middle of the eighteenth century. Where some 30,000 metric tons were extracted annually around 1570, that quantity had increased to 2,000,000 metric tons by 1800. The Tyne estuary was constantly filled with coal ships—inspiration for the term "sea coal"—which supplied the combustible rock to the Continent for use in industry, or to London, where it was mainly used for heating. Blacksmiths and locksmiths were, for some time, the only artisans who used coal. But furnace innovations expanded industrial possibilities. The reverberatory furnace was one such development. Invented in Wales at the end of the seventeenth century, it isolated the combustible fuel and the sulfur produced from the refined metal, to produce much purer forms of lead, copper, and tin. In 1700, one third of English coal was burned in industry, in lime factories, in salt manufacturing (which consumed 7 percent to 10 percent of all coal production), and beer breweries. In the eighteenth century, wood scarcity and its consequent higher cost made the exploitation of coal deposits even more profitable.[41] New extraction methods yielded greater quantities which, in turn, encouraged greater industrial applications: in glass, soap, and tile factories; brickworks; the treatment of alum; sugar refineries; and in specific textile operations (washing, drying, dyeing, etc.). Only the iron industry resisted coal, until the coking and puddling processes were developed; from that point it held a virtual monopoly as a combustion source in British industrial manufacturing. Increases in its production and consumption were considerable. The amount of coal consumed by the iron industry rose from 21,000 metric tons in 1750 to 5.5 million in 1830. Coal became the dominant fuel in Great Britain, as well as in a few pockets on the Continent. Its consumption continued to rise, from 2.7 million metric tons in 1700 to 5.2 million in 1750, to a staggering 15 million metric tons in 1800. Put into perspective, coal production increased 500 percent between 1750 and 1830.[42] This boom paved the way for British expansion, established the "great divergence" between Europe and Asia after 1800, and ushered in the age of fossil fuels.[43]

This progress also prompted coal burning to power steam furnaces. Based on Robert Boyle's scientific experiments on atmospheric pressure in England in the 1660s, and Denis Papin's theoretical work on pistons in France around 1680, the Englishman Thomas Savery put theory into practice by adapting their principles to enhance mining in 1697. Savery invented the steam pump, which was used to extract water from mines. Inefficient and dangerous, it was improved upon by Thomas Newcomen in 1712. His atmospheric engine proved popular: 300 were produced between 1733 and 1773. Newcomen's noisy pumps were contained in engine houses that dotted the landscape in Devon and Cornwall.[44] They were instantly recognizable by their signature smoke plumes. Newcomen's engine was made even more efficient by James Watt. In 1769, Watt separated the condenser from the furnace, which enhanced the power of the machine in wet mines. Working with Matthew Boulton in the Soho foundry in Birmingham, Watt found ways to increase power and efficiency in Newcomen's design before installing his first steam engine in the Wilkinson forge near Coalbrookdale in 1776. It was an instant success. Soon after Watt and Boulton filed their patent, their engines were supplanting Newcomen's pumps. By the end of the eighteenth century, more than 1,000 steam engines were in operation in the British Isles. In 1830 they began to replace both hydraulic wheels and horse power in the textile industry. Three quarters of the cotton produced in Great Britain was produced in steam-powered spinning mills.[45] The steam engine radically transformed transportation as well. Between 1802 and 1814, Richard Trevithick, in his mines in Wales, and George Stephenson at Killingworth experimented with steam locomotives running on rails. In 1830 the *Rocket* transported passengers between Manchester and Liverpool. It was the first railway to rely exclusively on steam power. In this manner, the railway age followed hard on the heels of the fossil fuel renaissance.

The combination of coal and steam freed Britain of the energy limitations imposed by biomass. Prior to 1830, renewable energy sources still dominated, even in Great Britain, and one could argue that rivers and hydropower had forged the path to the Industrial Revolution.[46] In hindsight, the steam engine's impact is nothing short of startling, but its growth masked the still minor role it played in many industrial sectors and the widespread resistance that rose up against its expansion. Nevertheless, coal-fired power, added to older forms of energy production, ensured an ever-growing industrial sector never dimmed for want of fuel. Although Watt and Boulton received only 110 orders from abroad between 1775 and 1825, there were nearly 60 steam engines in Belgium in 1790 (most of them for mining operations). In 1800, between 60 and 70 were in operation in France; this number rose to 625 by 1830. In the United States, although many engines were fueled by wood before 1820, the new nation could boast 500 in 1825, then 3,010 in 1838.[47] In other words, the steam engine's rise to glory in the 1830s coincided with the ascension of the railroad and the advent of big industry.[48]

As a new energy source, coal was flexible, reliable, independent of natural conditions, and applicable to many industries. But its emergence also upset numerous social and economic norms, and quickly provoked complaint. Pollution from mining meant that multiple court cases followed the opening of the first coal mines. Such was the case in Newcastle, in the first half of the seventeenth century, where 80 hectares of land had been devastated, the air polluted with sulfur dioxide and nitrogen oxide, and the waterways contaminated with heavy metals.[49] The emissions provoked fear. In smoke-filled London, the epidemiologist John Graunt drew a correlation between rising death rates and declining air quality.[50] In France, dense fumes—opaque, sulfurous, and bituminous—combined with foul odors to breed unsettled anxieties. Outbreaks of consumption in London and the Liège region were attributed to coal burning. Paris imposed a general ban on the burning of coal; only blacksmiths were exempted. In 1778, when a steam engine was built for the purpose of pumping water from the Seine, the king ordered that it be installed as far away as possible from the Tuileries Palace after the faculty of medicine warned of the sulfurous gases that it would emit.[51] In Japan, coal smoke was also considered noxious and a threat to human and cultural health. There were protests around the coal mines in Kyushu.[52]

The rise of coal as a source of fuel was contentious, but acclimatizing to the new reality was inevitable. During the 1780s, wood shortages in Austria and the region around Charleroi in Belgium—as in Paris and Barcelona—enabled the British coal boom to establish a beachhead across the English Channel and spread its tendrils throughout the Continent. Governments offered incentives to industry to use coal, with exemptions and special authorizations being afforded to those who complied. Many entrepreneurs falsely promoted coal as a benign energy source simply by ignoring medical warnings and asserting its safety.[53] To render coal burning more tolerable involved refining it into coke—as the British had learned and done—but this refinement was delayed in continental Europe until after 1820. As a result, coal's acceptance came only after a slow process of acculturation that met with frequent opposition.[54] Home to the oldest and most extensive coal-fired industries, Great Britain also suffered from the most polluted air. Black, smoky, and sooty Staffordshire received the dubious distinction of being nicknamed the *Black County* in the 1820s. Georg May described the county in apocalyptic terms: "The fields are covered with soot and the air polluted with smoke with little being able to grow here.... It is difficult to find a green leaf on any tree or bush. Some parts of the district look as though they have been laid waste by a destructive fire."[55] In 1828 the young writer and follower of Saint-Simonism, Gustave d'Eichthal, commented in Newcastle that "the sky is darkened from the smoke of the steam engines and by the coal dust, burned on tiles in the mines to eliminate it."[56] These plumes of smoke were the signature feature of industrial Great Britain.[57]

Grease in the Wheels

Industrial development also generated numerous byproducts; it multiplied traditional chemical operations and resulted in the advent of a more concentrated and expansive chemical industry. In general, organic products were replaced with more polluting mineral substances that had the benefit of rendering greater efficiency in industrial production. For example, gentler organic acids, previously used widely in tanning and dyeing, could not compete with nitric and sulfuric acids.[58]

A rise in demand drove this transition and intensified production. Traditional crafts increasingly adopted more and more powerful chemicals in their operations to accommodate this pressure. After 1750, crafts such as pottery, porcelain, wallpaper, gilding, and hatmaking acquired a more industrial dimension by using chemicals that had been previously prohibited. For example, commercial pressures dismissed concerns about using mercury nitrate in the drying of rabbit and hare skins for the purpose of making felt hats. While mercury use remained forbidden in France, competition with British hatmakers meant that French milliners circumvented these restrictions by employing mercury as part of a "secret" composition.[59] Mercury was also used in other craft workshops such as glass and window silvering, and metal gilding. As an obvious consequence, the quantity of mercurial gases emitted into the air increased markedly between 1780 and 1830 wherever this work was carried out.[60] The high rate of illness amongst watch and jewelry makers in Geneva led the city's Society of Arts to investigate if the cause was related to the amalgams being used in these crafts.[61] By the end of the eighteenth century, toxic minerals had found their way into pigments used to decorate porcelain, earthenware, and wallpaper. Reds were made with cinnabar (composed of sulfur and mercury), vermillion (from mercury) and minium (lead oxide); greens from verdigris (copper oxide); yellows were drawn from antimony or massicot (another lead oxide), or orpin or orpiment (from arsenic sulfide); while most whites came from ceruse (a carbonate of lead), bismuth, or tin. These colors were prepared for use first by crushing the metal composites before either adding acid to them or heating them over intense heat, which produced harmful fumes. The grinding processes produced a fine powder, which caused often lethal occupational diseases. During the 1780s, various learned societies took an interest in this sudden rise in illnesses among these workers. The French Academy of Sciences launched a competition to investigate the "insalubrious arts," which included metal gilding, hatmaking, tinsmithing, the grinding of colors, and tank and pipe cleaning. Studies were also initiated by the Society for the Encouragement of National Industry (SEIN) after 1803, as well as by the Institut de France after 1819.[62]

The most interesting transformation took place within the chemical industry itself. Alum is the obvious choice to demonstrate how the mining and chemical industries amalgamated from 1750 onward. Alum was used in metallurgy but was primarily used as a mordant for colors in the tanning and dyeing industries. It is a compound of aluminum sulfate and one or more other minerals found in veins of alunite or schists of pyrite. Alum was used worldwide, but up until the beginning of the nineteenth century, the Papal States dominated its production by controlling the Tolfa mine in Italy.[63] England also laid claim to deposits rich in alumina on the Yorkshire coast, which were mined heavily by order of Henry VIII after his break from the Vatican in the sixteenth century. The resulting industrial complex in the region, which became the birthplace of the English chemical industry, devastated the coastline over the course of the eighteenth century.[64] In a similar vein—and with comparable environmental consequences—Sweden concentrated alum's extraction industry in the southeast of the country. Alum production involved heating the mined rock in huge ovens to up to 600° Celsius, sometimes for months on end. Other operations such as maceration, washing, concentration, and crystallization emitted huge amounts of smoke that bothered the locals living nearby.[65] When schists of pyrite were involved, urine was employed in great measure (around 200 metric tons per year in Yorkshire in the eighteenth century), before being replaced by ammonia (a byproduct in the distillation of coal) after 1820. Because pyrite contains both sulfur and sulfuric acid, unwanted sulfates emerged from the process. Its acidic gases became a major source of pollution in nearby rivers. In Swedish factories, schists containing large amounts of coal were also used as a source of combustible fuel after 1810. They contained, however, a lot of sulfur and 5.8 milligrams of cadmium per kilogram. Almost 5,000 kilograms of cadmium were released into the environment between 1726 and 1840, traces of which can still be detected today.[66]

Alum and the pollutions it produced mark the birth of the chemical industry. During the 1770s, the French chemist Jean-Antoine Chaptal, building on the research of his German counterpart Andreas Marggraf, found a way to produce synthetic alum by mixing sulfuric acid with clay and potash. It was first produced in Paris in the Javel factory beginning in 1780, and subsequently in the Chaptal factory in Paris after 1797. At the end of the eighteenth century, France was importing 1,410 metric tons of alum and 970 metric tons of sulfates; by 1830, however, France was producing 7,175 metric tons of synthetic alum and 1,414 metric tons of sulfate, enough to make it self-sufficient on both counts.[67] At the beginning of the nineteenth century, synthetic alum was also produced in Yorkshire, most notably in Ravenscar, a village completely transformed by the new industry.

In the main, the chemical industry structured itself around the production of acids. They were necessary components in a great number of metallurgical operations, such as refinement in coin making, scouring, the scraping and cleaning of minerals, and so forth. Acids were also required in the textile industry, from bleaching to acid baths to aid in fixing colors on cotton.[68] Which is to say that acids became imperative— indeed, defining—contributors to industrialization. The intensity of their concentration was also increasingly important; countries that could produce stronger acids had an undeniable economic advantage.[69] These stronger acids were in stark contrast to the weaker, often organic acids that had been in use up until 1750. The more corrosive the acid, the faster it worked. Speed fueled market efficiencies, but at a cost. The harsher the acid, the greater damage it imposed on the environment and human health. In the eighteenth century, three acids were of particular importance: nitric acid (aqua fortis), sulfuric acid (vitriol oil), and hydrochloric acid (muriatic). Up until the 1770s, nitric acid dominated industrial operations, and because it was difficult to transport it was manufactured throughout Europe. Its manufacture was so polluting, however, that it was made in outlying suburbs, and only where demand was high. The chemical reaction of distilling clay mixed with saltpeter over intense fires took hours in sandstone pots, which were prone to cracking. The distillation process emitted thick clouds of gas throughout the neighborhood.[70]

In the first half of the eighteenth century, sulfuric acid was obtained from the combustion of sulfur or pyritic ores, and the resultant condensation of gas in glass globes. The process was costly and the equipment fragile. All of this changed in 1746, when John Roebuck from Scotland, an associate of Watt's in several industrial operations, succeeded in condensing the gas inside boxes lined with lead. These "lead chambers" meant that sulfuric acid could be produced in much greater quantities than before, and greatly increased its value and importance. The sulfuric acid produced after concentration was exceptionally strong, at 65 percent proof. This great innovation was made possible because of recent advancements in Great Britain in lead's use as a liner. At the same time, these lead chambers were rarely airtight and thus posed a threat to surrounding areas.[71]

Finally, hydrochloric acid was the least well understood of the three acids, before 1830. It was also the residual byproduct of one of history's most polluting processes: the manufacture of artificial or caustic soda (sodium hydroxide). Pivotal to soap making, where it was combined with fats, and in glassmaking, where it lowered the point of fusion of quartz glass, artificial soda underwent production mutations similar to those of alum. The similarities consisted of the transformation from a plant-based soda, derived from burning algae or saltwater plants, into an "artificial" or synthetic soda through a process of double decomposition of sea salt with sulfuric acid. Named after its inventor, the Frenchman Nicolas Leblanc, this Leblanc process was created in 1790 and used throughout the nineteenth century. The harmful effects of caustic soda's production were nevertheless quickly experienced in France, until 1825 the only country engaged in its manufacture. Quantity and severity of the pollution mixed in equal parts throughout the surrounding landscape: for each unit of caustic soda produced, a quarter unit of the hydrochloric acid used in its production was discharged into the atmosphere. Before 1830, little was done to recoup or repurpose this waste. All around Paris and Marseille, the two major production centers in the world, acidic vapors destroyed crops for a radius of several kilometers.[72] Hydrochloric acid was also used in producing chlorine, whose bleaching qualities were brought to the fore by the French chemist Claude-Louis Berthollet in 1785. Thanks to the effectiveness of Berthollet's methods, chlorine was adopted mainly in the bleaching or discoloring of textiles, wool, and cotton, as well as paper. In Paris, potassium hydroxide was added to the mixture in order to reduce the corrosive effects derived from the production of Javel acids. The result was the famous *eau de Javel* (Javel bleach), first produced in 1788. Consequently, France pioneered the way in manufacturing bleach, and British industrialists looked to France for guidance.[73]

In order to round out this picture of the birth of the chemical industry, some discussion of sal ammoniac (ammonium chloride), used in dyeing and metallurgy, is warranted. It was formerly imported from Egypt, but Britain and France became self-sufficient at the end of the eighteenth century. Sal ammoniac was obtained by the distillation of bones and rags with calcium sulfate. It produced some of the most intolerable odors of the new chemical age. Its pyroligneous residues infiltrated soils and contaminated water tables. In France, Antoine Baumé launched his sal ammoniac production (1766–1787) before another chemist, Jean-Baptiste Payen, founded a large factory in 1797 on the plain of Grenelle near Paris. For a century, Payen's factory was one of the world's major black marks in industrial chemistry: a standalone warning against the transgressions of the new industry.[74] Regardless, new industries fed each other and became entrenched by their necessity. Demand for sal ammoniac grew with the emergence of lighting gas production from coal distillation. The distillation of coal to extract the flammable gas produced a lot of highly polluting residues: not only coke of poor quality and ammonia, but also bitumen, tar, sulfur, volatile compounds, and heavy metals escaped into the environment. The technology, anchored in the interplay between scientific chemistry and coal energy, was acquired simultaneously in England and France. The first plants were established in 1812 and 1817 respectively. Immediately after the factories' installation, toxic residues—smoke, unpleasant odors, contamination of wells and groundwater—became the subject of recycling attempts, but most efforts had the effect of simply shifting the problem elsewhere rather than resolving it.[75] But again, industry was becoming entrenched. Because the supply of lighting gas can only be conducted through expensive pipelines, it became an exemplary sector, despite its multiple pollutions, that reflected the progressive insertion of industry into the city during the 1800s.

Making the Industrial City

Inasmuch as early industry occurred in rural areas—and prior to 1830, biomass and waterways were the two principal forms of power for industry—a new kind of city gradually appeared on the landscape. New, smoke-filled cities—designed specifically for industry—imposed themselves in northwest Europe. However, it was not until the middle of the nineteenth century that a veritable growth in these cities became noteworthy. In the United States, for example, despite the spread of steam power, the majority of factories continued to use hydraulic energy, conforming to the pastoral vision of production. Before 1840, industry remained fundamentally rural; the first great American complex, at Lowell in Massachusetts, relied wholly on hydropower in 1833.[76] The scene was quite different in Britain. During the 1780s, Anna Seward observed the growing problem of urban pollution in her poem, "Colebrook Dale":

> Grim Wolverhampton lights her smouldering fires,
> And Sheffield smoke-involv'd; dim where she
> stands
> Circled by lofty mountains, which condense
> Her dark and spiral wreaths to drizzling rains,
> Frequent and sullied; as the neighbouring hills
> Ope their deep veins, and feed her cavern'd flames;
> While to her dusky sister, Ketley yields,
> From her long-desolate, and livid breast,
> The ponderous metal.[77]

London had adopted coal in the seventeenth century and was without a doubt the most polluted city on the planet, due in large part to household use. In 1680, an anonymous author wrote a pamphlet in which he deplored the smoke, employing the words "polluted" and "polluting." While the reference was clearly directed at the city's air quality, it also implied connotations of the "stained and polluted" moral values and ideals of the elite who profited from the creation of such filth.[78] During the eighteenth century, London attracted more diverse workshops and factories to its surrounding neighborhoods.[79] It thereby tripled its use of coal between 1700 and 1800 and became known as the "smoky city."[80] On a visit to London in 1739, the historian and geographer Thomas Salmon commented that London "produces so much filth and so many disgusting vapors from artisanal furnaces that it is impossible that this air is actually breathable." He also identified numerous respiratory illnesses suffered by Londoners.[81] After 1750, there was an increase in the number of trials pertaining to the contamination of the air by factories. In 1772, John Evelyn's *Fumifugium* (1660) was republished, with the editor emphasizing the fact that there had been a sizable rise in the number of factories since its first publication. In addition to the preindustrial polluters who had sullied the landscapes Evelyn wrote about a century earlier—the brewers, the dyers, and the soap makers—glass works, metal foundries, sugar refineries, and fire pumps all along the Thames added to the litany and severity of environmental contamination afflicting London. Pollution was bad enough that the city's elite migrated to the West End, upstream, abandoning the dirtier industrial zones to the south and the east near the port.[82] Retrospective analysis of the atmospheric air quality in London revealed very high levels of pollutants. Levels of sulfur dioxide rose from 20 milligrams per cubic meter (mg/m^3) in 1575, to as much as 40 mg/m^3 in 1625, 120 mg/m^3 in 1675, 260 mg/m^3 in 1725, and 280 mg/m^3 in 1775. (The concentration in London between 2000 and 2010 was 2–5 mg/m^3, while current Chinese levels hover around 60 mg/m^3.) Similarly, levels of nitrogen dioxide rose from 7 mg/m^3 in 1625 to 40 mg/m^3 in the middle of the eighteenth century. (This is the current standard in Europe, already elevated due to automobile emissions.) Also, fine particle levels sat at 60 mg/m^3 in 1675 and 130 mg/m^3 in 1725. (Current maximum permitted thresholds are 40 mg/m^3, and London reached a maximum of 28 mg/m^3 in 2011.) Even if these measurements were not overly precise, there can be no doubt that the advent of coal meant London in the eighteenth century rivaled and exceeded the worst pollution levels in contemporary Asia.[83] At the end of the 1810s, tensions flared further when William Frend strongly denounced the harmful smoke which was poisoning the city. In *The Pamphleteer*, Frend compared factories to volcanoes, and lamented the metropolis's shift in priorities away from common sense and avoiding harm.[84] By 1819, enough criticism of London's air quality led to the formation of the first parliamentary commission whose job it was to attempt to tackle the hazard.

Manchester emerged as the prototypical British industrial city. It differed from London inasmuch as its growth depended entirely on the textile industry. In 1745, its population was only 17,000, but by 1816 it had grown to 100,000. Hundreds of spinning mills sprang up in rapid succession along the Irwell River and its tributaries. In 1782, the first steam engine was installed in one of the mills, and by 1794 four mills were running on steam. This move away from hydraulic power meant that factories moved to the city, which was quickly overcome with smoke and other air pollutants. The scientist Thomas Percival published a report denouncing the rise in air pollution, which led authorities to intervene. In 1816, 82 spinning mills powered by steam consumed 50,000 metric tons of coal. This did not take into account mechanical workshops, foundries, sugar refineries, or chemical producers.[85] On a visit to Manchester in 1835, Alexis de Tocqueville gave this eloquent description of the city:

Surrounding this asylum of misery, its fetid and muddied rivers ... run slowly, tinted myriad colours by local industry.... It is the Styx of this new hell.... Look up, look around and see the immense palaces of industry. Can you hear the furnaces and the whistling of the steam? These vast buildings prevent air and light from reaching human dwellings, over which they tower and which they surround in a perpetual fog.... A thick and black smoke covers the city. The sun looks like a disc without rays.... It is in the middle of this cesspit that the grandest river in human industry is sourced and will fertilize the world. From this filthy sewer, pure gold runneth forth. It is there where the human spirit attains perfection and goes to die. Oh, that civilisation can produce these marvels and that civilized man can be reduced to a savage.[86]

In Staffordshire, at Stoke-on-Trent, pottery took hold after 1750 to such an extent that the urban district between Manchester and Birmingham became known as *The Potteries*. By 1800 it was the largest pottery producer in the world. In order to solidify and finish the various clays, a lead glaze or enamel was applied. With more than 500 immense furnaces burning coal at more than 1200°C, combined with the use of lead in production, the area became a hotspot for lead poisoning.[87]

Comparable cases were found on the Continent in Belgium and France. In Wallonia, on the Sambre and not far from Liège, Charleroi headed down the path to industrialization in the 1730s. It had an abundance of coal to be mined, which resulted in the landscape becoming disfigured, the phreatic tables being contaminated, and plumes of smoke contaminating the air. A relatively small town in terms of population (30,000 in 1830), the number of factories in Charleroi turned it into a veritable industrialized district, where metalworking stood side by side with glass producers, soap makers, brewers, dyers, tanners, brickworks, and myriad other crafts. Authorities were thoroughly committed to the area's industrial growth, and the locals, enjoying not a little prosperity, raised few complaints. At the beginning of the nineteenth century the mayor, Barthélemy Thomas, was both a doctor and a factory owner. It was common at the time for medical commissions to agree on the benignity of industry, particularly the chemical industry.[88]

On the spectrum of cities undergoing industrialization, Paris offered a very different case. Before 1770, Paris was a densely populated "enlightened capital" that fostered a political, cultural, financial, religious, and commercial culture, rather than an industrial one. But in the eighteenth century Paris "became industrious and frenetic," "an industrial hub," according to the editor of *Annales de l'industrie* in 1817 and 1823. In 1822, in a medical topography of the city, Dr. Claude Lachaise described the prevalence of the city's smoke and acidic vapors. Gone was the romantic Parisian ambience countenanced by Menuret de Chambaud as recently as 1786. At the same time, Dr. Alexandre Parent-Duchâtelet reported on the death of the Bièvre River, which had been transformed into an urban sewer. In a short amount of time dozens of factories had sprung up along its banks, reconfiguring its natural course through the city, and ultimately poisoning it with the pestilential waste they emptied into it.[89] To properly understand the manner in which Paris, and many other cities besides, devolved into such dangerous and industrial places is to investigate the political mechanisms of this transformation.

*　*　*　*

The new, dynamic economies that profited from and invested in mining, steam power, and the chemical industry, brokered an inextricable bond between modern capitalism and hitherto unseen levels of pollution. Certainly these pollutions were still mostly limited to very specific areas, and often located on the outskirts of more populated cities. They did not, could not, and increasingly stopped trying to conform to the environmental standards of preindustrial France. This foretold a new normalcy, which was only accepted slowly and grudgingly in many quarters. However, as with all pioneering advances, the headlong rush to adopt innovation lead to rethinking contemporary inhibitions and no number of safeguards could stymie the deleterious consequences of progress. Under the pressure of economic development, attitudes changed and policies shifted to respond to new nuisances. The only option was to make way for the great industrial transformation: in a period of revolutions, legal and political evolutions rendered pollutions acceptable, and even desirable.

3

The Regulatory Revolution

Amongst the new products that transformed the pollution narrative, mineral acids played a pivotal role. These inorganic acids have received little attention from historians interested in the Industrial Revolution, who focused mainly on steam energy or the mechanization of textile manufacturing.[1] Yet, if the world of industrial pollution witnessed a major upheaval around 1800, it was largely because of the radical expansion of production of these acids. Thanks to the development of the lead chamber at the end of the eighteenth century, greater volumes of acids were produced: global production of sulfuric acid increased from 75 metric tons in 1770 to 800 metric tons by 1830.[2] As it was used more and more in the textile sector and in the manufacture of chemical products, sulfuric acid became a global marker of industrial and economic development.[3] As quantities grew, so too did sulfuric acid's propensity for leaking corrosive vapors out of factories and into the surrounding environment. Just as landscapes surrounding mines, coal-consuming cities, and other preindustrial manufacturing had become impoverished, the development of industrial chemistry produced new sites of localized pollution.

Before 1830, acid production was concentrated in Great Britain and France. The first lead chamber in the United States was operational in Philadelphia by 1793, but acids were not produced in any great quantities until after 1820. The Netherlands also witnessed early flirtation with acid manufacture (1774), but its production was sporadic. Elsewhere there was no activity at all. The German principalities did not begin production until 1837, in Berlin.[4] The controversy that emerged in the 1760s over the safety and legitimacy of its production primarily took place, therefore, on either side of the English Channel. French cities, less familiar with accepting the coexistence of industrial pollution and urban living, were particularly sensitive to this debate. Over the course of several decades, opposition to the new chemical manufactories challenged the safety and

Figure 3.1
Chaptal in front of his sulfuric acid factory, Louis Bouchet, 1801. Probably posing in front of one of his sulfuric acid factories, the chemist and industrialist Chaptal embodied the conversion of scientific and political elites to industrialism. Accused of pollution both in Montpellier and in Paris, Chaptal contributed to defining—as expert and man of the government—new environmental regulation favorable to the chemical industry. The French legislation (which he helped inspire) subsequently spread across continental Europe in the nineteenth century.

legitimacy of the affront to the environment and human health, both within the factories and in the surrounding community.

These debates were exacerbated by the revolutions on both sides of the Atlantic, which constituted especially rich moments for revamping the law, the economy, and the social contract—and which accelerated political economy transformations already underway. In France, where the institutional disruptions were particularly important, relationships between society and environment were reinvented. The problem of pollution was debated with ever-sharpening acuity, and new solutions were ever more extreme: the imperial decree of 1810 on "insalubrious establishments" marked a successful and radical transformation of the law. This was the world's first law on industrial pollutions, which sealed a new pact between industry and its environment, but also contributed to accepting and entrenching pollutions as an inevitable part of the new modernity.

New Chemistry and the Birth of Public Hygiene

"Capable of causing severe illnesses." "Dangerous: incites blood in the cough spittle." "Lethal." "Suffocating." Such late eighteenth-century descriptions of the acid vapors emanating from factories came from laborers, police commissioners, and the Parlement of Paris, the sovereign judiciary court. Similarly, in 1757 the British supreme court, the King's Bench, called the gas emissions from a sulfuric acid factory in Twickenham, the London quarter where Joshua Ward established the first such factory in 1736, "noxious." Like many other workshops that created nuisances, acid production was prohibited within the city limits.[5]

Whereas fear and reticence shaped the early modern construction of laws surrounding nuisances, a new perspective emerged during the final third of the eighteenth century. The decisive turning point in Europe emerged from the chemical studies of Louis-Bernard Guyton de Morveau. When Dijon's local authorities implored him in 1773 to disinfect one of the vaults in the city's cathedrals, Guyton tested out hydrochloric acid vapors. He reused this method a few months later in a prison. These experiments were regarded as major victories against the hazards of infection and had an immense impact. For Alain Corbin, this disinfection constituted a decisive step in the cultural transformation of olfactory smells: up to this point, the battle against miasma and pestilence was fought with vinegar-based and other aromatic compositions used against putrefaction; or fire to cleanse.[6] By the following year, the physician Félix

Vicq d'Azyr prescribed these kinds of fumigations to treat epizootic diseases in the south of France, while Étienne Mignot de Montigny and Philibert Trudaine de Montigny—both members of the French Academy—recommended similar practices to arrest the spread of contagious livestock diseases. In his two *Dissertations* on the waters of the Seine (1775 and 1787), Antoine Parmentier, scientific counsel to the lieutenant-general of police for Paris, asserted that acid and alkaline vapors contributed to clean air by neutralizing the miasmas dispersed in the atmosphere.[7]

Recruited to write the "Chemistry" section of the *Encyclopédie Méthodique* (1786), Guyton praised acids, "the key of chemistry," without bothering to pause and consider their more dangerous applications or inherent risks in the laboratory.[8] In 1801, having risen to the French Academy and become a public figure in the Consulate, Guyton published his *Traité des moyens de désinfecter l'air* [Treatise on Means of Disinfecting the Air], in which he attempted to demonstrate the superiority of his disinfection methods. His colleague and fellow chemist, Jean-Antoine Chaptal, at that point Minister of the Interior, was quick to distribute the work throughout the prefectures. Subsequently referred to as "Guytonian fumigations," Guyton's acid vapor methods spread throughout Europe. In Spain, the state encouraged their adoption in the fight against yellow fever in Cadiz and Séville.[9] In Great Britain, the Scottish physician James Smyth carried out nitric acid fumigations in 1780 in response to an epidemic. The Royal Navy repeated the exercise in 1796.[10]

Controlling pollutions through disinfection, rather than preventing them outright, marked a critical feature of the chemical revolution that crested in the 1770s. Throughout Europe its pioneers, from Henry Cavendish and Joseph Priestley in Great Britain to Carl Wilhelm Scheele in Sweden, and especially Antoine-Laurent Lavoisier in France, introduced the foundations of modern chemistry with their work on combustion, air, water, and the conservation of mass. In 1789, Lavoisier wrote his seminal *Traité élémentaire de chimie* [Elementary Treatise on Chemistry], two years after proposing the new classification of elements for the Enlightenment, alongside Antoine de Fourcroy, Berthollet, and Guyton.[11] The classification formalized a new language for understanding chemistry and its components. These new chemical discoveries—not least the new method of "seeing" chemical work—encouraged greater convergence and collaboration between chemistry, medicine, and pharmacology. As a result, Chaptal, Berthollet, and Fourcroy—physicians before converting to chemistry—became captivated by the new work's potential industrial applications and developed a number of acid- and chlorine-based substances in the interest of marrying manufacturing activity and public health.

A kind of pre-public hygiene movement—integrating neo-Hippocratic medicine and the new chemistry to clean up urban spaces—accompanied this transformation in medical and scientific knowledge. Paris, "la ville des lumières," fought to eliminate putrid miasmas that emanated from its environment, be it stagnant air or water, swamps, wetlands, riverbanks, narrow streets, and densely populated spaces: theatres, prisons, ships, workshops, and hospitals.[12] In order to eliminate bad odors, a key indicator of pollution, new artificial ventilation systems and methods to dry out humid spaces were devised. But chemical treatment became increasingly common. In Paris, the chemist Alexis Cadet de Vaux promoted quicklime to disinfect garbage dumps and horse slaughterhouses. He also experimented with sulfuric acid and hydrochloric acid as disinfectants in dumps and landfills. At the beginning of the nineteenth century, Parisian authorities encouraged the establishment of caustic soda plants next to organic matter waste sites so that the hydrochloric acid could annihilate miasmas. Increasingly, chemists started working with chlorine alone as a purifying agent; however, these processes were immediately met with opposition. In one of the first chlorine bleaching attempts in the textile industry, the industrialist O'Reilly testified "the poor workers suffered cruel pain from the powerful vapors. I saw them rolling on the ground in agony; frequently even serious illnesses followed from these first exposures to muriatic acid (hydrochloric acid)."[13] By the 1820s, the chemist Antoine Labarraque sought to diminish the product's more corrosive qualities by adding lye. Labarraque managed to remove the butcher's odors by immersing the unused animal entrails in sodium chloride tubs. The great disinfection campaign that followed was not exempted from controversies surrounding chlorine's purported innocuous qualities.[14] At the minimum, however, the debate supported the notion that chlorine and chlorine-related products be manufactured under the direction of eminent chemists, even if that work should take place in the heart of the same cities that had once evicted polluters to their outskirts.[15]

As disinfection became entrenched as a public health practice, chemists emerged as the unquestioned experts in evaluating sanitary hazards, whether the risks pertained to the copper pots used in distributing milk, lead counters used by wine bottlers, the varnishes that coated kitchen utensils, the effects of cemeteries on the still-living population, and especially emissions from manufacturing. In 1778, the French Royal Society of Medicine entrusted the chemist Pierre Macquer with the task of evaluating the ambient toxicity of an antimony factory, and in 1783 Fourcroy was asked to determine whether a Rouen sulfuric acid manufacturer's activities threatened to degrade the environment and pose a health threat to nearby inhabitants. Chemical expertise tended to favor industrial expansion, and chemistry bounded the debate in ways that opponents of the new techniques—lacking facility with this new language and practice—found themselves hard-pressed to make their objections heard. When a similar new factory was proposed in Paris in 1778, authorities could argue that "it is understood in medicine and chemistry that the evaporation of sulfur, rather than being detrimental to health, is in fact a boon. It purifies dirty air. It prevents epidemics. It even fortifies the lungs."[16]

Industrial operations throughout Europe found chemists indispensable. Chemists were inextricably involved in all manner of works to compose and transform matter in the factory, and in the combustion or disposal of wastes—or in creating new products and generating value from that waste. In Prussia, they assisted in porcelain manufacture, furnace construction, and in developing new dyes and pigmentations. In St. Petersburg, investments were made in military and medical institutions where new chemical research sites were built specifically for chemists to expand chemical knowledge, working independently from traditional artisanal knowledge. In the Habsburg Empire, they acted as prime movers at the mining academy as experts in copper works.[17] Absent any formal or professional accreditation, the title "expert" was loosely attributed to any individual on whom judicial courts could call to testify under oath. Increasingly, experts were designated as anyone who could make a technical evaluation of industrial practices or on their relationship to health.[18] Chemists and chemical experts came from a class of (almost exclusively) gentlemen of means—who else could afford to experiment with matter?—most of whom were also part of a moneyed elite. Their expertise and their economic interests were rarely far apart. Macquer and Berthollet were members of the French Bureau of Commerce. And while Guyton promoted his beloved acids in his 1801 *Treatise*, he was perfectly aware of their commercial value, independent of any sanitary concerns:

> But if it is possible, without danger, to restrict the prohibitions, to shorten the duration of the sanitary trials, how can we refuse to deliver trade from these obstacles? A few hours of fumigations [of the goods], by sulfur combustion, by the emission of simple oxygenated or muriatic acid gas, according to the circumstances, will infallibly give a more secure guarantee than a month of exposure to the open air, regardless of the nature of the contagious virus. What is the cost they would incur in comparison with the benefits that would result?[19]

Removing older regulations that had protected public health against epidemics (quarantines, isolation, destruction of infected goods), the new project of "pre-hygienism" encouraged the active pursuit of commerce and wealth in multiple countries.

The Art of Governing Pollutions

Political economy and civic administration actively promoted chemistry as a means of creating a new balance of power between industrialists and their environment. More than anyone else, Lavoisier embodied this convergence: the greatest chemist of his time was also an administrator and beneficiary of the Ferme générale, in charge of the collection of taxes (1768–1791) and gunpowder commissioner (1775–1792), where he encouraged new methods of manufacturing saltpeter with artificial nitrates. As a member of the Académie des Sciences, Lavoisier also participated in drafting multiple scientific reports, some of which were related to public health; "public hygiene" featured in the subtitle to volumes 3 and 4 of his complete works, published in the 1860s. More generally, Lavoisier was regularly consulted on the establishment of new factories. Allied closely with the new liberal economy in France, Lavoisier and other physiocrats countenanced a new epistemology that situated the emergence of public hygiene at the junction of chemistry, political economy, administration, and statistics.[20]

From the end of the seventeenth century, the emerging political economy was concerned with the development of resources and the well-being of the population. When the English economist William Petty formulated his "political arithmetic," he was dependent upon John Graunt's demographic and epidemiological studies. At the end of the eighteenth century, liberalism and the rise of statistics were the dominant lenses through which public health was understood. Social amelioration and economic benefit were concomitant goals in the new art of government. According to large surveys on population and wealth in France conducted by Turgot, Jean-Baptiste Moheau, Louis Messance, and Lavoisier, social and economic well-being were mutually constitutive.[21] The transition was even more notable in Great Britain. After 1760, doctors such as Thomas Short distanced themselves from Graunt's correlations between dirty air and mortality. Instead, they promoted the idea that morality was a more crucial indicator of mortality than stale air. More generally, with Adam Smith, who published *The Wealth of Nations* in 1776, and Jeremy Bentham's utilitarian thinking of the 1780s, the economic imperative began to assert itself as the most influential gauge of social progress.[22] Despite its diversity and weakness before 1830, liberalism accompanied a new theodicy of natural law and soft commerce. For the new physiocracy—this "government by nature"—as for other forms of emerging liberalism, free trade and private property ensured national progress, and the unrestricted circulation of materials established the most perfect order.[23]

This new art of governing obviously introduced major changes in the regulation of nuisances. First of all, regulation was regarded as a hindrance to the proper functioning of society. As the French trade intendant Isaac de Bacalan argued in 1768: "Regulations are almost always harmful. They interfere with the manufacturer's industry, hinder competition, stifle genius, enslave, and abase.... All regulations are absurd and harmful."[24] Critics of French regulatory control invoked the English model to demand more autonomy from systemic authority and oversight—police or judicial—and to establish a new social order.[25] The Turgot government (1774–1776) in France constituted the first large-scale attempt to free society from its privileges and routines, as well as a host of old regulations, including those dealing with pollution.

In addition, liberalism inspired a reflection on manufacturing techniques, and invited the opportunity to break down traditional artisanal practices and their institutions in the name of economic growth. The division of labor promoted by Adam Smith was bolstered by Enlightenment reason and permitted the conceptualization of a new manufacturing order that Diderot and Alembert's *Encyclopaedia* (1751–1772) illustrated by championing streamlined workspaces. Reason, efficiency, and technical proficiency were the key elements of the new manufacturing industry. In this vision of manufacturing production, work was classified and divided according to an impermeable mechanism that separated the artisanal world of labor from management driven by scientific expertise. From simple curiosity, scientific discoveries acquired practical utility.[26] Michel Foucault famously showed how much this new endeavor was at the source of a new biopolitics or an expression of power that no longer dealt only with land and territory, but also with the lives of populations, bodies, and the environment.[27]

The new industrial epistemology profited from discrediting the old regulations that were rooted in the artisanal world and by perceiving their enforcement as archaic and "corrupt" practices. Engineers and chemists contributed to the practical realization of the liberal project of remodeling the environment according to human needs under the control of the state.[28] Engineering and mining schools were created in every Western state; chemists were recruited to collaborate. Their task, for instance, involved optimizing the production of precious metals and mercury, with the Royal Academy of Mines and Metallurgy (1778) in Spain. In North America as in overseas empires, mining expertise marginalized indigenous knowledge. In Sweden, the Bureau of Mines brought together chemists in the service of industrial development, who found themselves at the end of the eighteenth century at the heart of the Enlightenment project. In Lyon, chemists provided support for administrative authorities determined to encourage industry.[29]

An historical paradox: liberalism strengthened the government through administration, because to dismantle the old regulations, voluntarist policies were necessary. Colbertism in France and cameralism in German principalities developed administrative practices that replaced, at the end of the eighteenth century, ancient local case law.[30] Scientific expertise was an essential cog in this transformation. The German doctor Johann Peter Frank, for example, determined a public health program, with its monumental *System einer vollständigen medicinischen Polizey* [Complete System of Medical Oversight] (1779–1827, in 9 volumes), which outlined state-controlled medical vigilance that maintained a discipline of the masses.[31] To regulate the urban environment, London and Paris moved gradually toward more centralized authority. After several unsuccessful attempts at administrative rationalization, London managed to create unified health infrastructures. In 1796, with the help of Jeremy Bentham, the Scotsman Patrick Colquhoun set up the Thames River Police, a management body designed to maintain order over port and river practices on the Thames, including nuisances.[32] In Paris, the desire for centrally administered social control was even more substantial. Beginning in 1777, the new Quarry Inspectorate, an administrative organ under the sole direction of the lieutenant-general of police, acquired the role of securing the urban environment from the collapse of quarries of stone and plaster, which removed the judicial courts from their traditional role of oversight. In addition, in 1780, the pharmacist Cadet de Vaux headed a newly created health inspectorate whose authority bypassed traditional police jurisdiction.[33] Government-led initiatives to improve the environment were rarely unconcerned with economic interests. For example, when water quality was fiercely debated after 1760 in London as in Paris, the technical priorities for drawing drinking water for residents turned toward the use of fire pumps, which were often financed by senior administrators or notables who sought to gain economically from the state-supported infrastructure. Public risk assessment became a source of private profit. In the name of a necessary capital-intensive modernization of the equipment structures, water pollution was exploited and, in so doing, largely marginalized as a key issue.[34]

Gradually, these kinds of developments resulted in a greater tolerance for pollution, justifying its ubiquity in the name of economic liberalization and expanding wealth. In 1772, the most important trial over industrial pollution of the Ancien Régime began in France. Neighbors of the manufacturer John Holker's vitriol plant in Rouen accused that acid gas emissions escaping from lead chambers were threatening their health and destroying the surrounding vegetation. After several months of proceedings, Holker, supported by Philibert Trudaine de Montigny, head of the Bureau of Commerce and member of the Académie des Sciences, obtained an evocation from the King's Council, a peculiarity of French law which moved judicial oversight of the case from the court to the central administration.[35] In the council, Trudaine argued on Holker's behalf against the minister and former lieutenant-general of Paris (1757–1759) Henri Léonard Jean Baptiste Bertin, who was a great connoisseur of police practices. Bertin's belief was that "One can hardly bring into the department of finance objects which concern only the police." But economic interests prevailed. In September 1774, the plaintiffs were dismissed and subsequent complaints against factory operations were forbidden.[36]

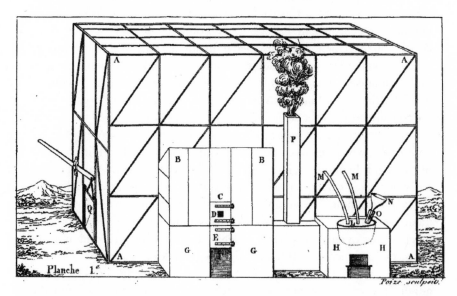

Figure 3.2
Lead chamber, 1812. *Source:* Rougier, "Mémoire sur la fabrication de la soude artificielle," *Mémoires publiés par l'académie de Marseille*, vol. 10 (1812): 177.

The decision marked an important precedent and encouraged the establishment of a second French sulfuric acid plant, in Paris, a few years later. Another Parisian trial relating to a nitric acid factory (1773–1774) testifies to the shift in regulations that occurred during this period. Breaking with custom, the lieutenant-general did not call on the police commissioners of the district, but rather turned to medical experts for testimony. In their report, the latter denied the insalubrity of the acid vapors and denounced the judicial "harassment" against artisans who they interpreted as the real victims. Similarly, in 1775, the Orléans police refused to establish a sugar refinery inside the city for health reasons; the Parlement of Paris ratified the decision, but the case was evoked in the King's Council, who decided, on the economist and statesman Anne Robert Jacques Turgot's recommendation, and in the name of remaining competitive with foreign industry, that the refinery should be established. The neighbors of the factory noted to this effect that the "dogmas of freedom" contradicted and superseded police regulations.[37]

Elsewhere in Europe, while the shift in jurisprudence was equally perceptible, it tended to be less radical. In Great Britain, in more than thirty years as Lord Chief Justice of the King's Bench (1756–1788), William Murray, Earl of Mansfield, took into account the changes brought about by urbanization and increasing industrialization; his decisions in favor of production imperatives increased over the course of his career. In a 1776 judgment against a sal ammoniac manufactory in London, Mansfield acknowledged the inconvenience posed, but argued in favor of inventive ingenuity and the importance of the manufactured products for the country and its independence.[38] In Brussels, disputes over acid manufacturing concluded in similar kinds of arbitration. In 1786, the police tried to close a factory after neighbors complained and the college of medicine confirmed its hazard, but the Privy Council of the Austrian Netherlands, which represented the central authority, opposed and overturned the police action. In Barcelona, the Royal Court of Catalonia faced many protests caused by nascent industrialization. In its arbitrations, it, too, became increasingly sympathetic toward the imperative economic activity. It advocated the introduction of coal, following the guidelines and recommendations laid out by French chemists.[39]

Environment in Revolution

The period between 1760 and 1840 is typically remembered as the "Age of Revolutions," or the wave of "Atlantic Revolutions." This revolutionary period built the foundation of the modern world: it ushered in newfound interests in and movements for civil and political rights, liberalism, and industrialization.[40] These major sociopolitical upheavals also had a significant impact on the environmental regulation of pollutions, because it established a new legal culture that consistently favored resource and territory expansion and exploitation at all costs, whether that meant in wetlands, forests, mining, or industrial manufacturing. Whatever the historiographic debates about the significance of the Atlantic Revolutions on subsequent relationships between industry and environment—and inasmuch as the period has undergone considerable and various scholarly efforts to reconcile radical economic development with protecting the public from harm and the struggle against an emerging environmental crisis—this period marks the undisputed and unmolested ascendance of the liberal and industrial worldview.[41]

In North America, the founding fathers of the United States engraved into their political project an agrarian ideal, but they framed that vision around an explicit notion that nature—all of it—was to be exploited and improved upon through the rational organization of society and work. In particular, Thomas Jefferson, a great admirer of Bacon, Newton, and Locke, designed his Monticello property with a symmetry that permitted the control of nature and its surroundings. The American Declaration of Independence (1776) constituted the very essence of this scheme: reason and natural law were beacons by which humanity could settle and cultivate the world. American naturalism subscribed to the notion that the environment should be tamed and improved upon through human work.[42] On pristine landscapes, American industrialization could develop along new, rational principles combined with an inherent pioneer spirit without the hindrances of Old World regulations, even if British common law served as primary inspiration. The American War of Independence (1775–1783) made industrial mobilization necessary, the momentum for which did not dwindle with the foundation of the United States. Reflections on wilderness were one of the consequences of the shock of human intervention and expansion into new territories. New opportunities, new resources, new wealth abounded. Bit by bit, as the frontier stretched westward, the pioneering, agrarian ethic of the Jefferson ideal shifted in favor of a world motivated by markets and industry.[43] The independence movements in Latin America (1808–1826) followed the same pattern: the emancipation from the Spanish Empire paved the way for new states and governments (based on the British model) to embrace liberal modernity based on the Enlightenment spirit. Nature needed to be seconded to economic need, which adapted as elsewhere to forms of slavery to produce vast quantities of coffee, sugar, and cotton.[44]

But institutional changes in France were unquestionably the most radical. Whereas the rural Cahiers de Doléances from the spring of 1789 suggested popular apprehension toward the expansion of industrialization, urban populations insisted that furnaces and workshops manipulating putrefiable materials be kept well outside the cities. In Paris, the new municipal authority tried to eradicate nuisances, but beginning in 1791 revolutionary legislation freed industry from the regulations that controlled its pollutions. The trade associations (thanks to Pierre D'Allarde's law), the Bureau of Commerce, the Office of Manufacturing Inspection, and procedures surrounding patent letters (authorizations that came from the King's Council) were all abolished. Preventive *commodo* and *incommodo* inquiries were also removed, and the Constituent Assembly decreed that tribunals were competent to rule on damages caused to property.[45] The result was that industries were thereafter free to establish their works wherever they pleased and employ whatever methods best suited them. However, the Declaration of the Rights of Man and Citizen upheld future legal intervention in arbitrating nuisance cases. "Freedom," it asserted, "consists of being able to do whatsoever does not bring harm to others. In this way, the exercise of an individual's natural rights is unlimited so long as it does not infringe upon another individual's freedom to enjoy the same rights. These boundaries can only be determined by the law" (Article 4). If the law was to ultimately decide, it was in accordance with the notion of public utility, associated with the growth in trade and production, meant to offer employment and resources to the populace. When in 1791, the Académie des Sciences inspected the pollutions emanating from a sal ammoniac factory near Valenciennes, the commissioners concluded that public interest in preserving the industry was worth some environmental sacrifice: "It is up to the local residents ... to determine if ... it would be more convenient to put up with the odor as a balance against the advantages of an establishment that can furnish work for many laborers and access to goods that benefit their country."[46]

At the onset of war in 1792, the European countries mobilized with an unprecedented ferocity that constituted a forced march toward rapid industrialization. Great Britain's production of coal, iron, and textiles all increased tenfold.[47] In France, military imperatives compelled a massive industrial shift to which the period's great scientific savants—Lazare Carnot, Fourcroy, Guyton, Claude-Antoine Prieur, Chaptal, Berthollet, Jean Darcet, Gaspard Monge, Jean-Henri Hassenfratz, and even Alexandre Vandermonde—were recruited. Hatters transformed their workshops to produce varnished helmets for the army; the need for uniform buttons marshaled copper acid works; textile manufacturing was recruited under the direction of the Supply Commission to make canvas, flags, and military uniforms. The Gunpowder and Saltpeter Agency (directed by Chaptal) centralized gunpowder production and hastily built a massive gunpowder factory in Grenelle, a district of Paris. In an even more direct manner, the arms industry flourished in the capital, where in the fall of 1793 the Manufacture of Paris constituted a vast, bustling, and nebulous collection of military factories working under Guyton's authority. Such were wartime preoccupations that, in all cases, preventing nuisances was relegated to an afterthought. Instead, the physical environment was conscripted to conduct war, to justify violence, and to engage in dangerous industrial production. On August 31, 1794, the gunpowder factory at Grenelle exploded, killing 600 laborers and injuring another 1,000.[48]

A few key forms of military production, such as in the use of leather, highlight the disruption of existing nuisance regulations. Beginning in the fall of 1793, as the revolutionary government searched for the fastest method of mass-producing materials for the rapidly growing army, it enlisted chemists to perfect the art of tanning. Armand Seguin, a longtime collaborator of Lavoisier's, proposed a "revolutionary" tanning method, which accelerated the process considerably by employing a concentrated sulfuric acid solution instead of the older, milder, organic tanning acids. After a laudatory endorsement of the practice to the Convention from Fourcroy, an industrial tannery was built just outside of Paris, but it released significant quantities of its acid into the Seine. Copper, used in making bronze cannons, proved to be another essential war material. A massive manufactory, which housed 1,000 laborers, was established on the île de la Cité right in the center of Paris, where it used copious amounts of acids to refine the metal under Fourcroy's and Guyton's oversight. Such was the scale of wartime industry and the quantities of new chemicals in use: never had so many acid fumes been discharged right in the heart of the city. One trial illuminates the spirit of the new, revolutionary regulations. In 1793, the National Convention was forced to react to the shutting down of a minium factory, which used lead tetroxide as a pigment. The municipality of Bercy sealed the factory's doors because of pollution emissions. But Fourcroy insisted that its production constituted a national utility, and Guyton was tasked with an expert mission to remediate the plant. The lead contamination of surrounding vegetables and livestock was incontestable; three cows had died as a result. The whole affair contradicted traditional law enforcement: whereas the municipal authorities demanded that the factory be distanced from populous areas, the government—supported by scientific experts—sought technical measures to maintain its place and practices. Even if the Convention finally concluded that the factory should be destroyed, the proprietor was generously compensated and encouraged to improve his manufacturing processes with the assistance of the chemist Nicolas Deyeux, which permitted him to rebuild his factory again in the center of Paris. Whereas technical improvement, under the cover of chemical expertise, was hitherto only intended as a strategy for eliding controls that inhibited sulfuric acid factories, such cases methodically paved the way for countering legal prohibitions.[49]

A Law for Protecting the Polluters

Once Napoleon Bonaparte came to power in 1799, France transitioned into a more authoritarian and centralized form of politics. The size and power of the state sought to eliminate obstacles that restrained economic growth, providing particular support to the chemical industry. In Paris, dozens of factories were opened both within the city and around its periphery. Chaptal's sulfuric acid and alum factory received considerable attention as a site of pollution. Since its construction in 1798, it provoked complaints and protests because of unhealthy fumes. Member of the Académie des Sciences, chemist, entrepreneur, and state councilor, Chaptal had in the meantime become minister of the interior. Year after year, he rejected every complaint against his factory. In 1803, local inhabitants called upon the justice of the peace to act, but Chaptal blocked even legal actions. He became a symbol of the unprecedented and inextricable connection between science and authority. The creation of new administrative structures based on these principles served as a critical historical pivot in French pollution regulations. Just before he became minister of the interior, Chaptal wrote an *Essay on the Development of the Chemical Arts*, which was both a treatise on industrial applications of the latest chemical discoveries and a guide for chemical entrepreneurialism, which he fostered and encouraged after he was appointed minister.[50]

Throughout France, resistance to polluting chemical factories threatened and compromised the march of industry. After Chaptal was replaced as minister of the interior in August 1804, authorities considered a national response to the recurring popular resistance. The new minister interrogated the Académie des Sciences "on the factories that emit nasty odors and on the danger they might pose to public health." The report was entrusted to Guyton and Chaptal. Predictably the two chemists concluded that public health complaints of polluting factories were not valid.

They distinguished between industries whose processes required the potentially harmful decomposition of organic materials and those that emitted fumes and gases from fire-based practices, particularly sulfuric acid workshops whose emissions they only considered bothersome but nothing more serious than that. Reversing the traditional order, Chaptal and Guyton called for central authorities to protect industrial works. Any constraints on their work would be, they claimed, "an unjust, vexing, and detrimental deterrent to manufacturing that would not remedy the harm derived from its operation." They defined instead a new approach: faced with the "arbitrariness" of a "simple police magistrate," the "prejudice" and "ignorance" of public opinion, or the "jealousy of competitors"— never mind the challenges inherent in the "chemical practice itself"— protecting public health was less important than protecting the chemical industry. "It is thus of the highest importance for the prosperity of chemical advancement that one finally poses limits which leave nothing more to the arbitrariness of the magistrate." How could an industrialist agree to invest significant capital, they reasoned, if he might have to stop production following a neighborhood complaint and a police order?[51]

In the short term, the report did not translate into new and coherent pollution legislation. Local structures continued to take charge of the management of industrial pollution at the turn of the nineteenth century, with a new orientation: the establishment of medical inspection committees. On the other side of the Channel, in Manchester, the British city most affected by industrialization, the municipal authorities changed the practices inherited from the ancien régime and created the Board of Health in 1796, and a new office of the police commissioner responsible for the health aspects of industrialization; a nuisance committee was also created in 1800 to manage steam engine fumes.[52] Following similar inspiration, Chaptal created a health council in Paris in 1802. Scientific expertise was paramount and directed the prefect to guide the local authorities. Almost every member of the Academy was a chemist or doctor who had previously promoted acids, chlorine, and artificial soda: Parmentier (first president), Charles-Louis Cadet de Gassicourt, Deyeux, Jean-Pierre Darcet; its main leaders were also members of the *Annales de Chimie* and the influential Society for the Encouragement of National Industry (SEIN), created in 1801 by Chaptal, and of which Guyton, himself vice-president, presided over the Chemical Arts Committee. A decree from the police prefect in 1806 reinforced the process through which scientific management rather than police authority administered the task of inspecting and monitoring polluting factories and workshops. Leaving oversight to the "chemical savants" gave authority to the scientists of the Health Council, leaving the police commissioners the simple task of conducting *commodo* and *incommodo* surveys, which were reestablished but reduced to simple neighborhood inquiries.[53]

Very quickly, the Paris Health Council took charge in resolving the pollution emanating from artificial soda factories. Since 1800, trouble persisted in the immediate vicinity of Paris, Gentilly, Belleville, Saint-Denis or Nanterre, but most of the works were headed by members of the Health Council, or their chemistry colleagues from the Academy. After 1809, the pace of production increased sharply in response to the British continental blockade, which cut off France's supply of vegetable soda. To stimulate the production of synthetic soda, manufacturers were exempted from the salt tax. The total destruction of crops and orchards for miles around each factory provided irrefutable evidence of the irreversible damage caused by hydrochloric acid vapors. The resurgence of pollutions forced the minister to order a second report from the Académie des Sciences. The composition of the new commission still reflected the face of manufacturing: beside Chaptal and Guyton, Fourcroy and Nicolas Vauquelin were also heavily invested in the chemical industry in the center of Paris. But, at the marked insistence of the minister who demanded some balance between the manufacturers' interests and the interests of local residents, the report's conclusion called for a compromise: it proposed establishing three classes of industry, based on their degree of hazard, and the chemists suggested instituting administrative authorization surveys before proceeding with upstream developments.[54]

These conclusions were reprised in the decree of October 15, 1810, which revisited the problem of "factories or workshops that spread an unhealthy or uncomfortable smell."[55] Before being able to operate, the classified industries were required to ask permission from the administration, which then triggered preventive surveys combining the consultation of *commodo* and *incommodo* with the scientific advice. The most harmful factories (first class) rested under the authority of the Council of State; the other factories were beholden to local prefectures. As a result, the new law confirmed the preeminence of administrative over criminal justice: police or correctional courts were not recognized as competent to judge the legitimacy of facilities already accepted by acts of public administration. Although Article 471 of the Penal Code reiterated penalties for environmental risks resulting from prerevolutionary policy (already mentioned in article 60 of the law of 3 Brumaire year IV), the decree bypassed it by its article 11, which left only the possibility of going to civil court and suing for damages. Procedures that could have been addressed by the criminal justice system were limited to litigation in the Council of State or prefecture councils, which could lead to the administrative closure of the factory (Article 12). The Restoration Regime reformed the law in 1815 but kept the essential provisions.

This legislation drew to a close twenty years of debate on the modes of nuisance regulation. It validated, ratified, and entrenched the move away from police authority in nuisance cases that manifested in the 1770s, elevating administrative oversight (grounded in economic liberalism) to protected status within the law. It also upheld the new regime's authoritarian politics, both internally and with an eye on expanding empire through conquest. If, between 1797 and 1800, the population could still gather in the street in demonstration against the nuisance-related improprieties of a local copper button factory and collectively call on the police commissioner to shut it down—a faint echo of the *sans-culottes'* protests from the Revolution—by 1811 the army was called on to intervene and protect industrial establishments against such citizen protests. This was the case at the Payen sal ammoniac factory, where neighbors complained that their wells had been contaminated by residues from burnt organic matter. Seemingly above the law, and sharing scientific and economic interests with industrialists, the Health Council's support for such interventions was unambiguous.[56] Until 1817, the Health Council was the only permanent forum for public health, which gave it a degree of national prominence. It served subsequently as a model for other councils created in the provinces to arbitrate industrial pollution issues: in Nantes (1817), Lyon (1822), Marseille (1825), Lille (1828), Strasbourg (1829), Troyes (1830), Bordeaux (1831), Rouen (1831), and Toulouse (1832), placing industrial and pollutions authority everywhere in the hands of scientists rather than police commissioners.[57]

The industrialist spirit of the Paris Health Council rested in its enthusiastic support for economic development. While its taste for economic liberalism was no doubt genuine, its collusion with the business world was a critical motivation in many of its decisions. The chemist Jean-Pierre Darcet is a good illustration of the invested interests shared between authority and industry. Himself an industrialist—and gold-medal awardee for the production of soda using the Leblanc process at the exhibition of industrial products in 1801—Darcet was also one of the chemists appointed to the Mint of Paris where he regularly met with Guyton and Berthollet, two of the three administrators of this laboratory where the alliance between science and administration became embedded.[58] In his factory in the Rue de Montreuil, Darcet produced soap and other chemical products, and in that of Nanterre, artificial soda; while after 1816 he was codirector of the Chaptal sulfuric acid plant. Darcet also held senior positions in the Society for the Encouragement of National Industry (SEIN), the General Council of Manufactures, as well as the Ministry of the Interior's Advisory Committee on Arts and Manufactures to advise the minister. Believing fundamentally in the benefits of industry for society but perceiving the incompatibilities between the harmfulness of certain industrial sectors and public health—which he did not deny—Darcet prioritized searching for technical improvements over pollution control or prevention. In one of his reports, in 1815, he anchored his decision to the decree of 1810: "by arranging the same factory in the first, second or third class, according to the quality of the processes employed therein, [the government] will undoubtedly contribute greatly to the improvement of our industry, and render the neighborhood of our great manufactories less unhealthy and less disagreeable."[59] His authority in the Health Council between 1814 and 1830 made technical improvements to the hygienists' credo. With the doctor Alexandre Parent-Duchâtelet, to whom he was a sort of mentor, Darcet guided Hippocratic medicine's mutation toward a social medicine that was attuned to the need for reform but captivated with the industrial world and its benefits. Henceforth, occupational diseases were no longer attributed to work. Instead, sex, age, low wages, or the immorality of the working classes became the primary indices for measuring health. These were markers to which the industrial system could respond.[60]

The Foundations of a New World

Like the Civil Code, Napoleonic legislation on nuisances spread across the European continent, albeit unevenly and with strong national variations. For example, in the annexed Netherlands (which included Belgium before 1830), it became legally binding in 1811. After the fall and retreat of French imperial influence, application of the decree became somewhat uneven. However, in 1824 a Dutch royal decree adopted the main provisions of the French legislation with a slightly less repressive scope, leaving local authorities with more power to order the closure of a polluting factory. In Amsterdam, for example, local authorities retained considerable power before the Council of State. Further, the legal reasoning and authority of municipal councilors remained marked by older practices: scientists played a much less important role than in France.[61] The law of 1810 did not cross the English Channel, however, as much for political and warlike animosities during this period as for distinct legal traditions. The British regulation of pollution continued to evolve from the common law, even if an attempt at revising existing legislation appeared when, confronted with the problem of smoke coming from steam engines, local police proved to be incapable of confronting the new hazard. In 1819, the House of Commons established a Select Committee to investigate industrial fumes, which was followed by a law (Taylor Act, 1821) requiring steam engine owners to adopt smoke abatement processes that controlled emissions from their furnaces. In spite of initially resisting the new legislation, some industrialists came to embrace it; the controls served as incentive to improve their boilers' combustion efficiencies, which increased productivity.[62] The law, however, failed to prevent the growth of air pollution in Britain. After a year of voluntarist action by the local authorities responsible for enforcing the law, a period during which a large number of industrialists were prosecuted and required to pay substantial damages, oversight and enforcement quickly subsided and the law fell out of favor due to a general lack of political will.[63]

In France, the application of the law experienced some adjustments that consolidated its intents and principles. In the 1820s, the frontiers of French justice and administration were fiercely discussed, as a result of the Bourbon Restoration's attempts to reconcile the Napoleonic Code with older juridical systems.[64] At the same time, however, more and more complaints from neighbors of polluting industries forced action. Two trials are particularly illustrative. The lengthier case dealt with refining precious metals in Paris. Despite prison sentences, fines, and damages awarded by the courts, industrialist interests prevailed: according to the law, the administration and its health council deputized the police prefect, who was beholden to their interest in matters of law enforcement.[65] The second important case pitted the soda manufacturers of Marseille against their neighbors, often landowners or big business bourgeoisie. By imposing penalties and indemnities on the manufacturers of the chemical industry that threatened their economic viability, the Aix Court of Justice revived the justice system of prerevolutionary France. In 1826, the Ministry of the Interior was forced to intervene and impose a compromise in order to allow the continuation of artificial soda production in the region: the intervention curbed judicial enthusiasm for pollution prevention, on the one hand, and constrained industrialists to use technical means to address the worst of their polluting practices, on the other.[66]

Justice was gradually being removed from the struggle over industrial pollution. Health councils were responsible for monitoring the regulatory operating conditions given to polluting establishments. But when industrialists did not respect these conditions, their noncompliance was not considered a criminal offense and could not be tried in police courts. Thus, industrialists often evaded paying fines, negotiating instead with the health council and the administration. The powerlessness of the public prosecutor's office was blatant, and in 1832 Alphonse Trebuchet, an expert on jurisprudence surrounding the law of 1810, was distressed.[67] Inasmuch as the law could be circumvented with ease, methods of seeking damages and protecting interests against industrial pollution were equally limited. The Council of State claimed that the civil justice system could not pronounce a verdict on the depreciation of neighboring properties (1824), nor on moral damage (1827): only materially damaged property could be compensated. In those same years, when the industrial risks associated with steam engines and lighting gas were being debated, the state councilor Héron de Villefosse justified this administrative approach and the marginalization of justice by a deliberate policy of favoring industrialization: it was "precisely because the application of the rules of the common law had presented enormous difficulties in cases of this kind that it has been recognized as necessary in France to lay down special legislation."[68]

In fact, increasing investments and certain technical devices rendered the more repressive legal approach that existed in prerevolutionary France impossible. The example of the distillation of coal by the gas industry demonstrates this. In 1817, when the industry was founded in Paris, the Health Council determined that its activities should have been classified in the most harmful first class of polluting industries, and theoretically forbidden in the city. But since the long pipes needed to transport the gas to places where it was consumed would compromise its viability, the council recommended second-class status and the furnaces and gasometers were established within city limits. The same desire to not curb industrialization explains why steam engines were permitted to operate freely. Competition with Britain, where smoke and dangerous establishments in the heart of the cities were tolerated, discouraged French authorities from imposing too many constraints on companies, which could weaken their viability against foreign competitors. Between 1815 and 1830, more than 100 steam engines were installed in Paris—far fewer than were built in London or other major British cities—while a handful of gas plants operated in both capitals. In both countries, scientists were consulted: in order to reconcile industrialization with concentrated populations, they defined the notion of a safety standard. In doing so, they entrenched the prospective industrial risk; in 1825, the new nomenclature for polluting establishments added the term "hazardous" to its title.[69]

Technical improvement became the modus operandi to manage nuisances. The Manchester doctor Thomas Percival praised the new regulations that relegated to technical innovation, and especially chemistry, the task of "attenuating or eradicating almost all the bothersome vapors."[70] In perfecting itself, "industry is like the Achilles spear; it heals the harm it does," wrote the Paris Health Council in 1821. In his influential work *De l'Industrie Française* (1819), Chaptal extolled chemistry's potential and associated sanitation with growth:

> In creating the art of manufacturing sal ammoniac, chemistry has given value to substances that had none before. Animal materials were rejected as being of no use for any purpose; today they are much sought after: gelatin factories and those of sal ammoniac fight over access to these raw materials. Thus chemistry, by extending its domain, creates value, and makes everything contribute to the general good.[71]

In this manner, the pharmacist Charles Derosne began to manufacture animal black, a material from animal residues, which resulted in a clarification product for beet sugar. Similarly, Payen recycled organic waste and ammoniacal water from gas factories in his large plant in Grenelle, where sodium chloride, lime chloride, animal charcoal, sugar, and other products were also made. Though propped up by the Health Council, these two manufacturers provoked several complaints and were subject to multiple administrative disputes. The Derosne factory was also the first in France to be finally closed (1834) under Article 12 of the law. Sanitation also involved reducing fumes and gases. But in the construction of "smokeless stoves" and devices for condensing and absorbing vapors, failures were numerous, even if those failures did little to temper hygienists' enthusiasm for the ingenuity of the industrial world.[72]

The hygienist credo, however, met with resistance from the people. Like some workers and artisans who denounced the social effects of industrial mechanics at the turn of the 1800s, simple citizens and residents strongly opposed the health and environmental risks posed by industrialization.[73] Of the more than 3,000 cases of unhealthy industrial establishments examined by the Health Council between 1806 and 1828, almost half were the subject of neighborhood complaints. The tendency toward local complaints and opposition increased in the 1820s. In the mid-1830s, Parent-Duchâtelet, then vice-president of the Health Council, articulated the nature of the confrontation as an epic struggle for power that he feared could become detrimental to industrial interests:

For several years, there has been a fierce struggle between private individuals and industrialists: the latter are pursued with ardor and activity which, so far, has not been justified; in a way, we dictate to authority the conduct that it must take.... Such a state of things cannot subsist any longer; because industrialists are indispensable to our social order, we must tolerate factories; and if so far individuals have needed to be privileged against the demands and carelessness of the manufacturers, the time has come when these must be defended against the ridiculous requirements and intolerable and tyrannical claims of others.[74]

Were the claims that were attached to the old legal regime of being able to live unhindered by others so intolerable? And the demands so ridiculous? Many doctors, practitioners rather than scientists, often supported the protests by neighbors of polluting industries, to the dismay of Parent-Duchâtelet. Some of these doctors also had honorable careers. For example, François-Emmanuel Fodéré, a physician in Marseille and a professor at the Strasbourg School of Medicine, published in 1813 the first *Traité d'Hygiène Publique* (*Treatise on Public Health*) in six encyclopedic volumes in which he outlined the toxicity of the new industrial world. Fodéré was also the first theoretician of acid rain; in his work he condemned the new regulations that favored industry and vigorously attacked industrial pollution and chemical agents, "destroyers of all organized beings, living or not." He strongly criticized the Health Council's practices, its methods and conclusions, as well as public health's subjugation to the requirements of industrialization.[75] Similarly, in Spain, the physician Juan Manuel de Aréjula, a supporter of Guytonian fumigations at the outset, ultimately concluded that acid gases caused "more harm than good." An eminent scientific authority who presided over the Supreme Court of Public Health since 1811, he was sidelined by his colleagues and the political powers after his change of heart.[76] Opposition to industrial progress was silenced, and these doctors were marginalized in their careers. Only legal recourse gave them a credible voice. For example, in Great Britain—where artificial soda production began after 1825—following the complaints of neighbors to soda factories in 1829, Scottish courts called on Edward Turner and Robert Christison, both of them doctors and chemists, to perform laboratory tests on the environmental effects of the factories' emissions. Turner and Christison concluded that acid fumes had a deleterious effect on plants, which had suffered considerable decline since the introduction of the new industry. They based their work on the methods conceived by the Swiss chemists François Marcet and Isaac Macaire-Prinsep, who for several years had been studying the effects of factory fumes on plants. In 1832, two botanists, the Englishman John Lindley and the Swiss Alphonse Pyrame de Candolle also confirmed, separately, that the harmful effects of acid gases on plants existed in and around manufacturing facilities.[77] The controversy that the authorities and the more prominent, industrially minded chemists wanted to silence was still enlivened by debate by the time industrialization witnessed rapid acceleration thanks to the advent of the first railways.

* * * *

Thus, the first stages of industrialization upset the old regulations by redefining healthfulness and unhealthfulness, and the boundaries between inconvenience and unwholesomeness. An all-conquering scientific chemistry transformed the industrial pollution paradigm and monopolized expertise to promote acids and their disinfecting properties. Local protections against polluting agents were disempowered: capital investment inherent in the new industrial system made any repressive attempt to control pollution incompatible with economic growth. The French decree of 1810 to govern unhealthy establishments serves as the keystone of the new regulatory system for pollutions. The first law on industrial pollution in the world, it was the result of a sudden increase in production coupled with the institutional reorganization of government structures. Public utility, inscribed in the spirit of the law, reorganized hierarchies of priorities in the name of economic enrichment. Founder of a new order, the law acknowledged pollutions as an inherent part of industrial society and as a new political project that imposed on society an industrialization yet contested for its deleterious effects.

II

Naturalizing Pollutions in the
Age of Progress (1830–1914)

If the eighteenth century witnessed the emergence of industry and its share of nuisances, the nineteenth century saw pollutions become constituent elements of modernity, and the word "pollution" gradually acquired its contemporary meaning. As the scale of industrial production continued to increase, so too did the number of toxic substances released. That satellite-view assertion, however, masks gaping contrasts and important redistributions in the allocation of production and its toxic discharges. At the beginning of the nineteenth century, industry in both Europe and Asia was primarily rural and widely dispersed. As the century progressed however, the disparity between the two continents increased rapidly. In 1870, Western Europe's industrial production accounted for 62 percent of global production. By the time of the Great War, Europe and North America accounted for more than 80 percent of global industrial production.[1] Put another way, if smoke superseded stench in the hierarchy of contaminants, this great phenomenon deeply affected the West first.

In Europe, progress was the capital idea of the century. Described in terms of growth and improvement, progress was a quest for power and a plundering of nature, which Europeans perceived as immense and hostile. The figure of Prometheus, the mythological titan who stole fire and gave it to humanity, became the ambivalent symbol of this age of potential progress. But Prometheus' suffering was no less symbolic in the period's miseries and its dreaded catastrophes.[2] Numerous accounts have depicted industrial expansion in the West as a race toward abundance and a manifestation of beneficial progress. Following postwar economic growth, referred to as the "Glorious Thirty" (1945–1975) period, historian David Landes likened this great industrial change to a liberation from the blockades and routines that had previously hampered Europe's rise.[3] For the most part, economic historiography has described industrial expansion in optimistic terms, listing myriad benefits, while also removing from the balance sheet the resultant ecological damage and social and spatial inequalities it produced. Statistics on the production and consumption of certain key products—such as cotton, coal, cast iron, and steel—were supposed to attest to the rising standard of living, which served as the foundational basis of the progressive narrative in the historiography.[4] In his book on the history of industry, Denis Woronoff noted that up until twenty years ago "the history of industrial pollutions and the sensibility of city dwellers to its nuisances had not been written." This point was reiterated by Geneviève Massard-Guilbaud, who remarked that in spite of an immense amount of documentation on pollution available in nineteenth-century archives, "particularly polluting industries, such as chemistry, have been studied without one word being mentioned about pollution."[5]

Pollutions were always an afterthought, however. Even if growth had not yet been subjected to theoretical rigor, pollutions were never used as a counterbalance to the exploitation of nature, which was treated as inexhaustible and apt to assimilate all waste from human activities. This fatalistic attitude meant that pollution was accepted as a regrettable, yet inevitable effect of global emancipation. The growing acceptance of toxic waste—its naturalization—fell in step with evolutionary and teleological readings throughout the century of progress. States legitimized a political economy based on production statistics, whose steady increase promoted a sense of well-being, pacification through soft trade, and political and social stability.

In order to follow the manner in which pollutions were naturalized in all their complex forms over the course of the nineteenth century, it is necessary to explore the principal sectors and territories involved, and to map out the sources of pollutants and their evolution (chapter 4). As pollutants increased exponentially, many sought to understand this phenomenon through all-conquering science; others protested—or even rose up when the harm or damage became all too evident (chapter 5). In the name of public interest and the greater national benefit, some states—those that considered themselves liberal and industrial—willingly sacrificed certain zones and populations while trying to contain nuisances. By means of regulatory laws and the promotion of science and its methods, pollutants were simultaneously fought against and accepted, denounced and permitted—resulting in a profoundly ambivalent relationship with industry and its achievements (chapter 6).

4

The Dark Side of Progress

In 1914, less than a hundred years after their construction began, railway lines extended more than one million kilometers around the world, predominantly in the United States and Europe. This radical innovation testified to the extraordinary transformation that resulted from advancements in mechanization and in the steel industry. Coupled with the growth of financial capitalism, preindustrial economies found themselves hurtling toward staggering expansion. However, apart from urban tramways, railway lines were not electrified until the early 1900s. The first significant electric line opened in 1905 and ran between New York City and New Haven. Instead, coal remained the key element in train development, giving the steam locomotive its characteristic plumes of black smoke, described by Victor Hugo, rapt with wonder: this "iron horse," which "tore at the trees as it passed," and spewed "burning coal waste and boiling water urine ... , and huge rackets of sparks."[1] The train unquestionably and unapologetically became the quintessential symbol of the industrial age.

Coal was described as "the principal resource of our industry," "a veritable philosopher's stone" of modernity: its influence conquered the world between 1830 and 1914.[2] Its significance grew in step with that of the steam engine and advancements in industrial techniques for metallurgical fusion. Even though hydraulic power and biomass continued to be used, coal grew in importance and became the standard pollutant of the industrial world.[3] The rise of coal was impressive. Between 1840 and 1914, the amount of coal mined in Great Britain rose from 30 million metric tons to 290 million metric tons. In German principalities, the production grew from 3.4 million metric tons to 190 million metric tons over the same period of time. A similar trend took place in France, with an increase from 3 million metric tons to 40 million metric tons, and in the United States, 2.1 million metric tons to 510 million metric tons. Also in this time period, the amount of cast iron produced per capita rose from

WHEN THE HEART OF THE POTTER REJOICES !   549.

Figure 4.1
"When the heart of the Potter rejoices!" Stoke-on-Trent postcard, undated (end of the nineteenth century). Coal fumes became the main symbol of pollution in the nineteenth century, especially in English cities. Stoke-on-Trent was the world's leading center for ceramics, to the point where the region was renamed The Potteries and included in the Black Country appellation. The back of the postcard reads: "This is the hole I am stuck in, very good for anyone with a bad chest."

5 to 130 kg in Germany, from 20 to 220 kg in England, and from 16 to 270 kg in the United States. The growth in cast iron production influenced many other industrial sectors; overall, global industrial production increased sevenfold between 1860 and 1914.[4]

If coal was the chief somber face of progress—with its black clouds and stench hanging over emerging industrial cities—there were many other processes that contributed to the rise in pollution. Urban growth reached a crisis with regards to the disposal of waste, while blast furnaces, nonferrous foundries, and chemical factories continued to grow in number and add to the increasingly worrying emissions in Europe and North America. This treadmill of production and waste management expanded into new territories at the end of the century.

Urban Metabolisms

Before focusing on coal's novelty and its many uses, it is important to first remember that for the majority of the world's population, the city was the primary producer of waste and nuisances (outside of large industrial sites) and that preindustrial city life had existed for some time. However, industrialization gradually changed the urban metabolism and its inner workings. The dynamics of its exchange of materials with countryside and mines, its access to resources, and the urban layout with respect to where goods were consumed and where they were disposed of were all altered.[5]

The impact urbanization had on society and natural environments intensified in parallel with industrialization, migration, evolving transport systems, and new modes of life. While there were only six cities in the world in 1800 with more than 500,000 inhabitants (Paris, London, Istanbul, Tokyo, Peking, and Canton), there were 43 cities of this size in 1900, most of them situated in Europe and North America. By 1850 there were 225 million people living in cities; the urban population in England had already exceeded the rural population.[6] Cities became ever more densely populated, and previously empty or agrarian spaces were rarely spared long from new construction. Despite the crowded nature of city living, a level of quality was sought for its inhabitants. One such example was the provision of water for Parisians, whose consumption increased thirty-threefold between 1807 and the 1930s.[7]

Population congestion translated into waste congestion, and the amount of waste matter requiring treatment and disposal became a major concern. In Paris the gross quantity of waste, not including the waste generated from industrial sites, jumped from 45,000 to 550,000 metric tons between 1815 and 1864. Animal waste was an added problem. While most cities in the west sought to place abattoirs far from city centers (British cities were the last to insist on this in the latter half of the nineteenth century, and London the last amongst them), the amount of manure to manage continued to increase.[8] In Paris, the horse population rose from 30,000 in 1830 to 80,000 in 1880. The English horse population peaked around 1850. In Germany, 3.5 million horses were divided between cities and countryside. A comparable number were found in American cities at the end of the century. As a result, vast amounts of organic waste were produced. In Milwaukee, a city of 350,000 people, some 12,500 horses produced 133 metric tons of manure per day. By 1900, the threat of infection was such that a push for motorization was seen as a means of reducing health risks associated with the presence of animals.[9]

The list of everyday wastes grew even longer when byproducts from artisanal and other diverse industries were also taken into account. The following is not a comprehensive list, but quotidian polluters included dyers, tanners, launderers and bleachers, brewers, and chemical product manufacturers. Most trades did not know what to do with the byproducts of their work. This was the case, for example, with the gas lighting industry. Gaslights had spread to numerous neighborhoods in Europe after 1850. The lights produced various residues from the distillation of coal (bitumen, sulfur, ammonia, tar, heavy metals), all of which were generally deposited into pits near the gas factories.[10] In 1873, Parisians noted the revolting appearance of the River Seine at specific times of the year, most notably when "the Saint-Denis factories took advantage [of a storm] to send all of the harmful residues they had in reserve into the river."[11] After 1880, Marseille, France's second largest city with 500,000 inhabitants, began stockpiling domestic and industrial sludge in depots, further aggravating the city's already critical sanitation problems. Interestingly, throughout the century in Europe, the urban mortality rate remained higher than its rural counterpart although the gap was tending to close somewhat. This fact was attributed by many to urban density, garbage, and sewage, all of which created veritable pits of humanity.[12]

As industrial systems and infrastructures took hold in both European and North American cities, they became undeniably more polluted than Asian cities. The contrast is really quite striking: in Tokyo the sanitary regime that typified the eighteenth century could persist with few modifications through the nineteenth century. The idea of soiling, with its religious connotations, lead to a very strict treatment of waste. Rather than getting washed away in the sewers amidst great quantities of water, the *excreta* were instead valued as fertilizer for farmland. Waste that was considered unclean, impure, or carrying disease was disposed of using rudimentary techniques. The mortality rate never exceeded that of European cities until the end of the century.[13] The only exceptions to the rule that Western cities were more contaminated than their Asian counterparts were the great colonial metropoles—Bombay and Calcutta, for example—whose pollution levels reflected those of European cities, where artisans and city dwellers lived beside sullied and insalubrious waterways.[14]

The 1870s saw a rupture in the day-to-day running of the average European city. Previously, human and animal excrement, butchers' scraps, mud in the street, wool rags, and other everyday waste products had in one way or another been reused or reappropriated. But the onset of industrialization and urbanization—combined with a higher standard of living—provoked a devaluing of scraps and dregs that were no longer deemed worthy of reuse. At the same time, industrial enterprises enjoyed access to new resources from further afield thanks to mining expansion, advances in chemistry, and the appropriation of colonial resources—further diminishing the need to reuse and recycle. Waste from this point forth was an end product, and nobody really knew what to do with it.[15] Numerous experiments, along the lines of London's *sewage farms* (in existence since the 1840s), were instigated to attempt to relocate excess sullied water to the cities' distant peripheries. However, many such urban efforts failed. In Paris, Vienna, Brussels, London, and Berlin, such ventures managed only to heighten tensions between the city and its suburbs.[16] This rupture in the workings of these and other cities also accentuated episodes of the *great stink* produced by masses of organic waste left in the heat, while toxic matter was released into local rivers such as the Thames or the Seine. London was particularly hard hit in 1858, and Paris in 1880 and 1895; the stench led to a public outcry and criticism in the press.[17] In Paris, the concern was over the industrial treatment of sewage from the manufacture of ammonia sulfate. Several ammonia sulfate factories had been built in the 1870s in the city's suburbs; they inspired the foundation of a municipal commission to examine the "*Odeurs de Paris*," but the commission never succeeded in resolving the crisis between urban dwelling and industrialization.[18]

The rejection of the preindustrial system that valued garbage varied greatly from country to country, and between larger and smaller cities. The shift away from older waste practices seems to have started earlier in the United States and England than in France, and in larger cities compared with smaller cities. In any case, by the end of the nineteenth century, various systems of garbage disposal and elimination replaced the older method of reappropriating garbage. The first choice was to eliminate garbage via a main drain, a system which the majority of European cities adopted. Alternatively, as was the case in New York from 1872, waste was simply carried out to sea. England created dumping grounds for its rubbish, and other systems such as incineration and recycling began to emerge elsewhere toward the end of the century.[19]

All these responses to the industrial waste upheaval contributed to pushing industry and its pollutions to the outskirts of cities; their relocation was made even more possible by access to railway lines. Great industrial cities like London, Paris, and Brussels attracted and welcomed many artisans and small decentralized industries. However, after 1860, industry tended to set up on the periphery, in abandoned zones, well away from bourgeois neighborhoods. In Haussmann's Paris, particular areas were known for their high levels of pollution. For example, the Villette district was one of the most polluted sites in the capital; its abattoirs, fuel depots, gas factories, and even a huge tar plant gave the district its notoriety.[20] The search for large, economical spaces for factories paved the way for the development of the first industrial suburbs, which were easily accessible by rail. In this manner, industrialization transformed urban landscapes as well as ideas about urban planning. In cities that were predominantly industrial and defined by the particular branch of industry that had long been established there—such as the textile centers of Manchester and Roubaix, or the steel cities of Detroit, Pittsburgh, and Le Creusot—the suburbs became hubs of industrial activity. The list of polluted areas sacrificed in the name of advancement grew ever longer. All major European and North American cities developed industrial suburbs, fashioning a familiar landscape of tall chimneys: those "long fingers, indicators of inconvenience." Each developed in keeping with the local topography, the type of industry, and social power relations.[21] In Paris, in 1912, there were 426 large chimneys spewing their noxious smoke within the city limits. But the increase in the number of chimneys was greatest in the city's suburbs, particularly in the northeast around Saint-Denis, Aubervilliers, and Pantin. A similar trend occurred in North America, where urban growth outpaced that of Europe.[22] For example, in New York in 1860 some 100,000 workers were employed in 5,000 factories, but a significant new source of pollution arrived after 1880 with the founding and expansion of petroleum distillation factories. Oil industry business magnates, like Rockefeller, chose to move away from the city, to Newton Creek on Long Island or to northern New Jersey. Longstanding rural areas were transformed into vast industrial estates. Waste and petroleum residues polluted surrounding water and soil, with little or no backlash from local authorities.[23]

In the nineteenth century, cities not only had to deal with new forms of pollution resulting from industrialization, but also had to keep in check preindustrial forms that continued to increase. In both cases the solution seemed to be to distance the source of pollution from the city. As Sabine Barles noted, a city "did not become cleaner and healthier except by unloading elsewhere the deleterious miasmas that sullied its soil." Thanks to the removal of pollutions—to the suburbs, underground, or further afield into the countryside—urban mortality rates declined at the end of the century to fall more into line with rural rates.[24]

Smoke-Filled Europe

Industry changed and developed at varying speeds and in myriad contexts over the course of the nineteenth century. Initially, steam engines and urban rhythms drew manufacturing industries into the cities. Subsequently, in a second phase of nascent industrialization, complaints by the urban bourgeoisie about pollutions (but also linked to complaints about high numbers of workers), pushed for industry's redeployment to the outskirts of the city. The expansion of railway lines accommodated this transition, just as it helped connect manufacturing with its raw materials—from mining, for example—brought from even further afield, from areas expressly dedicated to their extraction. Sometimes, manufacturing shifted closer to its resource base. In Europe, industrial zones were largely concentrated around coal deposits.[25] Around 1830, however, coal was little used outside of Great Britain, with many places continuing to use animals, water, or wind as principal sources of power.[26] In the 1860s, Great Britain extracted 100 million metric tons of coal every year. This amounted to more than half of the coal mined around the world, and four times as much as was mined in the United States or in Germany.[27] Before World War I, coal mining in Britain continued to expand, but its relative global importance decreased by 20 percent, ceding its lead in coal production to the United States.

The emergence of industrial zones meant that pollutions from industry were concentrated in and around specific areas. The scattered nature of small-scale artisanal production yielded to vast areas of land sacrificed to industry of unprecedented scale. Where small workshops had previously dotted the landscape, industrialization witnessed an agglomeration of manufacturing, inspired to maximize returns through the intense exploitation of raw materials, energy sources, and labor. As a result, coal mines became surrounded by steel, metallurgical, and chemical factories—and notably coking plants, which were necessary for removing the toxic impurities such as sulfur, benzene, and phosphorus from the coal before it could be burned as a fuel. Given the polluting nature of these factories (30 percent of byproducts were simply released untreated into the atmosphere), it is little wonder that they were required to set up farther from urban residential areas. For example, in London from 1870 onward, the municipality banished such factories from the metropolitan area.[28]

In the meantime, carbonaceous smoke became a major polluting source in industrial zones. The largest steel works were located in southern Wales, Yorkshire, and around Birmingham in Great Britain; in the Ruhr area in Germany; Wallonia in Belgium; and Lorraine, France. The textile sector also adopted coal as a fuel source, and smoke-engulfed Manchester accounted for 90 percent of the English cotton industry from the 1830s onward. Such was the scale of the industry, Manchester's coal consumption grew from 737,000 metric tons annually in 1834 to more than 3 million metric tons in 1876.[29] In 1852, steam engines provided

450,000 horsepower in Great Britain (six times as much as in France); by 1870 Britain could boast that there were 100,000 steam engines in operation. They were often defective and all of them emitted sulfur dioxide.[30] The extent of the pollution created by the great factory chimneys drew the attention of numerous travelers, social researchers, and authors. Amongst them, Charles Dickens's famous novel, *Hard Times*, described the sooty sky of Coketown (a fictitious city reminiscent of Manchester) in 1854: "monstrous serpents of smoke" trailed over the city. In another passage, the same serpents were "interminable."[31]

Devastated by acid rain, the English Midlands became known as the "Black Country," with a record mortality rate: 20 percent of deaths from bronchitis in Great Britain were the result of atmospheric pollution, not counting the rise in other pulmonary illnesses and cancers. Estimates suggest that one million Britons died prematurely between 1840 and 1900 due to air pollution.[32] Around this time, the first movements against industrial smoke began to avail themselves of available statistical data to correlate the rise in pollution with concomitant increases in respiratory illnesses. In 1873, the General Register Office showed a hike in the number of deaths following several dense fogs that paralyzed London in December. Coal and steam power were incorporated throughout much of Western Europe and the United States, but certain areas held out longer. The Mediterranean regions formed one example, where only Marseille and Barcelona made the switch to coal-steam power before 1900. Elsewhere in Spain and on the Italian peninsula—in Turin and Milan—the abundance of waterways, and the absence of coal seams, rendered the alternative hydraulic power much more profitable.[33]

After 1880, industrialization entered a new phase, often referred to as the Second Industrial Revolution. Driven less by textiles—and more by iron, steel, and chemistry—it was heavily reliant on coal. On the Continent, this "age of coal" rose to dominance first in the German states, with its large coal deposits in the Ruhr region, Westphalia, and Silesia.[34] In 1850, the Ruhr produced 1.5 million metric tons of coal, and employed 12,000 miners. By 1910, some 400,000 miners extracted 110 million metric tons of coal, making the region the capital of the Continent's steel industry. Coke, derived from coal, fed blast furnaces to produce cast iron and steel. Stimulated by railroad expansion and the race to produce a weapons arsenal, metallurgical industries in the Ruhr Valley grew to immense proportions: Krupp's factories at Essen, for example, employed 30,000 people around 1900. In 1912, a government inquest revealed that the bulk of the region was polluted with smoke, soot, and deposits of sulfuric acid, which impeded vegetable growth and contaminated local fruits.[35]

In Belgium, the Liège Basin remained one of the most important in Europe. The opening of new coal mines reinforced industrial power in Wallonia, which became the principal producer of zinc. An abundance of zinc ore, known as calamine, built the reputation of Moresnet, also known as the Vieille-Montagne deposit. But although this resource boosted the economy, it proved harmful for the environment, because zinc's purification process was particularly polluting. The numerous zinc metallurgical factories that were established in Liège and Angleur released metal oxides and other heavy metals, such as lead and cadmium, into the atmosphere. The refining techniques seem to have been imported from India after 1740, where sizable factories were already in operation during the early modern period.[36] Founded in 1837, the Vieille-Montagne Zinc Mining Company introduced a new refining method, which quickly set it apart as the first multinational in Europe after it proceeded to take possession of calamine deposits across the Continent.[37] Zinc became the flagship product, reflective of a new urban modernity. Its patina was used as a waterproof layer in the roofs of industrial cities. However, like many industrial products, effluents from its manufacture produced a high environmental cost.[38]

In France, a dispersed proto-industry continued to endure, which meant that no French equivalent—in size or scale—to the significant industrial areas of the Lancashire textile region or the steel manufacturing Ruhr Valley ever emerged. Consequently, the issue of smoke pollution never became as big an issue as it did across the Channel. Nevertheless, Le Creusot, which was home to the first French steam engine for industrial use in the 1780s, became the site of one of the largest factories on the European continent at the end of the nineteenth century. Nestled in a coal- and iron-mining basin, the region was galvanized by the development of railway lines and the creation in 1837 of the Schneider Frères company. Between 1830 and 1866, the city's population grew from 1,300 people to more than 23,000. Industry was everything. Le Creusot was entirely organized around an immense factory, which formed a vast city of iron and steel at the beginning of the twentieth century. The factory covered some 1,000 hectares, extending 4 kilometers, with a 300-kilometer network of railway tracks linking the various buildings together.

Coal smoke affected almost all industrial zones, but it was particularly noticeable in cities, where it came into direct and more regular contact with people. London—where coal was burned to heat dwellings, and industrial effluents regularly affected city life—remained the world's most polluted metropolis. In 1905, the physician Henry Antoine Des Voeux, member of the Coal Smoke Abatement Society, coined the term *smog*—a fusion of the words smoke and fog—to describe and draw attention to the new toxic fog.[39] This epitomizing term, coined to capture industrial smoke's devastating impact on the atmosphere, came at the beginning of a long and lethal history.

Drains and Outlets

More insidious than smog, because it was less immediately visible, water pollution also increased dramatically because of industrialization. Prior to industrialization, it was forbidden to dump industrial waste into rivers anywhere in the world, at the risk of legal prosecution. The new liberal industrial regime, however, freed manufacturers from these legal constraints. Under pressure to increase production, engineers redefined the perception of health risks as part of their vision that moving waters cleaned all manner of hazard. This new train of thought was inspired in Great Britain, where water-closets (WCs) were introduced around 1810 as a new system of eliminating excreta by means of a modern network of drains. The British model was adopted extensively in Europe and the United States. All drains led to natural waterways, which ultimately discharged into the sea. Moreover, air pollution control measures had the unhappy effect of exacerbating water pollutions as a means of disposing of unwanted waste. This was especially the case with the burgeoning chemical industry, which was responsible for releasing a large quantity of filtered and liquefied residues into river systems.[40] The result was a new, resigned attitude to water pollution, seen as a necessity for sustainable prosperity. As Antoine Poggiale, a member of Paris' Committee of Public Health and Sanitation, indicated in 1875 in reference to the Bièvre River:

> Could we remove from the banks of the Bièvre the industries recognized as unhealthy, such as tanneries, taweries, starch-work, and dyeing, as was done under Louis XIV for the factories established on the banks of the Seine? This would require a new law; but our morals, the respect of property, as well as our legislative principles do not appear to be equipped for or desirous of such a radical measure.[41]

Whereas certain types of rural waterway pollutions in Western Europe—such as those from the retting of plant textiles—were in decline and ultimately concentrated in just a few areas such as the Lys Valley, Russia, and the colonial world, newer forms appeared.[42] Urban rivers were the first to suffer from more intense pollution. In Great Britain, a parliamentary commission on water pollution (1866–1867) recognized the explicit link between industrial activities and waterway pollution; and it was at this time that the term *pollution* began to circulate. The Irwell River in Manchester, the Tees and Mersey in Liverpool, the Aire in Leeds, the Thames, and others: all became unsuitable as sources for drinking water. In addition, fish disappeared downstream from industrial installations.[43] Textile industries were particularly harmful to rivers, as was the case for the Espierre River at Tourcoing and Roubaix. But all types of industry discharged their effluents. In the 1850s, rivers in the industrial areas of northern France and in Pas-de-Calais became "black as ink" as a result of beet juice, and eau de vie emissions from distilleries, dyeing workshops, and other textile factories.[44] In Saint-Etienne, the foul smells emanating from the small industrial Furan River, which crossed the city, were accentuated by waste products from butchers and tripe shops, tanneries, and textile factories. The majority of city rivers that supported industry along or near their banks, could expect a similar outcome. With growing health concerns and a general inefficiency in municipal guidelines and laws, authorities decided to cover over these rivers, thus transforming them into sewers. Numerous city rivers became drains for sullied water, which of course encouraged the possibility of creating still more waste.[45]

Of all the numerous activities that sullied waterways, the principal culprits were those that involved the extraction and transformation of metals. Not only did they require a lot of water, but they also produced huge amounts of residual waste. In the Ruhr around 1900, all biological life was destroyed in a tributary of the River Rhine called the Emscher, which traversed the entire region from one end to the other. No restrictions on waste disposal, or proposals for any kind of treatment plant, were proposed before 1914.[46] In the Lorraine Region, the Hayange forge— already suspected of pollution in preindustrial France—developed into a vast complex of blast furnaces founded by the Wendel family. Here the Fensch, a small tributary of the River Moselle, was considered one of the most contaminated rivers in the region in 1900: fish had disappeared and animals could no longer drink from it.[47] In Great Britain, tin mines in Cornwall and Devon, which accounted for 90 percent of world production in the nineteenth century, also had a detrimental effect on local rivers with their arsenic emissions. Tin was an essential metal for the bronze, brass, and fine metalworking industries; the rivers in Cornwall were devastated by these mines. It is worth noting that if measures taken to control atmospheric pollution sometimes led to an increase in river pollution, those taken in order to limit the amount of waste poured into waterways meant that other means had to be found to dispose of byproducts. The need to collect arsenical compounds from manufacturing—preventing them from entering waterways—had the unhappy consequence of contributing to the wide diffusion of arsenic as a repurposed pigment in a large number of consumer goods (for example, wallpapers, artificial flowers, and toys).[48]

Despite the overwhelming number of waterways detrimentally affected by mining activities, it took a long time to address the problem. In copper mines, where smoke was the first thing to draw attention, water contamination was also a recurring problem at the end of the nineteenth century.[49] In the middle of the century, England was the world's leading copper producer; copper was extracted in Cornwall, before being transported to Wales to be smelted in factories, most notably in Swansea. New mines were opening in North America, Chile, and Japan. In the United States, production jumped from 8,000 metric tons in the 1860s to 367,000 metric tons in the early 1900s. Despite evolving techniques, the increase in pollution kept pace with that of production. Vast copper mines in Montana and Arizona devastated surrounding areas and were barely kept in check before the twentieth century.[50] Meiji-era Japan underwent remarkable industrial growth, establishing a modern economy within its archipelago and internationally. It became the second-largest producer of copper in the world; the Ashio mine alone, situated 100 kilometers north of Tokyo, provided 30 percent of national production around 1895. The strategic stakes were such that Japan ignored the potential environmental consequences of its industry. Besides sulfur anhydride, wastewater carried copper byproducts (heavy metals like arsenic, lead, zinc, and cadmium) into the river Watarase, which was also used to irrigate rice paddies. Some 100,000 hectares of soil were contaminated; the extent of the pollution threatened the health of the peasantry, who petitioned local authorities in vain.[51]

It is impossible to comprehensively review each type of mine everywhere. But consider petroleum extraction, still in its infancy mid-century, but which proved to be particularly harmful to water ecosystems. Because of the remoteness of the oil fields from the cities and the euphoria that accompanied the rush to black gold, the pollutions from the first oil industry did not garner as much attention as urban air pollution. However, during the 1860s, there were numerous ecological catastrophes in Pennsylvania, Texas, and in Baku near the Caspian Sea. Some rivers in the United States contained so much hydrocarbon residue that steamboats sometimes caught fire as a result. In 1886, when pipelines and the first tankers began transporting oil, 1.4 million barrels of crude oil were accidentally spilled into the Caspian Sea.[52] At the beginning of the twentieth century, Canada, the United States, Indonesia, Persia, and Mexico were the leading oil producers. In Mexico, north of Veracruz, British and American petroleum companies established wells and pipelines that sullied waterways. In December 1910 and January 1911, close to 6 million barrels of crude oil ended up in the Buenavista and Tuxpan rivers. The exceptional spill captured the imagination, but it cloaked the chronic spills omnipresent at every oil well.[53]

A Fin-de-Siècle Atmosphere

By the end of the nineteenth century, industrialization had experienced numerous reorganizations during the Long Depression (1873–1896), during which the leading industrial countries in Europe had faced a rise in unemployment, a drop in agricultural prices, and a deceleration in economic growth. This period of crisis and of the transformation of industrial capitalism has received considerable historiographic attention. It is largely understood in the context of the fears of decline coming from the British elites: those fears linked amplified pollutions with the degeneration of the race. At the same time, industrial hierarchies were in turmoil; the United Kingdom lost its industrial supremacy to the United States and Germany. In 1913, for example, it produced only 10 percent of the world's steel (down from 36 percent in 1875); the United States and Germany were responsible for 42 percent and 23 percent, respectively.

After the Civil War, American industry continued to grow at an impressive and steady rate. Between 1865 and 1914, production grew twelvefold; the annual growth rate was 4 percent, and the population more than doubled, from 31 million to 76 million between 1860 and 1900.[54] This golden age of "rampant capitalism" involved the construction of vast industrial zones in Pennsylvania, Ohio, and around Chicago. Within a few decades, these great industrial and urban conglomerations and the smoke they produced threatened a long-standing agrarian tradition. Entrepreneurs and authorities alike struggled to take effective measures, in the face of the industrial confidence and enthusiasm that had seized North American elites. According to businessman W. Rend in 1892, "smoke is the incense burning on the altars of industry," proof of man's greatness, his capacity to transform "the merely potential forces of nature into articles of comfort for humanity."[55] However, in 1911 a report by the United States Geological Survey revealed that smoke and its consequent damage cost the American economy 500 million dollars a year.[56] Pittsburgh, situated in the heart of the Appalachian coal mining district, became the world's leading steel industry center. By 1868, the *Atlantic Monthly* referred to it as the *smoky city*: "every object in it is black" and one saw "smoke, smoke, smoke—everywhere smoke!" Pittsburgh consumed 5 percent of the coal produced nationally and had 300,000 residents in 1880. It became the poster city for steel production, just as famous for its exorbitant pollution levels, which have received considerable scholarly treatment from North American historians.[57]

Other countries like Spain, Italy, and Russia also began their industrialization process. The Iberian Peninsula became a worldwide leading producer in the extraction and production of minerals such as mercury, lead, copper, and zinc; although most of the mining industry was controlled by large foreign companies. The resulting pollutions were significant and diverse: in Andalusia, lead poisoning extended from the Sierra de Gador mountain range to the Sierra Morena mountains. Hydrargyrism, or mercury poisoning, struck people in Almaden, in the Ciudad Real province, the site of one of the world's largest mercury deposits.[58] In 1873, a consortium of investors funded by British capital, purchased the vast Huelva mining complex in Andalusia from the Spanish government. The mines were situated along the Rio Tinto; this river would later give its name to a future giant in the mining industry. All the old mining installations were expanded, and operating methods modernized to make Huelva the world's premier copper mine: its once low-volume output increased to 4,000 metric tons between 1876 and 1878, and then to 15,000 metric tons at the end of the 1880s (which amounted to 1 million metric tons of ore extract). Copper was an essential raw material for new industrial trajectories at the end of the century. Particularly polluting methods of extraction were employed by the mining company, which consisted of burning iron pyrite in the open air to obtain iron, copper, and sulfur. Surrounding cities and agricultural land were literally devastated by sulfuric acid, resulting from the 270 metric tons of sulfuric anhydride and sulfur dioxide released each day into the atmosphere during the extraction process. The fumes were sometimes so strong that the Rio Tinto company had to lay off its workforce. Arguing that there were no alternatives to their modus operandi, the company's directors preferred to pay damages to local property owners, who were obliged to accept them.[59]

International trade spurred on rapid industrialization in particular regions in Asia. Meiji-era Japan, stimulated by the state and supplying the military, surpassed 1 percent of the world's industrial production in 1897, making the country a "veritable paradise for polluters."[60] Industrialization similarly intensified in British and Dutch colonies in southeast Asia and Oceania: one famous example was the exploitation of the Lyell mountains in Tasmania, where an extremely polluting copper mine was opened in the 1890s.[61] Other coal-based industries sprang up throughout the British Empire, albeit on a much smaller scale than in Britain itself. In the Commonwealth colonies, where populations were steadily growing (Canada, Australia, and New Zealand), immigration, interior exploration, and railway lines supported and maintained expansionist projects—and stimulated the opening of vast copper and tin mines. As mining declined in Britain, this prompted a labor migration. British engineers and several tens of thousands of out-of-work miners left Britain, and flocked to operate these new mines and lend their expertise.[62] In India, where Indian capitalists reclaimed the domestic textile market previously under the control of foreign industrialists, the first large-scale factories opened in the 1860s in Bombay (cotton) and Calcutta (jute). In 1914, with 6 million spindles and 260,000 workers, the textile industry in India was ranked sixth in the world. The Tata family had also established an impressive steel company. Once the Suez Canal opened in 1869, Bombay became a key port in the British imperial system and international trade with London, Manchester, and Glasgow. At the turn of the last century, cotton manufacturers employed almost 80,000 workers in Bombay, or 10 percent of its population.[63] Industrial smoke affected the already vulnerable working class, as well as the not-insignificant class of untouchables, made up of a majority of rural migrants who returned to their villages for seasonal work, and at the end of their lives.[64] According to British experts convinced of their own superiority in these matters, the extent of the smoke observed in Calcutta was in the first place the fault of poor quality coal in India, and in the second the incompetence of local builders.[65]

In the peripheral spaces of capitalism's world-system, pollutions varied greatly depending on local economic structures and how well each country managed to integrate new global dynamics. In Latin America, industrialization first expanded to meet the needs of Europeans. Hence the development of the mining sector and the rapid growth of the food-processing industry between 1880 and 1914 (sugar refineries and packaging plants in Cuba, flour mills and refrigeration plants in Brazil and

Argentina).[66] European investments also extended into the Ottoman Empire, with France financing operations at the Eregli coal mine along the coast of the Black Sea, while numerous large spinning factories equipped with steam-powered looms appeared in the 1890s. In Salonica, industrial fumes became the symbol of a contested European presence, believed to be the root of the decline of ancient artisanal and cottage industry crafts and nature's degradation.[67] In Africa, although some initiatives were boosted by various states—as was the case in Egypt—they remained precarious. Only South Africa, as part of the British imperial system, saw rapid growth following the gold rush in the 1880s. It became the world's leading gold producer; vast mines contaminated the environment with mercury.[68]

Changes at the end of the nineteenth century signaled the emergence of new trajectories and profound spatial rearrangements of industrial dynamics on a global scale. Industrial firms were also reshaping themselves as a result of these transformations, becoming more international in their outlook as they broke into foreign markets. This new context reconfigured territories of industrial pollutions and shaped a fin-de-siècle atmosphere in which intensifying international competition served as backdrop. New materials emerged, like specialty steel and nonferrous metals (zinc, lead, aluminum). Innovative sectors, such as electricity, which began to remodel how energy was organized, stimulated production. And chemistry reconfirmed its place as a vital sector of the economy.

The Chemical Industry's New Frontier

The latter half of the nineteenth century marked the consolidation of the chemical industry. From 1830, organic chemistry no longer referred to the chemistry of animal and plant substances, but rather to that of carbon and its compounds, as a result of coal's dominance. The boundary between organic and inorganic compounds became blurred to the point where it was possible to produce both through synthesis. A historic turning point was reached when German chemist Friedrich Wöhler synthesized urea in 1828. Thereafter it was possible to supplement resources found in nature, particularly where natural supply was limited.[69] Research by French chemist Marcellin Berthelot furthered advancements in synthetic chemistry when, between 1850 and 1865, he reconstructed methane, methanol, and benzene from their elemental building blocks. In 1860 he wrote *La Chimie organique fondée sur la synthèse*, considered one of the bibles of this new discipline. Between 1860 and 1910, the number of known organic compounds grew from 3,000 to 140,000, offering a substantial quantity of new substances to the industry, most of them carbon-based, but also introducing new forms of pollution.[70]

The second half of the century saw the introduction of chemical treatises, university chair positions, and laboratories. Famous chemists such as Justus von Liebig in Germany and Berthelot in France trained future generations of chemists in an environment that worked at the interstices of university research and the industrial world. Science mingled with industrial and commercial interests. Chemists were also strategically linked to public policies to ensure the survival of the production model, where industry's financial stakes would always outweigh questions of public health or environmental protection.[71] Even though this alliance between industrialists and public bodies differed from country to country—chemical engineering, for example, had more trouble establishing itself in France than in the United States, or in Germany—it was always a key point. Chemists turned professional and quickly organized themselves into learned societies such as the Chemical Society of London (1841), the Société de chimie de Paris (1857), and the Russian Chemical Society (1868), all of which devoted much of their efforts toward industrial practice. Manufacturers also created their own societies, like the Manufacturing Chemists' Association in the United States (founded by a group that produced sulfuric acid in 1872), the Verein Deutscher Chemiker in Germany (1877), and the Society for Chemical Industry in the United Kingdom (1881).[72] In France, however, the Société industrielle de chimie was not established until 1918, motivated by the Great War and the mobilization of science in the service of the nation.[73]

The chemical industry's centers of gravity shifted toward Germany and the United States. When it came to squeezing value out of industry's unwanted wastes, the German companies reigned supreme. Major companies such as Bayer, Hoechst, and BASF were all founded at the beginning of the 1860s.[74] In 1913, for example, Germany was producing 85 percent of the world's manufactured dyes. Across the Atlantic, 170 small chemical factories in the United States only employed 1,000 workers in 1850. By the advent of the First World War, its chemical industry had become the second largest producer; it boasted 67,000 employees, and trumped even Germany in certain sectors, such as sulfuric acid production.[75] The Belgian company, Solvay, expanded to the United States, initially opening a soda ash plant in Syracuse (the first factory to integrate industrial chemistry), followed by another in Detroit in the 1890s. Some of the biggest American enterprises, many of which would become the worst polluters of the twentieth century, were established near the end of the nineteenth century. Dow Chemical opened its doors in 1889; Monsanto followed in 1901. A third conglomerate, DuPont de Nemours, founded in 1802 as a gunpowder mill, began to branch out into industrial chemistry at the end of the century.

This background explains the rapid development of industrial chemistry and the devastation it wrought on the environment. Coal chemistry offered particularly significant opportunities because of its varied compositions and the potential to exploit it for financial gain. The dyeing industry did not delay in taking full advantage of this.[76] Mauveine was the first synthetic dye and was discovered serendipitously in the 1850s when sulfuric acid was mixed with aniline taken from coal tar. However, these new dyes, which were replacing plant tinctures and metal oxides, were also the origin of many contaminants. For example, one of the first aniline dyeing factories was built near Basel around 1860, and in 1863 the owner was brought to trial over polluted waterways caused by industrial pollution. He was forced to sell his factory and leave the city. As a result of this incident, the Prussian government required manufacturers to dispose of their waste in the North or Baltic Seas, to the great displeasure of Dutch authorities who complained about this practice. The Rhine, whose flow and depth were thought sufficient to dilute the remnants of chemical residues, consequently became one of the most important sites for the chemical industry in continental Europe, and one of the most polluted rivers on the continent. In 1875 more than 500 factories lined its banks. Most were German or Swiss; no other river in the world had previously been so overrun by the chemical industry.[77]

The great range of chemical products used by the processing industry spread to control economies the world over. A diverse range of manufacturers including but not limited to glass, soap, textile, metallurgy, and wallpaper, as well as tanneries, became dependent upon sulfuric and hydrochloric acid, as well as caustic soda, chromium, arsenic, mercury, borax, and other substances. This guaranteed an indefinite demand. Sulfuric acid remained the key product; it was the most powerful and stable of mineral acids, and as lead chambers became more efficient, more of it could be produced. Thus it was produced in huge quantities: 90,000 metric tons in France in 1867, and 1 million metric tons by 1913. France was the world's fourth largest producer behind the United States (2.1 million metric tons), Germany (1.7 million), and Great Britain (1.1 million), but well ahead of Italy (410,000 metric tons).[78] Production increases were a good indicator of the chemical industry's growth globally. It was needed to manufacture artificial soda (Leblanc process), without doubt the most polluting sector of the nineteenth century with its significant emissions of hydrochloric acid, whose vapors were extremely corrosive to the skin, eyes, and respiratory and digestive tracts. The chemist Anselme Payen, an enthusiastic promoter of this sector during the Second Empire, conceded that the acidic vapors "attacked buildings' metal hinges, [and] condensed in leaves' stomata, which immediately dried up and fell off. Traveling through the air, [they] exert a pernicious influence on the health of populations living in their path."[79] As the LeBlanc process spread after 1825, Leblanc soda factories reached their apogee in Europe in the 1850s. Great Britain dominated the market, employing close to 20,000 workers near Bristol, Birmingham, Glasgow, and of course in the Liverpool area—in St. Helens, Widnes, and Runcorn—where they literally devastated the surrounding areas.[80] Along with hydrochloric acid, the waste produced was impressive. Around Liverpool in 1862, for every 280,000 metric tons of soda produced, soda factories emitted almost 4 million metric tons of gaseous, solid, or liquid residue, 500,000 metric tons of which were toxic sludge containing, amongst other things, calcium sulfide, calcium hydroxide, and other bituminous products. The countryside was desolate and for the first time forced the British government to legislate on chemical pollution (*Alkali Act*, 1863).[81] Similar issues arose in Belgium. In 1855, a government inquiry commission measured 850 cubic meters of hydrochloric acid emitted each day from factories near Namur and noted the detrimental effects on workers and locals alike.[82]

Although caustic soda factories were built away from city centers in Widnes, Namur, and Marseille, pollution was such that the industrialists were forced to consider other modes of production. The Belgian chemist Ernest Solvay derived sodium carbonate from ammonia through a simpler, more cost-effective method, which released far fewer pollutants into the atmosphere. It replaced the Leblanc process during the 1890s and completely reconfigured the chemical sector. In 1913, around 1.8 million metric tons of caustic soda were produced, much more than the 120,000 metric tons produced using the former method. By 1901, there was only one Leblanc soda ash factory left in France, near Marseille. By the eve of the Great War, the Solvay Company was the largest chemical group in the world, with a network of 32 subsidiaries in the United States and Europe, employing 25,000 people.[83] Even though the Solvay process did not emit foul odors or acidic gases, it did produce large quantities of sodium chloride, which was often dumped into rivers, creating a new pollution cycle. In Lorraine (which became part of Germany in

1871), Dieuze (1879), Sarralbe (1885), and throughout the Meurthe Valley, caustic soda factories quickly contaminated waterways, rendering the water unfit for human consumption in Nancy and Sarreguemines. In Sarralbe, the Solvay Company was forced to pay damages in the amount of 100,000 marks to the city to finance the drilling of new wells after its factory contaminated the water supply.[84]

Another branch of the chemical industry was the manufacturing of new metals, of which aluminum was one of the most important. With large deposits in the earth's crust, aluminum does not exist in its pure form naturally; its refinement involves a complex series of chemical reactions. Aluminum's emergence reveals the unprecedented alliance that developed during the second half of the century between the state, scientists, and industrialists. The first step in aluminum production involved extracting aluminum oxide from bauxite, and then transforming the oxide into aluminum by means of electrolysis. Examined in great detail in the 1850s, this new metal became—even more so than zinc—the new, modern version of semiprecious metals that were stable, pure, light, and malleable. In 1854, the chemist Henri Sainte-Claire Deville, with support from Napoleon III, began production of aluminum at one of Javel's factories in Paris. Another chemist, Henry Merle, owner of a soda ash factory in Salindres in the Gard department in southern France, extended production at his establishment to include luxury items.[85] In the 1880s, the electrolysis process—patented simultaneously by Paul Héroult in France and Charles Martin Hall in the United States—prevailed. However, the Bayer process, named for the German chemist Carl Josef Bayer, offered an easier means of production from aluminum oxide extracted from bauxite, by treating the oxide with caustic soda. In 1889, the French Electro-metallurgical Society (SEMF) began to manufacture pure aluminum and various alloys. Production was energy intensive, and as a result was often located in Alpine valleys in order to take advantage of more economical electric power—first in Switzerland, and then in Froges in the Isère department, the Maurienne and Tarentaise Valleys, as well as the southern Alps, and the Pyrenees. In 1900, 5,000 metric tons of aluminum were produced worldwide, provoking concomitant pollution on a much larger scale than before: 35 to 50 kilograms of fluorine, a highly toxic gas, were released per metric ton of metal produced.[86] The Parisian manufactories (Javel, Gentilly, Nanterre) quickly provoked significant complaints pertaining to the proximity of factories to hydroelectric power stations and bauxite mines. The history of aluminum factories in the Maurienne Valley, led by Pechiney, is entangled with environmental disputes by farmers and foresters whose livestock and trees were affected by fluorosis after 1895; bee and fish populations were also decimated, while corrosion of factory windows was also notable. In 1905, farmers were awarded substantial compensation.[87]

In 1839 a new chemical process, the vulcanization of rubber, was introduced and was equally responsible for many future environmental contaminants. Vulcanized rubber was used in myriad products (shoes, waterproof clothing, lab equipment, and others), and became increasingly important with the advent of the automobile. Michelin (1889) in France and Goodyear Tire & Rubber (1898) in the United States launched their competition for worldwide supremacy in tire sales. In 1909, Bayer began to produce a synthetic rubber. In fact, the list of extremely polluting manufacturers continued to grow; the list would not be complete without mentioning the synthesis of nitrocellulose (1846) and celluloid (1870). Both were highly flammable; the latter served as a staple for photographic and cinematographic film. Also worthy of note was Payen's work on the extraction of cellulose from wood, which revolutionized the papermaking industry, but at a high pollution cost. Similarly, after 1900 the first petrochemical products—for example, petrol, kerosene, phenol, formalin, and Bakelite—challenged and overwhelmed coal-based chemistry's stranglehold on industrial dominance.[88]

The Agrochemical Industry

Industrial chemistry also began to transform agricultural practices. Growth in city populations and resulting stresses on urban systems meant that traditional, organic fertilizers such as manure, liquid manure, and urban waste were insufficient in accommodating growing demands for food. Moreover, agronomic science cited nitrogen, phosphorus, and potassium (the famous magic formula NPK—the acronym representing the chemical symbol of each of the three elements respectively) as the best solution for the earth's needs. Ancient practices were first usurped by the provision of natural fertilizers, such as guano—excrement of seabirds comprised of the aforementioned three key elements—imported to Europe after 1840. After 1870, Europe doubled its imports of fossil fertilizers, such as sodium nitrate from Chile and Peru, phosphates from Florida and the Maghreb, and potassium from mines in Alsace. Worldwide trading of these fertilizers is an example of one of the many one-sided ecological exchanges that enriched the emerging industrialized powers, while at the same time inflicting massive environmental damage in the countries that provided the resources. Between 1840 and 1880, Peru exported close to 13 million metric tons of guano to Europe and North America; around 1900, Chile produced two-thirds of all fossil fertilizers consumed on the planet.[89]

These substances paved the way for the introduction of manmade fertilizers by virtue of having already replaced traditional methods of recycling organic matter locally. With the rising cost of guano and other mineral-based fertilizers—combined with diminishing reserves—a need arose for the chemical industry to produce fertilizers. Agricultural science was at the helm of this movement. In the eyes of numerous agronomists, primarily Liebig in Germany, chemistry was supposed to create an agriculture based on the capitalist manufacturing model.[90] For agronomist and chemist Pierre-Paul Dehérain: "agricultural science's mission should be to increase output, an honorable goal worthy of the greatest efforts, since its success will reduce the number of human creatures who suffer from hunger."[91] But the addition of chemical products, such as fertilizers, was more widely accepted in Great Britain than in France, where from the outset its advances were challenged. Country farmers were very wary of these expensive products, and in 1847, the *Annales d'hygiène publique de médecine légale* mentioned an example of cows having been poisoned after consuming clover in a field treated with ashes from metal factories.[92] In 1858, the press reported that in some departments in the north of France "farmers were being harassed by fertilizer merchants who claimed that their chemical concentrations were more efficient than mere manure. The Société impériale d'agriculture that had carried out trials to investigate such claims, warned against the use of these concentrated fertilizers, which according to the Société would in no way replace manure."[93] Confronted with popular doubt, chemists and industrialists promoted their new products by disparaging farmers and their so-called archaic and conventional practices. In France, the agronomist Louis-Nicolas Grandeau decided that one way to overcome farmers' reluctance was to educate them about the products in simple terms. In Germany, experimental agriculture stations served as part of this acclimatization.[94] Many publications also denounced the farmers' obscurantism. The director of the *Journal des fabricants de sucre (Journal of Sugar Manufacturers)* wrote in 1886: "Chemical fertilizers permit us to control farming the land, while manure allows the farmland to control us."[95]

Of all the artificial fertilizers, phosphate, rich in phosphorus, was the most important. In its natural state, phosphate was found in bones, and was difficult to use in agriculture due to its insoluble nature, not to mention the limited supply of bones. In 1842, the English chemist John Lawes treated bones containing phosphate mineral with sulfuric acid to obtain a high performance fertilizing agent, superphosphate, which he developed at his experimental farm at Rothamsted Manor, and then at his factory at Deptford Creek, a suburb of London.[96] Prior to 1880, superphosphates were predominantly used in Great Britain (600,000 metric tons). In comparison, only 100,000 metric tons were used in France due to a popular mistrust of superphosphates, as well as a recurring problem of fraud surrounding the quality of these products.[97] Even if the amount of chemical fertilizers used globally in the last twenty years of the nineteenth century remained comparatively small to that of the twentieth century as a whole, their use was already wholeheartedly entrenched in the United States and Europe. Companies like Saint-Gobain invested in an American model of superphosphates production. In 1892, a plan was established to build fifteen factories and, by 1906, the factories in Bordeaux and Montluçon were each producing 60,000 metric tons annually.[98] In Marseille, the Schloesing Company produced chemical fertilizers, and in order to promote its business it launched a newspaper called the *Gazette des champs*, intending to convert the agricultural world to "the ideas of progress." Its salesmen also carried jars of its products with them to give demonstrations.[99] Thus, the movement to promote chemical fertilizers was launched. Controversies surrounding their use were quelled by *La Culture intensive illustrée (Intensive Culture Illustrated)*, an educational newspaper published by the Société nationale d'encouragement à l'agriculture (founded in 1880 based on an equivalent publication for industry). The newspaper was in print between 1901 and 1914.[100] Germany, Denmark, the Netherlands, and Belgium applied the greatest amount of fertilizer per hectare; the United States was Europe's main provider.[101]

Synthetic nitrogen dates from the beginning of the twentieth century. The synthetic production of ammonia through the fixing of nitrogen in the air provoked a race and lively competition between European and North American chemists. The famous Haber-Bosch process—named for its two inventors and developed between 1909–1913—finally won out in Germany. It provided an artificial nitrogen fixing process, which was cost effective in producing not only nitrogen fertilizers but also various explosives, which became more important with the march toward war. This decisive innovation was industrialized in Germany during the Great War, before it was eventually used to develop a new, intensive, agricultural model based on chemical inputs.[102] By introducing these into the

Figure 4.2
Poster for the promotion of chemical fertilizers, Society Saint-Gobain. *Source:* F. Hugo d'Alési, 1896, Paris, Bibliothèque nationale de France. During the nineteenth century, the Saint-Gobain company, a former glass and mirror factory, became a major producer of new chemicals and fertilizers. Mixing the sight of peaceful countryside and smoking plants, perceived as a pledge of prosperity, the advertising poster tries to reassure the public and promote the new products for agriculture, which were initially perceived with suspicion.

soil, with only some being absorbed by plants, these fertilizers resulted in a steady source of new pollutants, which would not be properly understood until later in the century. Nitrogen dissolved into nitrate in water. Excess nitrate polluted waterways and groundwater tables through eutrophication; the process was further exacerbated with the addition of superphosphates.

In addition to this artificialization of soils came the first insecticides, another unwanted byproduct of the chemical industry. The word insecticide appeared in France in 1838, and was then applied to a product in 1858, before being adopted into English in 1866.[103] Arsenic and sulfur compounds received the most attention, and agronomic research on insecticides enjoyed support from the highest spheres of state and academic science (the chemist Jean-Baptiste Dumas was the best example). Production was stimulated by the phylloxera crisis, which decimated European grape vines during the 1870s. Several synthetic products were recommended in vain, including naphthalene (a product of carbon synthesis), carbon disulfide, and even potassium sulfocarbonate.[104] However, it was not until the First World War, or even the Second World War, that insecticides were put into widespread use.

*   *   *   *

Between 1830 and 1914, pollutions increased considerably while changing shape as industrialization progressed. Spread less thinly, concentrated more in cities and industrial basins, or pushed to the outskirts of town, pollutants provoked more complaints locally as they posed more of a risk. Coal and chemistry were the two pillars of the new industrial world, and even before the Great War—which accelerated chemistry's rise to dominance—chemical industries had become powerful multinationals with the ability to impose their definition of what constituted toxic products, and to suppress protests against them. However, in those countries still undergoing industrialization, where nuisances continued to accrue, the general population, experts, and authorities were hardly passive. As industrial acclimatization won over the elite and their experts, complaints became widespread and debates on the risks of contamination increased. Even though many scientists and economists promoted technical solutions with a view to reducing the number of nuisances, civil societies and various states sought ways in which to limit and control their effects.

5

Expertise in the Face of Denial and Alarm

The responses to the pollutions of the nineteenth century—the uncertainty surrounding physical and chemical phenomena; the close relations between experts, industrialists, and public authorities; and the growing public acceptance of industrialism—all seemed to contribute to playing down their potential harm. The contaminations produced by lead and its carbonate, white lead, a common white pigment used in paints, perfectly illustrates the difficulties of experts and populations in understanding the period's risks. Lead poisoning, from the solid form, dust, or steam manifested itself in several different industries: mines and foundries, to be sure, but also glassworks, printing, and building. Although lead production was a very old activity, controversy about the harmful effects of white lead intensified when it enjoyed a crescendo in popularity and became a fundamental feature of the new industrial society. The nineteenth century witnessed a growing cult of white: a symbol of cleanliness to cover the walls, if only to hide the black soot encrusted there. After 1830, Holland lost its monopoly: Great Britain, Belgium, and France, and later the United States—gradually became primary manufacturers and the number of white lead factories increased. For example, while its production was virtually insignificant in 1830, French manufacturing rose to 10,000 metric tons in 1850 and 25,000 metric tons in 1914, mainly in Lille, where about twenty factories were in operation. On the eve of World War I, nearly one third of the lead extracted or imported in France was used to make white lead.[1]

The danger of white lead is well known—a less toxic substitute, zinc white, was developed at the end of the eighteenth century—but its harmfulness was denied and it was never banned because certain authorities and industrialists were intent on protecting their lead-producing activities.[2] Expert authority, the legitimacy of which was built up in industrializing Europe in relation to public regulations, could no longer be

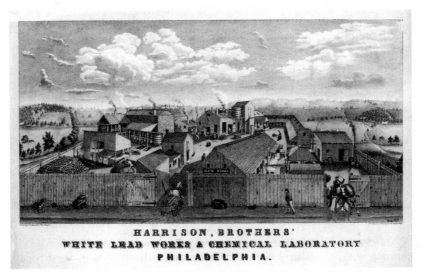

HARRISON, BROTHERS'
WHITE LEAD WORKS & CHEMICAL LABORATORY
PHILADELPHIA.

Figure 5.1
White lead factory in the United States (lithograph by William H. Rease, 1847). The deleterious effects of white lead on workers have been hotly debated by health experts since the nineteenth century. After 1850, the United States asserted itself as the leading producer of lead carbonate. The Harrison brothers' plant in Philadelphia produced a very diverse range of chemicals, from metal oxides to acids. This type of plant was at the root of the chemical contamination of environments.

treated simply as the rising force of liberating knowledge: scientific knowledge was an interested actor. Many recent studies reveal the links between expertise and the invisibilization of risk. The analyses developed by North American historians of science shed light on the complex chronology of this story, which encompasses economic, scientific, social, and political issues. They demonstrate that environmental debates were rarely independent of economic and market interests; while technical and legal expertise was recruited to solve the economic development versus pollution equation, the result invariably meant subverting the latter's importance.[3]

Even if the local, temporal, and sectoral situations varied, the role of expertise—whether of a physician, economist, scientist, engineer, or administrator—remained consistent. The expert promoted the idea of a benevolent and universal knowledge detached from local secrets and arcane artisanal practices. Such applied knowledge was designed to serve the benefit of the state and of industry. In concerns over contamination, the dominant medical expert was the hygienist, who completely exonerated industrial emanations from any responsibility in health crises. Meanwhile, the political economy encouraged the continuous development of those same productive forces. But the industrial era was deeply ambiguous, and many people challenged the apparent absurdity of promoting economic and industrial growth over human welfare. Pollution became an aporic symbol of big industry's excesses. As environmental warnings and protests arose, new knowledge emerged with them, notably an incipient form of ecological science.

Hygienism and Expertise

Faced with pollution's scientific unknowns and uncertainties, hygienists became the main actors responsible for assessing their effects and working to contain them, at least until the last third of the nineteenth century. Public health recruited from various disciplines in the process of responding to the new threats. Medicine, pharmacy, chemistry, public administration, and statistics were brought into collaboration to make up modern public health, even if its institutionalization as a field of knowledge operated slowly and very differently from country to country.[4] In the 1830s, a community of hygienists organized itself in France around the *Annals of Public Hygiene and Forensic Medicine* (AHPML). Thanks to the prestige of its members and the consistency with which it produced new findings, the journal acquired a considerable influence in Europe, putting France at the new discipline's vanguard before a public hygiene movement spread internationally.[5] In practice, public hygiene in France remained the remit of health councils—renamed hygiene and public health councils in 1848—and from then on were ensconced at the departmental level under the prefect. These were responsible for conducting the investigations of polluting workshops and factories, as well as monitoring and overseeing them. The Paris council was home to the highest academic authorities, and dominated the architecture of French public hygiene discourse. Such familiar luminaries as Parent-Duchâtelet, Louis-René Villermé, Etienne Pariset, Jean-Pierre Darcet, and Payen were members. In fact half of the members of the drafting committee of the AHPML were on the Health Council.[6] Hygienists influenced health policy throughout Europe: for example, in 1842 in England, the reformer Edwin Chadwick, former secretary to Jeremy Bentham, wrote his *Report on the Sanitary Condition of the Labouring Population*. This work was influenced by the French hygienists and served as the inspiration for the General Board of Health (1848), the importance of which he managed to impress upon the British government, as a means of better organizing local health authorities.[7]

The learned community mixed doctors, pharmacists, chemists, industrialists, and administrators, without really identifying disciplinary boundaries. For example Payen, who ran the large chemical plant in Grenelle just outside of Paris for more than fifty years—where he produced, for example, sal ammoniac, sodium chloride and lime, sugar, bitumen, which made the site one of the most important foci of industrial pollution in the capital—was an academician, renowned chemist, and professor of industrial and agricultural chemistry at the École Centrale, the most prestigious engineering school in France. In 1842 he joined the Hygiene and Public Health Council of Paris where he sat until his death in 1871. As such, Payen was one of the most influential counselors in hygiene and embodied the proximity of industry and chemistry to health expertise. Even if public hygiene promoted a more objective kind of expertise—freed from economic interests—that gained power and momentum over the course of the nineteenth century, when asked to report on pollution, hygienism was unmistakably a form of industrialism. The paradox was apparent: as an instrument of government committed to industrial novelty, population growth, and technical progress, public hygiene tended to acclimate nuisances by reassuring public opinion. The real modus operandi of the new expert councils was primarily, according to Alain Corbin:

> to reassure, to defuse anxiety aroused by stench.... The optimism they evinced toward harmful practices was based on confidence in the advance of chemistry ... By skillful propaedeutics of technical progress, the experts on the councils succeeded in getting neighborhood industry accepted.[8]

Elsewhere in Europe—in Italy, Belgium and England—similar processes were taking place, inasmuch as doctors were becoming more specialized in hospital practice or as general practitioners, leaving room for chemists to participate in hygienism.[9] Of course, the hygienist *doxa* remained tempered by numerous local contingencies. In many instances, doctors managed to direct the hygiene councils and proved to be more reluctant in reconciling public health with the new industrial realities. Where doctors and not chemists retained control over health councils, there was a tendency in their recommendations to curb the establishment of particularly polluting or dangerous industrial establishments.[10]

A brief investigation of white lead illuminates the archetypal ambiguity between knowledge and power in nineteenth-century pollution concerns. In the 1830s, the Clichy factory—founded in 1809 by the well-connected chemists and colleagues Jean-Louis Roard and Louis-Jacques Thénard—dominated national production. Regularly alerted by the hospitals that received the factory's workers suffering from lead poisoning, the Paris Health Council visited the plant several times between 1830 and 1836, drafting reports and instructions prescribing the ventilation of the workshops and better hygiene for the workers. The reports, however, never attributed a causal link between lead manufacture itself and illness. In 1835, Henri-François Gaultier de Claubry, one of the council's chemists, criticized the negligence of workers rather than the danger of lead:

> A large number of attempts have been made to reduce such accidents. Preservative masks and a host of other similar means have produced no results, because of the discomfort they pose to the worker, and the laborers' obstinacy against taking any precautions which are the most in their interest.[11]

The Dutch manufacturing process, though less dangerous, was denigrated by the French experts, mainly out of sheer patriotism; denial of the problem prevailed.[12] Having studied this controversy closely, Judith Rainhorn shows how the invisibility of the white lead problem and its harmfulness depended on the economic market and the industrialist paradigm in which hygienists were entrenched. In 1845 *L'Atelier*, a workers' paper, and other critics castigated the Clichy factory as a human abattoir. Convicts, the needy, the hungry, and the elderly without resources were its primary victims: pushed into what amounted to a death sentence amid the cry of "Lead!"[13] A prohibition decree was signed by the French state in 1849, but it was quickly abandoned; the interests of the manufacturers outweighed the protection of the workers. An administrator from the Seine department briefly outlined the stakes in 1852: prohibiting white lead would be "wrong" because it ran "contrary to science." Further, a ban would favor "fraud and immoral smuggling," which would result in "a considerable deficit in the public treasury."[14] In the name of science, which cannot be restrained, and imperatives of competition, the state must not intervene by prohibiting a product certainly dangerous, but above all very useful. Until the 1900s, the white lead industry flourished in Lille, founded on the promise of technological change and the lack of regard shown for the labor force.[15]

Based on statistics, public hygiene matured during the 1840s, under the influence of the physician Villermé.[16] He resisted environmental factors as the primary cause of disease or mortality. Rather, he focused on social conditions, concentrating on the misery and immorality of the working classes. Despite the debates it provoked, social inequality theories in the face of death and disease dominated. It raised demands for higher wages and shorter hours of work (especially for children), but without questioning the organization of work and its influence on health. The transition from medical topography to hygienist inquiry—that is to say, the shift from environmental etiologies to the examination of social conditions and forces—reinforced the link between industry and health progress. Living conditions were the new foundations of health, and these conditions were supposed to improve with industrial development. The workers became in some way responsible for their misery and their poor health. In his famous investigation of the textile industry, commissioned by the Academy of Moral and Political Sciences, Villermé observed that "the workshops are not exposed to these alleged causes of insalubrity."[17] This new etiology subsequently spread throughout Europe. In Belgium, it was universally accepted by the medical elite, even though still in the 1840s "industry was far from being recognized as that token of health defended by the Parisian hygienists." During the second half of the century, however, Belgian industry hewed more clearly to this vision.[18]

Medical and chemical perspectives on pollution remained all the more ambiguous as industrial emanations were often regarded as having disinfecting properties; some were even deemed beneficial for public health. Many testimonials attributed antiseptic virtues to factories' fumes. Even at the end of the century, it was not uncommon to find authors affirming that the smoke was "rather more useful than harmful" for the exposed population, because it reduced in particular the risks of malaria. For example, the Catholic University of Lyon professor, James Condamin, praised the hygienic virtues of the waters from the nearby Gier River, polluted by the dyeing industry. These waters "saturated with acids and tannic materials," he proclaimed, would prevent "typhoid fever" and contribute to sanitizing the river. Industrial smoke promised similar benefits: "if it darkens our houses and fills our apartments with dust," he argued, "at the same time it saturates the air with tar atoms, the inhalation of which is good for health."[19]

However, the neo-Hippocratic model of linking diseases to miasma and environments remained a powerful concept. The 1832 cholera outbreak that shook the European continent helped to reactivate interest in this older notion for a time. In Manchester, faced with concerns over smoke deaths in the 1840s, doctors were quick to launch investigations. Local organizations that opposed smoke also used the available statistical data to correlate the rise in pollution with increases in respiratory diseases. In 1866, Manchester's Medical Officer of Health, John Leigh, asserted that the excess mortality observed in the city was the result of its "flawed atmosphere."[20] Many testimonies from the Victorian period involved air pollution in the development of respiratory diseases and rickets, but also some mental diseases. Although these assertions, which were difficult to prove, were generally rejected by industrialists and authorities anxious not to slow down industrial progress, arguments warning against the dangers of smoke were increasingly common and consistent.[21] Amid ubiquitous fears of race degeneration across Europe, experts were more concerned with the role of water pollution in the spread of diseases after 1850. Workshop hygiene and water and air quality surveys gained in significance in the emerging international debate: the scientific standards for distinguishing polluted water from drinking water appeared at international hygiene congresses, the first of which was organized in Brussels in 1852. Between 1876 and 1907, thirteen international congresses on hygiene and demography were held in European capitals, followed by one in Washington in 1912. Doctors, engineers, and chemists debated various topics, including assessments of the severity of pollutions; in this context the physicians André-Justin Martin and Jules Arnoult proposed in 1889 measures to reduce "pollution from industrial residues."[22]

Around 1900, many doctors and biologists more explicitly denounced the dangers of fumes to human health. In England, Albert Rollo Russell called into question the relationship between smoke and the high urban mortality rates in the British capital in his book, *London Fogs* (1880).[23] Similarly, Parisian doctors frequently evoked the specter of fog in Manchester and London. A few years later, in France, surveys were launched to "measure the degree of air pollution in Paris." These studies marked the introduction of "pollution" as a term used by chemists in discussing the atmosphere.[24] As a prerogative of the state, public health was transformed into health science: in 1879, the new *Journal of Hygiene and Sanitary Inspection*, published by the new French Hygiene Society, testified to the rise of health concerns. A decree of the young French Republic of 1884 entrusted the Advisory Committee on Public Health, set up in 1872, with the task of studying the urban water regime; a laboratory was established in 1889. The body was charged with overseeing water quality in order to limit such epidemics as malaria, typhoid fever, and cholera.[25] After 1890, municipal health offices were established, while in the United States physicians also gradually acquired more authority over public health issues.[26] Results from the German doctor Louis Ascher, who investigated the increase in lung diseases in the smoky districts of Stuttgart and conducted laboratory experiments by subjecting animals to industrial fumes, circulated widely. In 1906, the *Journal of the American Medical Association* reported on his work. In 1907, two American doctors simultaneously published the results of their investigations proving the links between industrial fumes and health conditions. The works included compiling a large body of data collected in Europe, and concluded by asking the federal government to intervene to reduce pollution.[27]

At the same time, however, hygiene was also undergoing a radical transformation due to the microbiological—or "Pasteurian," according to the name of its French champion—revolution.[28] The Frenchman Louis Pasteur and the German Robert Koch discovered and explained a microbial world, in which tiny organisms propagated and affected the body. They also found that it was possible to control microorganisms and considerably reduce human mortality by means of prophylactic precautions. These new understandings dramatically reshaped previously understood links between health and the environment. By implicating these small, invisible organisms as prime disease vectors, medicine diminished the importance of environmental factors and the impact of pollution. After 1890, international health conferences promoted the principle of bacteriological evidence in epidemiological investigations. The International Office of Public Hygiene (OIHP), established in 1907, ratified the creation of a permanent network of hygienic surveillance. "Microbial pollution"—the expression first appeared in the 1890s—restructured the hierarchy of concerns about industrial nuisances; the new microbiology seized more and more expert attention, occupying a central place in the "hygiene battle." The function of Pasteur Institutes merged research laboratory and training center, but they also became factories manufacturing chemical products required by doctors. The lab at Lille was a good example. Inaugurated in 1898 and directed by Albert Calmette, it participated in improving sugar and beer manufacturing processes. At the same time, it supported the application of chemical fertilizers to agriculture and engaged in water purification in response to pollutions derived from industrial discharges.[29] The microbiological turn altered the role and nature of expertise in the treatment of pollution. But still, at the end of the nineteenth century in industrial circles, the hygienist remained arbitrating counsel balancing several interests more than a true defender of public health. To wit, in 1883 the physician Léon Poincaré, father and uncle of the famous mathematician and president of the Republic, began a report on caustic soda manufacturing with the following declaration: "If it is a duty for ... hygienists to protect public health against the dangers created by industry, it is also for them to protect the industrialists from the unfair attacks that sometimes public passions mount against them."[30]

The Political Economy of Nature

Where doctors and hygienists sought to reconcile the preservation of public health and the imperatives of production, engineers, harnessed in service of the political economy, were exclusively focused on encouraging industrialization. The economic sciences, charged with thinking about manufacturing activity and guiding the action of governments, explored methods of stripping economic growth away from its natural constraints and taking unprecedented possession of the world through science and technology. This industrialist obsession concentrated its enthusiasm for increasing wealth, largely overshadowing its negative effects. A transition was taking place. In the eighteenth century, physiocracy read the land and its resources as the foundation of economic wealth. Over the course of the new century's first decades, classical and then neoclassical economists transferred notions about the wealth of nations to work and capital. The models of David Ricardo and Thomas Malthus at the beginning of the nineteenth century extended the work of Adam Smith and insisted on natural limits to the expansion of production. But after 1830, further justification for industrialization demanded an escape from the "Malthusian trap" through the development of international trade and the establishment of a new energy system based on the increasing use of machinery and fossil fuels.[31] This was an important conceptual revolution. The new, steady-state spectrum focused on nature's ability to provide sufficient resources and materials to allow the expansion of production, but the issue of environmental consequences—in the form of waste and discharges—received little consideration; pessimistic theories about the inescapable depletion of land and energy were continually marginalized.[32] Dominant in the West, the new political economy recruited physics to solve the study of material flows devoted to the optimization of utility. According to the new paradigm, this was the only way to improve the human condition. It reflected a worldview that contributed to rendering pollutions invisible and altogether inconsequential, as well as eclipsing the sufferings inherited from the aristocratic past.[33]

More than ever, from 1830, liberal economists and social reformers, such as Saint-Simon's followers in France, enthusiastically lauded the Promethean mastery of nature. This constituted nothing less than a new era of continuous improvement:

The object of industry—so claims the work, *The Doctrine of Saint-Simon*—is the exploitation of the globe. That is to say the appropriation of its products to the needs of man; in accomplishing this task it modifies the globe, transforms it, gradually changes the conditions of its existence. It follows that through it man participates, apart from himself in some way, in the successive manifestations of divinity, and thus continues the work of creation. From this point of view, industry becomes a form of worship.[34]

This *episteme* deeply permeated nineteenth-century world visions. The glory of industry—a term that now referred to manufacturing's transformative activities—triumphed in different schools of thought; the word "industrialism" was also spreading, alongside "socialism." The latter also promised material progress and praised manufacturing work. In spite of certain doubts expressed by Charles Fourier as well as by Karl Marx, the machine became a source of hope, which contributed to the concealment of the predations engendered by industrialization. In both languages of progress, the environment was converted into an indefinitely reproducible commodity.[35]

Far from appearing as threats and sources of nuisance, the large industrial automatic machines and their chimney stacks that emitted fumes evoked prosperity. Paradoxically, they were the guarantors of health and sanitary cities and countrysides, because they eliminated dirty pockets of antiquated and broken artisanal craftworks. In comparison, the major steam and chemical industries, as well as their new products (including rubber, zinc, aluminum, white lead, and synthetic substances), were valued as hygienic. The grime and the stench from the older practices of working with organic materials were regarded as archaic and far more polluting. At the dawn of the 1830s, the Scottish economist and chemist Andrew Ure offered a remarkable illustration of this point of view. In his apologia for large factories, described as rational and clean spaces where everything is dedicated to order and productive efficiency, Ure argued that the enormous automatic manufacturing "[has] become an object of the first desire to mankind all over the globe." He did not hesitate to write that "it is no uncommon thing in Manchester to see engine-boilers equivalent to the force of from 200 to 300 horses generating their steam without any sensible smoke," while the mismanagement of heat in smaller and older breweries of London frequently "dissipate[d] the fuel in such black clouds."[36] In the rhetoric of most economists, social observers, and many workers, industrial fumes were celebrated as a sign of progress and employment. In 1841, the Irish essayist William Cook-Taylor exclaimed:

> Thank God, smoke is rising from the lofty chimneys of most of them! for I
> have not travelled thus far without learning, by many a painful illustration,
> that the absence of smoke from the factory chimney indicates the quenching
> of the fire on many a domestic hearth, want of employment to many a willing
> laborer, and want of bread to many an honest family.[37]

In fact, blue sky over Manchester became synonymous with a general
strike, putting the proletarian population in an awkward precarity. This
tension highlighted some of the city's contradictions: although an Association for the Prevention of Smoke emerged in 1842, it struggled to
mobilize the workers in its support of better air quality.[38] "What can we
do?" the young journalist Angus Bethune Reach asked about Manchester's smoke-caked skyline. In his investigations published in the *Morning
Chronicle* in 1849, he noted: "Purify the air ... and you will deprive the
inhabitants of bread. The sinister machine must continue to function,
otherwise hundreds of thousands of people will starve."[39]

Throughout the industrializing world, such nuisances were perceived
as a necessary evil while the resolution of industry's social implications
became a universal preoccupation. At the end of the nineteenth century,
Eugene Lebel, the head of a movement of fishermen opposed to pollutions in the Somme, recorded the response of an industrialist to a mayor
who had decided to file a complaint against the pollutions emanating
from the industrialist's sugar refinery. "If I hear of you," the industrialist warned, "if a complaint is made to the Administration, I will send
all the workers of your commune home immediately—employees at my
factory—and I will tell them that it is your fault the measure has been
taken."[40] Such employment-related blackmail and infatuation with technological progress frequently prevented any hint of intervention. Geneviève Massard-Guilbaud has called this representation of the industrialist
world "a smoking chimney culture," which captivated and dominated
Europe and North America and helped make invisible the damage
imposed by progress.[41]

Engineers, who gradually replaced hygienists as the primary experts
and managers of pollution, were even greater champions of the industrialist imagination. Already a powerful body at the beginning of the 1800s,
they continued to professionalize at the intersection between industry
and the state. In France, the École des Mines, the École Centrale, the
Conservatoire, and the École des Arts et Métiers trained the technicians
of industrial development. The *Annales des Ponts et Chaussées* [Annals of
Bridges and Roadways] (1831) became one of the new and most important sounding boards for the new engineering profession, alongside the

Annales des Mines [Annals of Mines].[42] Throughout Europe, engineers were initially responsible for the safety of steam engines and gasometers, and then by extension the many mechanical textile industries. The pollutions emanating from the mechanized factories led to the foundation of industrial hygiene after 1860, a discipline that appraised the internal health of workshops and factories. The engineers' confidence in techniques never seemed to have been shaken by protests and complaints. They sought first to adapt the workforce to industrial processes and to play down workers' pathologies.[43] The number of engineers increased rapidly in Germany as in the United States, where the number grew from 7,000 in 1880 to 136,000 in 1920. Their influence was reinforced by their numerous publications, their professional associations, and their technical skills—all proof of a credible expertise. For them, the reduction of fumes and the treatment of water were first and foremost a technical matter and an engineering problem. The development of pollution control devices, in comparison, offered financial opportunities for the business community.[44]

Economists and engineers were instrumental in naturalizing pollution and creating a worldview that valued industrial landscapes and their aesthetics. In Belgium, between 1850 and 1855, a team of lithographers gathered by the publisher Jules Géruzet made more than 200 images of the country's main industrial sites.[45] These lithographs of the "new world" offered a characteristic representation of the cult of progress and hope stoked by industrial capitalism's emergence. Similarly, in France, the engravings of illustrated periodicals—*Le Magasin Pittoresque* (1833), *L'Illustration* (1843), *Le Monde Illustré*—and, later, the popular works such as *Les Grandes Usines* by Julien Turgan (beginning in 1859) or *Les Merveilles de l'Industrie* by Louis Figuier (1873–1876) spread the industrialist creed. All of these representations in the technical and mainstream presses praised the machine, the ordered spaces, the factories' large interiors, and their placement in the landscape: all from flattering, almost idyllic, perspectives. In these press and publishing images for the general public, chimneys often emitted smoke, and black plumes were sometimes haloed with the words "work" or "prosperity" in business letterheads.[46] How could it be otherwise? Far from lamenting nuisances and damage to landscapes, travelers visiting Le Creusot, like the engineer Gaston Bonnefont, extolled the dynamism and benefits of these omnipresent fumes:

> This visitor is impressed with the heaps of smoke that spread into the atmosphere—polychrome smoke intertwined with red and blue, yellow, white, light gray and dark gray, and rising to above the city, to whom, distributed in bizarre patches, it forms like a halo which has its eloquence, and which even

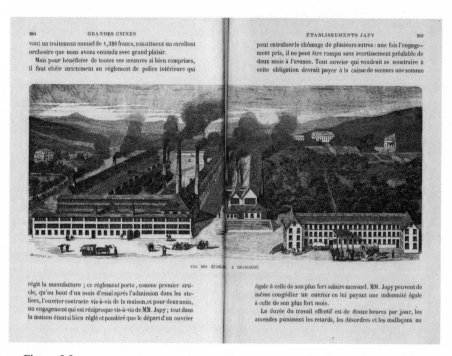

264   GRANDES USINES

vant un traitement annuel de 1,320 francs, constituent un excellent orchestre que nous avons entendu avec grand plaisir.

Mais pour bénéficier de toutes ces mesures si bien comprises, il faut obéir strictement au règlement de police intérieure qui

ÉTABLISSEMENTS JAPY   265

peut entraîner le chômage de plusieurs autres : une fois l'engagement pris, il ne peut être rompu sans avertissement préalable de deux mois à l'avance. Tout ouvrier qui voudrait se soustraire à cette obligation devrait payer à la caisse de secours une somme

VUE DES ÉTABL. A BEAUCOURT.

régit la manufacture ; ce règlement porte , comme premier article, qu'au bout d'un mois d'essai après l'admission dans les ateliers, l'ouvrier contracte vis-à-vis de la maison,et pour deux mois, un engagement qui est réciproque vis-à-vis de MM. Japy ; tout dans la maison étant si bien réglé et pondéré que le départ d'un ouvrier

égale à celle de son plus fort salaire mensuel. MM. Japy peuvent de même congédier un ouvrier en lui payant une indemnité égale à celle de son plus fort mois.

La durée du travail effectif est de douze heures par jour, les amendes punissent les retards, les désordres et les malfaçons au

Figure 5.2
"View of the Japy metallurgical factorie at Baucourt, Haut-Rhin." *Source:* Julien Turgan, *Les Grandes Usines* vol. 7, 1888, 264–265.

has its own poetry. Chimneys everywhere, of all heights. Industry is alive and breathing, powerful and continuous, testimony of a ceaseless and prodigious work. This stranger expected to find the city sad and black, a population bearing the marks of exaggerated toil and insufficient wages; on the contrary, he finds a clean and cheerful city, a population whose appearance proclaims ease and moral tranquility.[47]

Toward a Science of Pollutions

Faced with pollution, the industrialist ideology mixed denial with scientific uncertainties. Rather than orchestrating a conspiracy of silence, industry practiced a longstanding refusal to build knowledge. At the beginning of the nineteenth century, sensory experience—through smell and sight—and direct observation were the extent of evaluating pollutions. In the 1840s Chadwick considered, for example, that the speed with which laundry hung outdoors was dirtied was a good indicator of the degree of urban pollution.[48] However, as the century progressed, tools for measuring air and water quality improved, and devices for estimating the degree of contamination of a particular environment were disseminated. In 1846, two chemists carried out measurements of toxic gases over the course of several days and at different distances from the Dieuze plant in Lorraine, one of the largest chemical complexes in France. Some 28,000 metric tons of salt, 4,000 metric tons of sulfuric acid, 6,500 metric tons of soda, 1,000 metric tons of lime chloride, 3,000 metric tons of hydrochloric acid, and various other chemicals were refined and manufactured annually. The scientists collected dew, placed litmus paper (which changes color according to acidity) and wet glass plates at different points, and analyzed harvested products using chemical reagents. Their conclusions unequivocally stated that abnormal acid compounds were present throughout the industrial plant and its surrounds, noting that these "unexpected results have contradicted generally accepted opinions."[49] Their methodology was sound: in Belgium, it inspired the "war on chimneys" of the 1850s.[50]

While measurement is a tool for liberal societies, the purity or harmfulness of substances is the subject of permanent controversy. Christopher Hamlin has studied the developments of this "science of impurity" in Britain, where water analysis techniques testified to new relationships between science, administration, and public health.[51] The supply of clean water was based on scientific experience at a time when urbanization and rising consumption by city dwellers required that pipelines and tanks be built and maintained by municipalities—and monitored by a complex battery of sophisticated instruments. In 1841, for example, the Scottish chemist Thomas Clark developed a measure of "water hardness" by foaming an alcoholic solution of soap in water in order to evaluate its composition in mineral salts. This so-called hydrotimetric process was subsequently adopted in continental Europe, but the measurement of organic matter present in water remained the subject of lively debate among chemists.[52] In France, a vast survey in the 1850s, ordered by the Minister of Agriculture and Trade Jean-Baptiste Dumas and led by the geologist Sainte-Claire Deville, evaluated the quality of the waters of the country; that of urban waters in particular was of great concern.[53]

After 1880, a strictly chemical analysis of water was conducted through bacteriological analysis, and the Advisory Committee of Public Health came to define "drinking water" as that which contains "no mineral, organic or artificial, capable of harming any organism that absorbs it."[54] However, to counter the heroic narrative that interprets the rise of bacteriology and lionizes Pasteur and Koch as pioneering a new era of science—succeeding a dark, prescientific age—it is important to recall the many uncertainties that persisted in the study of components and contaminants. How do toxic products transform and dissolve in air, soil, or water? How do they interact with plants and other living things? The scientific literature on the effects of industrial pollutants on environments remained limited. Expertise was, at best, incomplete and partial. In 1883, the doctor Poincaré still denied the responsibility of the Solvay soda makers for the fish kill that struck the Meurthe River. Instead, he mistakenly attributed it to small parasites coming from the sewers.[55] Be that as it may, toxicology and bacteriology became professionally established disciplines in the second half of the nineteenth century. This was equally and importantly the case in the colonial world, where public health dictated the actual beginnings of the chemical analyses of elements imported from the West. As a result, after 1870 chemists became regularly consulted experts in India.[56]

At that time, German scientists were also beginning to discuss pollution thresholds and the capacity for rivers to purify themselves spontaneously below a certain level. This was an essential notion with a normative function, since it made low-dose pollutions acceptable rather than having pollutions completely banned. Moreover, with the chemical industry dominating global production, more industrialists were infiltrating the field of scientific expertise to counter disputes and promote doubt and uncertainty. This was particularly the case in the synthetic dyes industry, where industrialists encouraged multiple investigations that only muddied the debate and benefited their interests. In 1886, the Verein zur Wahrung der Interesse der Chemischen Industrie Deutschlands (German Chemical Industry Advocacy Organization) was founded. It emerged as a powerful lobby that promoted the idea of industrial self-regulation in water pollutions and challenged—often successfully—the severity of the nuisances its production efforts released. They absolved manufacturers of any responsibility for pollutions, even in cases where zoologists noticed the destruction of all organic life near the factory discharge areas.[57]

In the global and dynamic understanding of the functioning of potentially altered natural environments, what about "ecology"? The term was coined in 1866 by the German biologist Ernst Haeckel, and its first institutionalized forms were in place around 1900.[58] The first work of scientific ecology was published in 1895 by the Danish botanist Eugen Warming. Translated into English in 1909 as *The Oecology of Plants: An*

Introduction to the Study of Plant Communities, it summarized the main traditions of nineteenth-century botanical geography, and examined the factors that influence plant communities such as heat, light, humidity, soil composition, and animal presence. Warming's efforts laid out the research program that influenced the first generation of ecologists, and modeled the emergence of new concepts for understanding the living world. However, ecology's emergence was considerably slowed by the Linnaean tradition, which stressed that nature was the product of divine wisdom and providence that human activities could not alter. As one Aveyron botanist observed at the end of the 1830s, "nature's admirable foresight ... restores the equilibrium every day, and constantly fights against human industry, which tends to destroy the wild species."[59] Over the course of the century, other models—transformism then Darwinism—competed in positing phenomena and dynamics of species adaptation according to the principle of "adaptation to circumstances." In particular, Lamarckian transformism placed environmental influence at the center of its model of analysis, making his followers the first scientists to develop experimental ecology research. This new approach stressed the influence of the physical environment on plants, as practiced at the Fontainebleau plant biology research station, created in 1890.

If, by environment, the naturalists were more interested in altitude or climate rather than pollution, some nevertheless observed the environmental devastations caused by industry, and sounded the alarm. In 1886, English botanist Mary Eliza Joy Haweis published *Rus in Urbe, or Flowers That Thrive in London Gardens and Smoky Towns*, a book that meticulously recorded plants that resisted the worst smoke pollutions the industrial cities could offer, and those that on the contrary suffered, such as roses.[60] Atmospheric pollution mobilized naturalists and lovers of birds and plants, first in Great Britain where the cult of "Green England" appeared quite early, but also in France, where the Society for the Protection of Landscapes was created in 1901.

The issue of climate imbalance caused by carbon emissions was also raised by the Swedish physicist Svante August Arrhenius in 1896. Widely translated and discussed in Europe, his work anticipated reflections on the greenhouse effect and the accumulation of carbon dioxide in the atmosphere, and pointed at coal combustion as a major source of the problem.[61] The international scientific circulations of the 1900s definitively show that the idea of global warming produced by industrial wastes was a topic of discussion before the Great War. Louis de Launay, a

French mining engineer and member of the French Academy of Sciences, concluded one such article on global coal reserves with a warning:

> To produce some 8,000 billion mineral fuels, how much plant matter needed to be left to decompose, very accidentally preserved from combustion over the course of geological time; the day when this carbonic acid has been restored to the lower layers of the atmosphere by our factory chimneys, what changes (of which we already have the early symptoms over the large industrial cities) will not fail to be realized little by little in our climates?[62]

These early scientific reflections on the consequences of industrial pollution were the more localized birthright of the territories that endured the pollution. But scientific knowledge from the West filtered throughout the rest of the world thanks to imperial expansion. In the colonies, environmental degradation was blamed on neglect by the native populations—on the archaic nature of their production processes more than on the effects of a still embryonic industry. After 1830, the military and some colonial administrators set up a narrative attributing the destruction of the environment to the uncivilized and backward "native" and to their harmful practices rather than to new industrial rationales. In the colonial rhetoric, which required that nomads become sedentary to better assist their integration into the market economy through the establishment of private property and rational farming methods, the civilizing process aimed to restore the environment, to protect it from practices deemed archaic.[63] This imperialist vision associated industry with modernity and hygiene, but it was hardly a universal mandate. The concepts of pollution and purity were still understood quite differently across space and cultures. In India, for example, European hygienists promoted a water "purity" regime based on their secularized science, but it contradicted the local and longstanding perspectives shaped by moral and religious conceptions, and far removed from European technics and major infrastructures.[64]

Crisis of Industrialist Expectations?

The intimate and inextricable marriage between industrial production, political economy, and engineering and scientific expertise was also accompanied by multiple warnings and denunciations, which swelled in number and severity over the decades. The apocalyptic descriptions of smoky cities and the condemnation of industrial waste were common across the political spectrum. Romantic and conservative authors both castigated pollutions, and so did many republicans, radicals, and socialists, not to mention the residents of industrialized areas, the first victims of polluting discharges. They frequently excoriated not only the new world's ugliness, but also its social and environmental ravages.

Following Tocqueville's eloquent descriptions of the new industries, horror intertwined with fascination in many of the travel stories from writers of the Romantic Age. For example, in *The Rhine* (1839), Victor Hugo was struck by the industrial landscape of Liège and its monstrous activity:

> At the foot of the dark and wooded hills, towards the west, two balls of fire glare and glitter, like the eyes of tigers; while from an orifice, eighty yards above your head, issues a fierce flame, which glances over the neighboring rocks and forests. A little further on, at the entrance of the valley, is a yawning furnace, which, when occasionally opened, sends forth volumes of flames. These are the forges…. The whole valley [is] filled with what appear to be the craters of volcanoes in eruption. Some emit immense clouds of red vapor, glittering with sparks. Others define upon their reddening glow the dark circumference of an adjoining village; in other places the flames are distinguished through the aperture of some mis-shapely edifice…. This warlike spectacle, seen in time of peace, like a frightful copy of devastation, is illustrative of the progress of industry, and the vast enterprises of Cockerell…. The spectacle is indeed striking…. The scorched and smoky workmen howl like hydras and dragons at their terrible occupation, as if tormented in that heated atmosphere by the demons of hell.[65]

In England, John Stuart Mill painted an equally overwhelming panorama of Yorkshire's industrial works, where chimneys spewed "disgusting" substances: "every volcano erupts its odious contents."[66] In France, the feminist and socialist Flora Tristan denounced London as the "monster city" and the "necropolis of the world." In her travelogue, she closely associated environmental and social damage:

> In London, one draws gloom with every breath; it is in the air; it enters at every pore. Ah, there is nothing so lugubrious, so spasmodic as the look of the city on a day of fog, rain or bleak cold! When one is in the grip of such influences, one's head is heavy and aching, one's stomach has trouble functioning, breathing becomes difficult for lack of pure air.[67]

Though it only existed on the fringes, because the subject matter rested outside dominant areas of interest and expectation, painters also engaged in portraying industrial civilization. In England, William Turner was one of the first to exalt the whirlwind of industrialization. But if such paintings as *Rain, Steam and Speed—The Great Western Railway* (1844) are conflicted though energetic representations of the steam civilization, his work was hardly a denunciation of pollution. Rather it was more an exaltation of Promethean techniques.[68] During the century, the motif of the factory in the landscape became commonplace, especially with the Impressionists after 1870. Between 1899 and 1901, Claude Monet traveled to London three times to seek the play with light in the pollution fog of the city.[69] The chimneys of suburban factories merged happily into the landscapes of Monet and Alfred Sisley, and countenanced scenes of elegant promenades along the Seine. Although Vincent Van Gogh's *Factories at Asnières* (1887) extended beyond the painting's frame, they remained carefree and their fumes were light and unforeboding. Georges Seurat's dark pencil drawings (1882–1883) offered a more circumspect view of the industrial suburbs in formation. As a counterpoint, Ernest-Jean Delahaye's *The Gas Factory at Courcelles*, exhibited at the Salon of 1884, testified to the grueling nature of the work. The care devoted to the depiction of noxious vapors does not in any way tarnish their suggestive force.[70]

Across the English Channel, the rejection of harmful vapors was common in nostalgic Tory circles that had championed Green England, but also amongst certain workers and radical quarters that included the denouncement of environmental degradation caused by smoke with their social criticism of mechanization. For example, in the publications of the Chartist movement—which mobilized in favor of universal male suffrage—mention of nature polluted by smoke was a recurring theme. From 1838 to 1852, essays and worker poems published by the Chartist newspaper *Northern Star* denounced such nuisance:

Even the very stars seemed troubled
With the mingled fume and roar;
The city like a cauldron bubbled
With its poison boiling o'er.[71]

After reform efforts failed, the Chartist movement championed a return to the land and the promotion of nature as an escape from the insalubrity of the factories. The rhetoric was profoundly democratic in its politics, but Andreas Malm also identifies gestures of a proto-environmental movement.[72] Among workers, pollution issues were closely linked to the types of organization surrounding production and labor's inherent occupational health risks. In 1846, *L'Atelier* published a petition of 200 foundry workers denouncing the "miasma-laden atmosphere" in which they worked, and demanding a reduction in the hours of work in order "to breathe two hours of clean air."[73]

Faced with industrial transformations and their impacts, the ruling classes' enthusiasm for industrial progress did not completely override social anxieties, nostalgia, or the revolt of the middle classes and certain elites. Many stakeholders feared the new threats and dangers of science-based industry. In the 1850s in France, the pamphleteer Eugène Huzar announced "the end of the world by science."[74] The physician François-Vincent Raspail, an ardent republican and hostile to the cloistered and exclusive world of the scientists, was equally vehement. He was also relentless. From 1845 to his death in 1878, he railed against the industrial poisons and castigated the "cursed factories":

> Trade plays with arsenic and corrosive sublimate, as it does with molasses and potash; ... they are dissolved in vats and dye boilers, which boil over to flow into streams, wells, sewers, and rivers. Through evaporation they corrupt the atmosphere up to ten leagues; and the poison thus arrives as a public health issue by dint of drinking water, food or digestible drugs, breathable air; and finally there are public calamities in all forms of death .... And industry brings back, with brilliancy, the most disastrous epochs to public health, but the most flourishing for the commerce of medicine, pharmacy, ... and the administration of funeral convoys.[75]

By the end of the century on the other side of the Atlantic, in New York, as in dozens of other industrial cities in the United States, middle-class reformers mobilized to fight against pollution and to purify urban air.[76] Associations—frequently led by women—emerged, such as the influential St. Louis Women's Organization for Smoke Abatement or the Chicago Citizens' Association for the Prevention of Smoke. They collected funds to pay lawyers and inspectors to investigate factory emissions, fight against unhealthy industries, and put pressure on industries that resisted regulation. In this so-called progressive era, these middle-class urban women confronted pollution, which they argued contravened the Victorian values that intrinsically linked cleanliness with prosperity. By mixing aesthetic, moral, and hygienic considerations, the reformers worked to obtain the support of the local authorities. The decades leading up to the Great War were marked by an increase in alerts and warnings, a proliferation of disillusionment with industrial developments, changes in science, and new products. In England, growing disenchantment with industrialization seized the intellectual elites, inspiring a new literature infatuated with disaster and the end of the world.[77] The socialist and anti-industrialist tradition prompted by William Morris, John Ruskin, and Edward Carpenter condemned the evils of industrialization, and especially the fumes that vitiated the air of cities. Nostalgia for the rural England of old—now an endangered place—was rooted in an explicit rejection of the proletarian and miserable city. The writer Edward Forster defended a naturalistic romanticism that resisted the modern impulse toward over-artificialization. At risk was the critical link between man and the natural world.[78] Some members of Parliament expressed indignation that "the countryside surrounding the [soda] factories is destroyed for miles and becomes a desert."[79]

The expansion of industrial capitalism into the twentieth century—with its new technologies (including electricity, chemistry, telephone, wireless telegraphy, and the internal combustion engine)—and the upheaval of national hierarchies they inspired only caused further disconcert. Such a brave new world, often seen as polluted and unhealthy, generated further anxieties about alienation and loss of control. Fears surrounding race degeneration became a concomitant obsession that nourished xenophobic discourse and jingoistic nationalism. Ultranationalist and Catholic writers often linked the rejection of immigrants and the denunciation of big industry's polluting nuisances, considered to be the two great threats to the nation's purity. In his novel *The Invasion* (1907), which described the influx of Italian immigrants to Marseille, the writer Louis Bertrand gave free rein to xenophobic stereotypes while strongly condemning the pollution that corrupted the landscape—these "industrial hells … which burn and which incessantly puff under the eternal fog of the asphyxiating fumes, vomited in big broths by the glassworks, the lead smelters, and the sulfur refineries."[80]

Dystopian stories also proliferated in this context, often making industrial fumes an irresolvable contradiction and typic marker of the industrial age. In France, for example, the novelist Anatole France published *The Penguin Island* (1908), an apocalyptic satirical narrative in which the future is dominated by "artificial air" in unlivable cities, and "perfected machines" serve as the cause of an alienated population. In the British novelist Richard Jefferies's *After London, or Wild England* (1885), a cataclysm brings about the collapse of modern industrial society and returns humans to a pre-pollution state of wild nature.[81] As a counterpoint to the ever-increasing industrial influence, the theme of returning to nature—a melting pot for environmentalism—subsequently spread throughout Europe. In the German Empire, this phenomenon was particularly poignant and structured itself around the so-called Life Reform movement (Lebensreform, 1892), which attracted artists but also followers of vegetarianism, naturism, and natural medicine. In 1900, above Lake Maggiore in Switzerland, a small cooperative of vegetarians created Monte Verità, a community that focused on alternative lifestyles, promoting especially a future based on simplicity and free from the toxic artifacts of modernity.[82] In France, this aspiration appeared in a brief anarcho-naturist movement, which emerged in the 1890s, composed mainly of artisans and craft workers whose industries were in crisis, and united around harsh criticisms of the manner in which the industrial world alienated workers and destroyed nature. Henri Zisly, one of its leading figures, damned "the fumes of the factories, … the continuous deforestation, and the infected air, stained by foul smells coming from the factories; these are the causes of the disturbances of the atmosphere."[83] In many pamphlets and articles, Zisly continued to denounce the impoverishment of "Mother Earth," who "can no longer sustain us without the help of chemical fertilizers." For these authors, human history was a story of decadence, where pollution and waste were among its most telling symbols.[84]

Complaints and Social Struggles

The dangers of pollutions also provoked more violent mobilizations from workers and the factories' neighbors. These social struggles were rarely well coordinated and largely invisible because they were localized and usually relegated to peripheral areas. In France, the regulatory framework put in place in 1810 and the *commodo* and *incommodo* procedure favored the expression of these complaints, which the administration archived with care, and which Geneviève Massard-Guilbaud has studied closely.[85] The individual letters or collective petitions sent to local authorities expressed fears over bad odors, signs of rot and disease, and concerns over the questionable color of water contaminated by industrial residues. In the north of France, fishermen and brewers who depended on clear water frequently complained about the effluent discharged from potash factories and sugar refineries. The waters, they claimed, "could only be compared to shoe polish."[86] Quarrels between industry and its neighbors were ubiquitous. An example among a thousand others: in Dijon in 1859, a confectionery manufacturer wishing to install a steam engine faced opposition from a neighbor who demanded that this type of equipment be kept outside the city. A few years later, while a mining engineer reassured the public that the inconvenience would be minimal, the inhabitants of the district—owners, coffee makers, pork butchers, carpenters, clerks, vintners, manufacturers of vinegar, watchmakers, and even the Sisters of the Visitation—protested collectively to the Prefect against the damage caused by burning coal:

A stinking and thick smoke, as well as an enormous quantity of dust and dark matter, which ruin the atmosphere and get into the respiratory system, are detrimental to public health. The same dust, carried away by the wind, continually penetrates homes, even when all openings are closed; it dirties the goods and furniture of a whole popular district and causes considerable damage to the neighbors. Often even the residues are carried by the air before their complete combustion, fall all inflamed in the attics which surround this manufacture, and can become a cause of fire.[87]

Such protest rhetoric was based on subtle strategies aimed at generating the compassion and interest of the authorities by establishing legitimate demands. The rhetoric often articulated a defense of health, appeal to the public good, risk of financial harm, and even calls for the respect due to honorable inhabitants. In 1848, the inhabitants of Le Rouet near Marseille opposed a plan to install a chemical plant that threatened to transform an entire district into a "desert" by destroying a "neighborhood covered with gardens and pleasant homes." They denounced "the factory that had fought its way into their midst despite their complaints, their alarms, their opposition." This was a problem: "industry tramples on the laws of the nation."[88] Even at Le Creusot, a space entirely dedicated to the Schneider industry, residents in 1908 complained to the prefecture about the extent of "black or yellow, stinking smoke" that made the air unbreathable, tarnished the silverware and caused "vomiting," and "respiratory tract disorders."[89] This case would probably have been quietly ignored and lost to history if it had not mobilized the physician Félix Martin, both senator of Saône-et-Loire and resident near the plant. In Belgium, similar mobilizations were common in 1848 against the zinc factories, for example, and their "harmful emanations ... which were already killing the trees and creating massive wastelands."[90] In England, the protests of the landed aristocracy, gentry, and municipal authorities were heard at the opening of the first soda factory in Liverpool in 1823, and then increased to become a national issue during the 1860s. In response to these protests, the British Parliament created the Alkali Act of 1863.[91] On the other side of the Atlantic, pollution opponents argued that bad odors constituted a threat to the nation.[92] Despite many similarities, strong national specificities typified the course of action taken. While the movements that burgeoned from civil society in England and the United States played an important role, such mobilizations remained weaker in Germany. Such distinctions could be attributed to the attitude of the authorities and to the different structuring of relations between state and civil society in each country.[93]

Economic context and social standing of both protesters and industrialists also influenced the intensity and character of disputes. The organization of stakeholders also mattered. For example, fishermen spearheaded numerous challenges to pollution; in protecting their own economic interests, they became both watchmen and sentinels of water quality.[94] In 1888 in Belgium, fishermen's associations were born simultaneously amongst the working classes in Ghent and Liège. The number of such groups quickly multiplied well beyond the country's two main industrial centers, and a central society was created as early as 1890 in order to "establish remedies to water pollution." Where the initial concern focused on engaging workers to protect local food resources, the threat to fish created an ever-broader context or opportunity for fighting against the pollutions themselves. For example, in 1890 the Liège fishing group mobilized against the alteration of the waters of the Vesdre by the wool industry. The struggle became more radicalized in the following decade. In 1905, a new society was created in Brussels to fight "against unscrupulous industrialists who dump the residues of their plants and factories in rivers, canals and tributaries, poisoning not only thousands of kilograms of fish but also on occasion bringing death to local residents and livestock."[95] Doctors and scientists were often in the vanguard of these associations, which tended to broaden their objectives, following the model of the Dutch Association Against Water, Soil, and Air Pollution. In 1908, the former Angling Society of Aerschot and its surroundings became the League Against Water Pollution, before morphing in 1913 into the Association Against Water, Soil, and Air Pollution. The issue received considerable visibility at the Universal Expositions in Brussels (1910) and Ghent (1913), during which the Central Fishermen's Society published a general map of polluted rivers in Belgium, where the streams of the Scheldt and Meuse basins were marked in black, indicating a permanent contamination.[96]

The working classes—laborers and peasants alike—had neither the means nor the resources to oppose industrialists. In Great Britain, for example, costs of legal action, the only remedy, remained prohibitive.[97] More generally, it was difficult to obstruct pollutions when livelihoods depended on the preservation of industrial activity, however toxic. This was especially difficult in a hypermasculine labor culture that delegitimized complaints. Except on rare occasions, labor had no audible voice—there were no codes or negotiating contexts in which workplace nuisances could be reduced—which further internalized popular acceptance of pollutions as a part of everyday life. In the Ruhr, the working population was largely immigrant; no mass protest movement developed against an evil that appeared in equal parts necessary and fatal. As one inhabitant in 1919 recalled: "We were totally fascinated by the size of the big industry and by the gigantism and the genius of human efforts ... we accepted the poisonous fumes of the iron and steel factories as something inevitable and almost no one complained."[98]

For all that, the rise of unions gradually provided workers with some political power and a platform to air their concerns. In the 1860s, workers' delegates at the World Expositions began to discuss the risks of poisoning, especially from the deadly dusts in which workers were immersed. In a popular conference devoted to hygiene in 1867, the worker L. Barbier beseeched his companions to "guard against the organic disorders so often produced in factories and workshops, against the poor condition of the premises, and against the use of unhealthy substances."[99] As in the trade unions, occupational health was frequently invoked at socialist congresses devoted to hygiene issues. The 1889 Workers' Congress of Paris demanded "the prohibition of certain kinds of industries and certain methods of production prejudicial to the health of workers," while another Congress organized at the Lyon Labor Exchange in 1894 required "the prohibition of the use of any industrial process recognized as irreparably harmful to the health of workers."[100] These demands aroused rebuke from some doctors, such as Henry Napias, who highlighted the irony: to "remove the dangers of an industry by removing the industry itself is a somewhat crude economic process, even a little childish! ... The railways cause accidents; must we return to the stagecoaches?"[101]

Workers' organization deepened against such attempts at dismissing labor's interests. At the end of the nineteenth century, the painters' union in France demanded a complete ban on white lead. Similarly, the National Federation of Factory Workers in the Match-Making Industry of the State succeeded in prohibiting the use of phosphorus—which caused necrosis of the jaw—in the manufacture of matches. Subsequently, phosphorus was banned throughout Europe between 1898 and 1906.[102] The General Confederation of Labor (CGT, 1895) inscribed the "suppression of professional poisons" on the agenda of its first congresses and recruited the Bonneff brothers—celebrated journalists sympathetic to labor—to conduct an investigation into the subject to convince public opinion.[103] In Parliament, the socialists tried, however unsuccessfully, to introduce a law on occupational diseases that would complement an 1898 law on industrial accidents. The Great War and its imperatives introduced a sharp inflection in the positions taken by unions over health in the workplace. Reform legislation dealing with reparations for occupational diseases prevailed after 1918; the fight against industrial poisons subsequently disappeared from the General Confederation of Labor's agenda.[104] Nevertheless, while the question of pollution at work undeniably emerged in the claims and the trade union movement before 1914, it remained secondary in relation to those of wages and hours of work.

In some situations, however, the extent of pollution sometimes resulted in mobilizations turning violent. In August 1855, popular unrest broke out in the Lower Sambre region in Belgium against caustic soda factories accused of decimating the potato crops. Since their establishment in the 1830s, farmers and landowners had been wary of the new industry, which they feared would upset their crops, livestock, and health. Regularly dismissed by industrialists, mining engineers, and doctors convinced that the acid discharges were innocuous, village complaints became increasingly vociferous alongside increases in emissions. A riot eventually broke out during the summer of 1855, spurred by scientific debate and severe socioeconomic tensions that accompanied the collapse of the old, proto-industrial activities in the Belgian countryside. The insurgency against pollution was akin to food riots: the protesters denounced chemical factories controlled by local dignitaries and accused them of starving the people by destroying crops. Placards in various parts of Namur proclaimed "Death to the hoarders" and "Demolish the chemical factories." The scale of these protests turned what had been a local dispute over a chemical nuisance into a vast, national controversy that forced the state to intervene. At the dispute's boiling point, hundreds of rioters stormed the factory before being violently repressed by the army, killing two people.[105]

In 1910, local residents of the calanques inlets near Marseille also mobilized in defense of the "natural" beauties of the Provençal coast, which were under assault from the activities of the multinational chemical company Solvay and its soda factory. Social networks merged around currents of regional identity to confront the faceless industry. The writer Frédéric Mistral supported the protest. Indeed, some fifty cultural and sports associations organized a mass demonstration of 2,000 people on March 13, 1910. The event would serve as a "place of remembrance" for subsequent mobilizations during the twentieth century.[106]

Another riot, of greater magnitude—without a doubt the most significant "ecological" revolt of the period—broke out on the site of the copper mines in the village of Río Tinto, near Huelva, in Andalusia. As early as the 1840s, sporadic complaints about mineral processing methods, which destroyed vegetation and affected public health, were registered by the Spanish state. The situation worsened with the arrival of British capital in 1873, which altered the scale of production. Opposition to the copper mines organized: its strength lay in its capacity to unify a diverse array of actors around the same protest. Big landowners gathered under an "antismoke" league (*liga anti-fumista*), alongside peasants and immigrant workers. The charismatic anarchist militant Maximiliano Tornet succeeded in mobilizing the workers by linking the protests against pollutions to more usual wage demands. Confronting upper and lower classes in opposition to the mine, the Spanish state and the medical elite of the country joined forces to defend industrialist interests. This only crystallized agitation against perceived social injustices and the iniquities of the powers having made the choice to sacrifice the general interest for the benefit of a few. Moreover, the English presence united the disenfranchised and the landowners against the foreign capitalist aggressor by playing on the sensitive chord of patriotism. At the beginning of 1888, the situation intensified, and on February 4, during a demonstration which consisted of nearly 15,000 people, the army opened fire. More than 100 people were killed (48 officially), making this dramatic event—all the more exceptional because pollution had been its catalyst—one of the most violent antiworker repressions in nineteenth-century Europe.[107]

On the other side of the world, workers and peasants in Meiji period Japan were galvanized by the intense industrial modernization and the pollutions of the Ashio mine, one of the biggest copper mines in the world. The rising anger manifested itself in the outbursts of violence against the "monopolists" during food shortages and famines. Villagers defended their rights by seeking to protect their inheritance and land ownership against big industry. In a sense, protesters were trying to scale back industrial progress by attempting to preserve older community social relations, which were threatened by toxic mines. For a long time, the miners had been asking for improvements in their working conditions. In the 1890s, these popular complaints were relayed by the deputy Shōzō Tanaka, local representative of the devastated regions, on behalf of small village chiefs and middle peasants. Tanaka emerged as spokesman and champion for the peasant victims of polluting mines. In 1891, peasants organized themselves into "Forum for the Salvation of Ashio," but they only obtained compensation. The fight continued at the local level and in the Chamber of Deputies; in 1896, 2,000 peasants marched on Tokyo, 3,000 the following year, and 11,500 in 1898, but they encountered police violence and dozens of them were brought to justice. In 1901, Shōzō Tanaka resigned his mandate and appealed directly to the emperor by handing him a petition in which he virulently denounced "the toxic leaks [which] are increasingly serious because of Western-type mechanization." Against the repressions inflicted upon the protesters, the petition beseeched the emperor to work for the purification of water, to "eliminate highly polluted land" and "to revive the extraordinary natural wealth of the coastline," "put an end to a mining industry that poisons—and permanently stop the flow of polluted water and toxic waste."[108] The mobilization reached a national scale: 800 academics from across the country traveled to Ashio to support the protest. Opposition to the mine increased after the Russo-Japanese War, which temporarily muted the protest. In 1907, a riot broke out. For three days, the workers dynamited and torched the industrial complex, and were severely repressed by the military.[109] As in many cases, the productive imperative of modernizing the country outweighed any other considerations. Despite their resistance, most local victims were forced to leave their polluted homes.

* * * *

Between fatalism, resignation, complaints, and insurrections, pollutions were hardly accepted in tacit silence and the general denial of hazards that sometimes received mention. A body of knowledge and practical application was gradually and unmistakably being put in place, even if it did little to interrupt or even slow the increase in pollution. Over the course of the nineteenth century, the negative impacts of industry on the environment became increasingly well known. The word and modern meaning of "pollution" entered the popular lexicon during this period, but a political process of acculturation was also established. If the learned and intellectual elites, engineers and industrialists, chemists and economists tended to diminish the hazards and accommodate pollutions so as not to slow down the advances of progress, which came to be exclusively identified with big industry, other segments of society saw in the phenomenon of pollutions the manifestation of the ambivalence of a development fraught with hazard. Between these deeply ambivalent and antagonistic interests, the state and local authorities tried to reconcile differences and pacify the deployment of large-scale industry and neighboring populations.

6

Regulating and Governing Pollution

The tragic events at the Ashio and Río Tinto mines revealed just how difficult it was to effectively manage pollution and its consequences. Copper's extraction and production are particularly illustrative. Thanks to its physical and chemical properties—most notably its resistance and good thermal and electrical conductivity—pure copper and its alloys enjoyed multiple applications; its production grew exponentially. At the beginning of the 1840s, Europe produced 35,000 metric tons of copper per year. Most of it came from Great Britain (25,000 metric tons); Wales was the center of the copper industry: it accounted for 90 percent of all British copper. On the eve of World War I, production changed scale. With even more applications—particularly in the electrical sector—demand grew, leading to the development of mines and foundries in Spain, Chile, Japan, North America and, later, the Congo. Worldwide copper production in 1890 reached 280,000 metric tons. It rose to 500,000 metric tons ten years later and surpassed 1 million metric tons in 1914, the same year production ceased in Great Britain.[1]

Copper ore needed to be cast and refined, usually in reverberatory furnaces. These complex techniques were energy intensive and polluting. After 1850, copper foundries worked closely with sulfuric acid factories, from which they obtained critical supplies. In Wales, Swansea—nicknamed *Copperopolis*—provides an example of how difficult it was to reduce the environmental impact of these processes, and the panoply of available regulations. Up until the 1880s, 75 percent of the world's copper was refined in Swansea, including ore that came from Chile and the United States. Its factories released 92,000 metric tons of sulfuric acid into the air every year. Very early on, this situation caused residents to complain, forcing authorities to intervene and manufacturers to search for technical solutions to limit discharges and avoid government intervention.[2]

Faced with mounting pollution, numerous legal, regulatory, and legislative responses emerged: lawsuits, municipal bylaws, and state intervention. Blackmail—to enforce work and competitiveness—was the authorities' credo, designed to legitimize the sacrifice of certain zones and people in the name of public advancement and national need. Fundamentally liberal and industrialist, the regulatory regime was regularly subject to various modifications in response to the political and economic changes of the period and the construction of nation-states and empires, as well as differences in legal traditions. The first point of contention was the relative importance accorded to justice, which always dominated in *common law* countries, while administrative intervention was preferred on the European continent. Whatever the elements of law, controlling pollution had to happen on a national scale. Given the vague interventionist hopes of the authorities, specialized solutions took hold in the form of recycling, calls for innovation, and improvements in manufacturing processes. In all cases, a veritable naturalization of pollution was taking place. As and when pollutions became unbearable, the ultimate solution was to banish them into the atmosphere through tall chimneys, into the sea, or into unpopulated areas.

Common Law Regimes

After 1830, industrialization continued to shake up the rules of law. In the Anglo-American world, common law remained the principal method of recourse to protect citizens, property owners, and their neighbors against industry's nuisances. It is easy to explain how this judicial system worked in Great Britain and the United States. According to the liberals, pollution was simply a collection of negative externalities created by some and tolerated by others—with the "invisible hand" of the market ready to amicably restore equilibrium between polluters and victims through compensation or litigation.

But as industrialization became a fundamental part of society over the course of the century, case law had to develop in tandem with industry. Judges' decisions varied depending on complex sociopolitical power relations. In litigation cases, if the defendant was an industrialist of moderate standing from a nonstrategic sector, with a small workforce and low capital investment, he would most likely lose to larger property owners. On the other hand, a powerful industrialist with significant capital, confronted by a more popular and poorly organized complaint, could generally expect to win.[3] This logic was demonstrated by a comparative study of the American states of New York, New Jersey, and Pennsylvania. In Pennsylvania, where heavy industry based around coal and steel dominated in Pittsburgh, judgments rendered between 1840 and 1906 found predominantly in favor of industrialists, in spite of significant pollutions. Conversely, in the other two states, where industry was more fragmented and less powerful, rulings were more varied and more in step with the traditional doctrine of common law, sometimes finding for the plaintiff, despite less intense pollution. Therefore, after the Civil War northern industrial regions, where the pro-business Republican Party held sway, tended to find in favor of advancing heavy industry. The majority of judges were affiliated with the party. Thus, when New York became a Republican state in 1893, rulings there suddenly mirrored those in Pennsylvania.[4] Elsewhere in the United States, judicial decisions depended upon the pollutants in question and the types of industry under examination. If judges were severe in cases of ancient artisanal nuisances relating to putrefaction techniques (tanneries, animal fat foundries, and the like), they proved to be more lenient toward pollutions emanating from modern industrial practices and those sectors—such as steel or chemistry—that symbolized the new economic preeminence of the young country.[5]

A similar trend of partiality toward significant economic interests was also common in Great Britain. Influenced by American jurisprudence, English civil law abandoned the principle of precedence for property owners in favor of a reasonable use policy, particularly in relation to running water. In other words, in the case of pollutions, a waterside resident would no longer have recourse to historic rights, but would have to instead prove unreasonable use of the water by an outsider.[6] The Barlow Judgment, passed in 1858 following a court case near Birmingham, precipitated this pivotal change. However, as in the United States, it had less to do with the internal evolution of the law, or the application of a particular doctrine. Rather, the primary impetus for change rested with how a particular judge interpreted and understood the social and cultural contexts in which these environmental disputes took place, and from which they were brought before him.[7]

Many debates were raised amongst legal scholars to discuss whether or not industrialization had altered the course of common law's adaptation and evolution. That is to say, disagreement abounded whether common law was a hindrance to industrial progress, and whether it was capable of solving pollution problems, or if it needed bolstering with legislative and regulatory acts. Many thought pollution law should be based on some kind of cost-benefit analysis—the balancing doctrine—in order to manage the nuisances and disagreements that industrialization provoked, and that, from this point of view, it corresponded to the general interest.[8] But as it was practically impossible for judges to calculate real costs and benefits—this was never actually done in the nineteenth century—judicial decisions had more to do with social power struggles.[9] This explained why, from a certain mistrust with regard to industry and its nuisances, the law turned progressively—albeit in fits and starts—toward a judicial system that became much more tolerant of polluters.[10] Judicial regulations depended on such diverse reasoning and circumstances that complete generalization was evidently difficult. In Great Britain, industrialization was grafted onto a traditional, urban, and aristocratic society. By contrast, in the United States, it constituted more of a pioneering occupation. As a result, judicial treatments varied widely. In cities built as a result of, or specifically for industry, there existed little regulation. Conversely, in cities that permitted only certain types and sizes of factories—and zoned their activities outside of the city limits so as not to upset the vacationing bourgeoisie—disputes and litigation were more commonplace. Thus, in spite of an overall movement to create case law that would favor industrialists, common law—up until the early 1900s—still played a role in safeguarding neighborhoods. Judgments were on occasion severe with regard to industrial polluters; some even included prison sentences.[11]

Common law's incapacity to fight pollution on a global scale can be explained by factors linked to the practice and organization of differing justice systems. First, in the Anglo-American world it was costly to go to court, and so preferable for the ordinary citizen to come to some sort of an arrangement with the industrialists, either by mutually agreed upon private contracts or clauses to compensate for damages or inconveniences, or purchase of properties, or some other measure. Second, fines were always minimal and thus had no impact whatsoever. Finally, judges struggled with attributing blame. In the case of smoke, for example, how could they decipher the guilty party when tens, or even hundreds of factories were in operation? Justice could only rule on more specific cases and rights. But even then it was still difficult to calculate damages—or to adjust them according to the number of guilty parties and victims to compensate.[12]

Regulation or Legislation

Faced with such legal impasses and a continual rise in pollutions, the justice system became ineffective and incapable of managing these new troubles. More and more frequently, the search for solutions turned to public authorities and the development of local or national regulatory controls. This phenomenon was common in Great Britain from the beginning of the nineteenth century, but only became an issue in the United States at the end of that century. Even if liberalism slowed the process and favored local over national initiatives, the move toward state administration was evident. As such, from the beginning of the 1840s, industrial smoke reached such unbearable levels in Great Britain that particular cities took matters into their own hands.[13] The *Municipal Corporation Reform Act* of 1835 reorganized and standardized municipalities' powers. Heretofore each municipality had functioned independently. Amongst the newly laid prerogatives was the guarantee of public health and safety for all citizens, street sanitation, fire prevention, provision of drinking water, and waste management.

In Manchester—the best-known case—local courts of justice, the Court Leet, deteriorated, inflicting fewer and fewer fines and, through collusion of interests, proved especially lenient toward one industry considered a pillar of local identity. The Court Leet was abolished in 1846 and replaced by municipal regulations, overseen by police commissioners. Nuisance inspectors were employed from the beginning of the century, and Manchester experimented prematurely with a local policy to regulate pollution without judicial oversight. It was no success.[14] The Manchester Association for the Prevention of Smoke (1842) pushed the city to pass the *Manchester Police Act* in 1844, which enforced fines for industrialists, and in 1847 the municipality created a smoke inspector position to collaborate with manufacturers to improve the situation. Other *Improvement Acts,* intent on reducing the level of smoke, were adopted during the 1850s in Birmingham and Sunderland (1851), Newcastle (1853), Leeds (1856), and eventually by most other large cities in Britain.[15] In 1846, and again in 1854, Liverpool passed a *Sanitary Amendment Act,* which hardened the fight against smoke. A health council was formed to monitor the measures put in place to control industrial pollutants. A nuisance inspector gave out 653 fines between 1854 and 1866. In Birmingham 360 were issued, and 445 in Sheffield in the same time period.[16]

In London the situation was particularly complex due to the convolutedness of metropolitan jurisdictions. In 1841, the Common Council requested a report on industrial smoke which led to a parliamentary interrogation. The idea to regulate smoke emissions took shape and, in 1851, the City of London was granted the right to fine manufacturers for smoke emissions, but only within its local districts, which excluded a large part of the metropolitan area. The 1853 Smoke Nuisance Abatement

Metropolis Act ordered all owners of industrial furnaces (and not just those run by steam engines) to burn their smoke. By 1887, some thirty inspectors were employed full time to oversee compliance with the 1853 Act. In 1857, the metropolitan area also created the Thames Conservatory Board, charged with monitoring the purity of the river. In 1870, the Metropolitan Board of Works ordered all offending manufacturers to move their businesses outside of city limits.[17]

In spite of all of these efforts, municipal policies were often limited in their effectiveness due to a lack of means for in-depth investigation or intervention. Typically, fines were set so low that they hardly acted as discouragement; there was a £5 maximum in Liverpool, even less in Manchester. Authorities were reluctant to hinder industry; rather they practiced persuasion and encouraged technical changes. In London, after a few months of zealous enforcement of the 1853 Act, during which 124 convictions were handed out, industrialists succeeded in reinterpreting the ruling to benefit themselves. After three decades of experimentation, one can only remark on the failure of local regulations. By 1860, pollution levels had reached unprecedented highs.[18]

These local failures forced the state to intervene. If the state intervened early in overseeing labor in Great Britain—with the creation of the post of inspectors of manufactures in 1833 and the vote of several Factory Acts regulating the dangerous activities like the mines or the railways—environmental regulation came later.[19] The Smoke Prohibition Act of 1821 for smoke emitted by steam engines rapidly fell by the wayside. However, with pressure from various associations and members of parliament, a parliamentary Select Committee on Smoke Prevention was finally created in 1843. Upon investigating pollutions in Manchester, Birmingham, and Liverpool, its members agreed on the necessity of national legislation, without actually referring to the French model. Drawing on the works of Chadwick, and with the creation of the General Board of Health, several laws gave local authorities increased power to control pollution, such as the Health of Town Act (1847), the Public Health Act (1848), or even the Nuisance Removal Acts (1846, 1848, 1855). Above all the 1875 Public Health Act (from which London was excluded) put imperative structures in place (such as health officials), which allowed private individuals to have access to a justice of the peace. But, despite these unifying sanitary measures, these laws did not directly address the problem of industrial pollutions, and thus did nothing to lessen growing anxieties.

After the 1858 Great Stink of London, royal commissions were put in place to improve the rivers' water quality. Virulent gas emissions from the chemical industry added to the black, coal smoke. In 1862, at the request of Lord Derby, a conservative parliamentarian, two-time prime minister in the 1850s, and property owner near Liverpool, the House of Lords established a Select Committee charged with investigating caustic soda factory pollution and ways to reduce it. Evidence collected by the upper house—a space where landed property owners addressed their concerns—revealed the extent of the problem: raw waste, burned fields, and sullied waters, most notably around Widnes and St. Helens, near Liverpool. Caustic soda manufacturers were accused of harming the country's agriculture and menacing private property. Public authorities could not ignore these results and passed an *Alkali Work Act* in 1863, which would endure for five years and be the first British law to specifically target industrial pollutions. Certainly audacious in a country where the new elite class was largely liberal, this law only applied to the manufacturing of caustic soda.[20] It severely limited toxic emissions (95 percent of hydrochloric acid was required to be condensed), and for the first time ever a body of state inspectors, under the direction of the Home Office, was mobilized to monitor its application. Under the direction of Robert Angus Smith, four inspectors were charged with inspecting factories in order to enforce the new rules.[21] Smith's approach was more preemptive than punitive, which might explain why British soda manufacturers ultimately chose to obey the law, rather than not. They saw some advantages to this legislation as a means of protection from lawsuits and mounting pressure by locals. Manufacturers were all placed on an equal footing thanks to emissions standards that applied to everyone and went beyond local regulations. Moreover, this legislation, much like the 1810 law in France, protected industrialists from being accused of *public nuisance,* which kept them out of the penal justice system.[22]

After the *Alkali Act,* there was a shift toward technical improvement in the fight against pollutions. This approach, even with limited resources, was successful and in 1868 resulted in the renewal of the law for an unspecified amount of time. In 1874, its scope was broadened to include other noxious gases as well as other industrial sectors that emitted acids and sulfur. However, after the emphasis on technical modifications had been in operation for several years, Smith noticed some adverse effects. The condensed gases had been emptied into rivers around caustic soda factories instead of into the atmosphere, and leftover sulfur from chemical factories had amassed. For each metric ton of soda produced, one and a half metric tons of waste heavily laden with sulfur was generated. Smith expressed concern over porcelain manufacturing, cement factories, gas lighting works, salt manufacturing, and copper metallurgy, all of which emitted a great many toxic composites (including acids, arsenic, and antimony). In 1859, particularly concerned with copper factories' pollution, Smith coined the term *acid rain*. The copper industry's powerful magnates resisted accusations about its deleterious effects (all the while paying compensation to their neighbors). In 1878, members of a royal commission created to report on the subject said that "although the amount of acidic gas emissions from copper factories was impressive ... , there is no need to impose overly stringent measures which would negatively impact the future of these factories" and "harm numerous people."[23] In 1881, the *Alkali Act* encompassed seventeen different types of industry for inspection; subsequent *Alkali Acts* in 1891 and 1906 expanded its reach. Within the scope of half a century, under Smith's direction (he was chief inspector between 1863 and 1884) and that of his successors Alfred Fletcher (1884–1895) and R. F. Carpenter, national legislation gradually replaced legal precedence of common law and municipal regulations, thereby demonstrating a change in scale of pollution.[24]

For a long time, water pollution was not regulated under national legislation. Indeed, the *Public Health Act* of 1848 encouraged cities and industrialists to dump their waste into rivers or the sea. Chadwick's famed sanitation system also encouraged this means of disposal. Moreover, the *Alkali Act* and its call for industry to condense acids before disposing of them, further aggravated the problem. In 1876, however, the *Rivers Pollution Prevention Act* put the project of finding a resolution squarely in the hands of the justice system and local authorities. It also confirmed that residents had the right to pursue upstream polluters in a court of law, reinforcing common law and ensuring undeniable recourse. But its results were limited; the law failed to appoint competent authorities to police sources of pollution at the level of local watersheds, which consequently left authorities powerless.[25] In 1893, a committee was appointed to examine the pollution of the Mersey and Irwell rivers near Liverpool. The commissioners could only attest to the deplorable state of both.[26]

State intervention in Britain, the most polluted country of the nineteenth century, had little influence on nonintervention policies in the United States.[27] From the 1880s through the Progressive era (1890–1920), municipal laws began to dictate acceptable atmospheric pollution levels and fines in larger cities. Unlike the various groups in Great Britain who battled against smoke pollution, similar associations in the United States seemed to work with more gusto and intent. Each city had its own institution to address the smoke problem; in Baltimore, Boston, and Chicago, the Health Department took charge, while the Department of Public Works handled that responsibility in Pittsburgh. Philadelphia had its steam engine inspector do the job; Cleveland its Police Department; the construction inspector in Milwaukee; the Public Health Commissioner in Rochester; and the Fire Department inspector in Cincinnati. The one fundamental principle shared by these various public health officials was an adherence to nonintervention. As such, when St. Louis introduced a stringent antismoke measure in 1893, the Supreme Court of the State of Missouri declared it illegal and unconstitutional: the city could only intervene against emissions that caused actual damage to inhabitants or property.[28]

In 1899, however, two Acts of Congress on waste and smoke reinforced municipalities' authority in their fight against pollution.[29] This new framework favored conciliation, with Chicago proving to be a model in the matter. In 1905, a federal Supreme Court judgment rejected a plea by sugar refiners, one of the biggest industrial polluters, against municipal police proceedings. In 1907 the mayor, Fred Busse—a coal merchant elected with the support of the business sector—created a smoke inspection service for the city, not only to reprimand, but also to counsel manufacturers on how to reduce pollution by improving their technical installations. Between October 1, 1909, and September 30, 1910, no less than 1,040 proceedings were filed against industrialists for smoke infractions, and fines increased to unprecedented amounts of up to $500. This new framework fostered collaboration among engineers, public services, and municipal associations, in order to encourage technical improvement. In 1907, the International Association for the Prevention of Smoke was founded in Pittsburgh. It became the Smoke Prevention Association in 1915, and in 1916, seventy-five American cities adopted antismoke ordinances. In 1917 the Smoke Abatement League in Cincinnati confirmed that the city's industrial emissions had been reduced by 75 percent. The combination of militant actions, municipal arbitration, and cooperation between the engineering and business worlds resulted in a notable reduction in atmospheric pollution in American cities before the Great War, at least in more affluent parts of these cities.[30]

Law and Regulation

On the European continent, regulation models differed greatly from those in Britain and the United States. Generally, regulatory bodies had more power; the justice system's only task was to set the amount of potential compensation for damages to property. Against a backdrop of latent nationalism, legal experts in France exacerbated the contradiction between British common law and specific legislation on industrial pollution, like the 1810 law. In 1854, the jurist Désiré Dalloz, a great admirer of French administrative organization, criticized the blatant injustices of the common law model:

> He who wishes to found an industrial establishment usually makes an informal request in the neighborhood where he wishes to build, and rarely is the result of this request unfavorable. Confident in this result, he builds his factory, installs its machines, and commences work; but soon demands for damages are made by numerous parties. Neighboring property owners speculate for the most part on the sort of damages they can claim, work to make a case, and pursue it on behalf of one amongst them before the courts. If he is rich, and if his establishment is productive, he may emerge the winner of this clash. Conversely, if the case is more expensive, he may be ruined by legal fees and high damage costs, forcing the closure of his enterprise.... The abuse is no less serious if an industrialist, with large amounts of capital, opens a factory in a poorer district, and therefore need not fear any legal proceedings being brought against him.[31]

During the nineteenth century, the administrative and preventive logic of France's 1810 imperial decree (to control installations that were sources of nuisances and risks) was used as the benchmark for the majority of European countries. In France, regulations were not altered until 1917. Regional prefects and the Minister for the Interior (sometimes the Minister for Commerce) retained full authority, and preserved the integrity of the law based on the industrial circumstances which had presided at its creation. As a result of changing times, its application adapted through legal precedent at the discretion of those involved, through industrial changes, and as a result of new industrial sectors. But the premise of the three classes of factories and workshops remained unaltered. In this way, the nomenclature attached to the law was always expanding: it was completely rewritten in 1866 and 1886, with additional tables added in 1872, 1879, and 1881.

This system of updating the law meant two things. First, French pollution regulation was constantly out of date, and reacting to change. Industrial innovations and emerging sectors were up and running long before they were documented on paper, often leaving public authorities to hastily scramble to resolve new sources of pollution. It also meant that industrialists often sought to use machinery that would classify their factory as a level 2 or 3, rather than 1, because public officials' surveillance of the former classes was less stringent. The decree therefore implicitly favored

NOMENCLATURE DES ÉTABLISSEMENTS INSALUBRES, DANGEREUX OU INCOMMODES
ANNEXÉE AU DÉCRET CI-DESSUS.

DÉSIGNATION DES INDUSTRIES.	INCONVÉNIENTS.	CLASSES.
Abattoir public[1]. . . . . . :	Odeur et altération des eaux.	1ʳᵉ
Absinthe. (Voy. *Distillerie*.)		
Acide arsénique (Fabrication de l') au moyen de l'acide arsénieux et de l'acide azotique :		
1° Quand les produits nitreux ne sont pas absorbés.	Vapeurs nuisibles	1ʳᵉ
2° Quand ils sont absorbés.	*Idem*.	2ᵉ
Acide chlorhydrique (Production de l') par décomposition des chlorures de magnésium, d'aluminium et autres :		
1° Quand l'acide n'est pas condensé.	Emanations nuisibles. . . .	1ʳᵉ
2° Quand l'acide est condensé.	Emanations accidentelles. .	2ᵉ
Acide muriatique. (Voy. *Acide chlorhydrique*.)		
Acide nitrique.	Emanations nuisibles. . . .	3ᵉ
Acide oxalique (Fabrication de l') :		
1° Par l'acide nitrique :		
a. Sans destruction des gaz nuisibles.	Fumée . . . . . ·. . . .	1ʳᵉ
b. Avec destruction des gaz nuisibles.	Fumée accidentelle	3ᵉ
2° Par la sciure de bois et la potasse.	Fumée	2ᵉ
Acide picrique :		
1° Quand les gaz nuisibles ne sont pas brûlés. . .	Vapeurs nuisibles	1ʳᵉ
2° Avec destruction des gaz nuisibles	*Idem*.	3ᵉ
Acide pyroligneux (Fabrication de l') :		
1° Quand les produits gazeux ne sont pas brûlés.	Fumée et odeur.	2ᵉ
2° Quand les produits gazeux sont brûlés.	*Idem*.	3ᵉ
Acide pyroligneux (Purification de l').	Odeur.	2ᵉ
Acide stéarique (Fabrication de l') :		
1° Par distillation.	Odeur et danger d'incendie.	1ʳᵉ
2° Par saponification.	*Idem*.	2ᵉ
Acide sulfurique (Fabrication de l') :		
1° Par combustion du soufre et des pyrites. . . .	Emanations nuisibles. . . .	1ʳᵉ
2° De Nordhausen par la décomposition du sulfate de fer.	*Idem*.	3ᵉ
Acide urique. (Voy. *Murexide*.).		
Acier (Fabrication de l').	Fumée	3ᵉ
Affinage de l'or et de l'argent par les acides. . . .	Emanations nuisibles. . . .	1ʳᵉ
Affinage des métaux au fourneau. (Voy. *Grillage des minerais*).		
Albumine (Fabrication de l') au moyen du sérum frais du sang.	Odeur.	3°
Alcali volatil. (Voy. *Ammoniaque*.)		
Alcools autres que de vin, sans travail de rectification.	Altération des eaux	3ᵉ
Alcools. (Distillerie agricole).	*Idem*.	3ᵉ
Alcool (Rectification de l').	Danger d'incendie	2ᵉ

[1] La création d'un abattoir public entraîne la suppression des tueries particulières établies dans les localités que dessert cet abattoir. (*Ordonnance royale* du 15 avril 1838, art. 2.)

Le décret du 1ᵉʳ août 1864 a conféré aux préfets le droit de statuer sur les demandes de création d'abattoirs publics : ce droit de décision avait été réservé à l'Administration supérieure par l'instruction du ministre de l'intérieur du 22 juin 1855.

Figure 6.1

The nomenclature for polluting industries classed in the first order. *Source:* Henri Naplas, *Manuel d'hygiène industrielle* (Paris: Masson, 1882), 34.

technical improvement, but nonetheless was rarely coercive. Once a polluting establishment was authorized (there was rarely an issue with authorization), it was practically impossible to shut it down, save in a few exceptional cases where a change in the nature of the enterprise was duly noted. But procedures in this particular instance took so long and were so complex—going as far as the Council of State—and industrial strategies and pressures to preserve their practices were so strong, that this rarely happened. In this way, permission to open a factory essentially equated with permission to pollute. In addition, many industrialists avoided the required authorization procedures and drew up mutual agreements with their neighbors. Local mayors possessed little power in the face of this complex administrative machine. Initially hesitant to have their staff get involved in the process, they rarely objected to polluting industries establishing themselves in their communes, even toward the end of the century when growing worries about health led, on occasion, to contemplating a form of urban planning.

But was the French public regulation system totally powerless to protect the population from industrial pollutions? The answer depended upon context, in particular reactions from neighbors (often property owners), and those of public health boards. In certain instances, the plaintiffs found sufficient resources to rally and force the administration and industrialists to reduce pollution levels, by moving installations, introducing filtration and condensing equipment, or reducing production volume. Some health councils could be quite averse to and harsh on polluting industry, and property rights were, in some circumstances, sufficient to protect property income not involved in industry.[32] Overall, however, the law did not aid in drastically reducing pollution; the means to do so were ignored, leaving civil justice to regulate any financial damages owing.

Established under Napoleon and the First Empire, the basis of France's legal system was adopted in many European countries, although incompletely and with very different national nuances. After Belgian independence in 1830, the royal decree of 1824 was confirmed. The decree was heavily influenced by the French decree of 1810, although for a time the judicial courts regained importance. In 1847, for example, the Court of First Instance in Liège declared itself competent to judge the request for the closure of the zinc factory at Vieille-Montagne. The following year, the Court of Appeal in Anvers ordered the closure of a brickworks because of damages it had caused to locals. Thereafter, case law took a different direction and the Court of Cassation finally reaffirmed the primacy of administrative authorization for judicial decisions.[33] Under the guidance of new liberal powers who expressed confidence in science and industry's capacities to resolve its problems, the law was reformed in 1849, and again in 1863, much to the benefit of industrialists. The state's attitude demonstrated a certain pragmatism, and the system evolved in conjunction with the political balance of power. To wit, in 1886 and 1888, the Belgian Socialist Party gained enough ground that it succeeded in bending the legislation to protect the health of factory workers.[34] In the Netherlands after 1830—the 1824 decree was also passed there by the Supreme Court in 1846—the state's role was also altered, and municipal administrations were given the power to retract establishments' authorizations and, in time, to order the closure of manufacturing plants and workshops. The power to grant permission to set up establishments was completely given over to municipalities under the nuisance laws (*Hinderwet*) of 1875 and 1896, with recourse reserved for the State Council. There as elsewhere, the law favored industrialists, who often circumvented the regulations by setting up establishments without permits and operating with an air of fait accompli.[35]

In the German states, pollution legislation remained marginal. In Prussia, for example, the 1832 and 1848 ordinances for steam engines were supposed to bring them into line with technical standards designed to limit smoke emissions. These were hardly enforced. From 1845, if insalubrious establishments required state authorization, authorities seemed less inclined to obstruct industrial development there than in France. Following the formation of the Reich in 1871, Prussian administrative services spread throughout the German Empire, and commercial law ordered that no factory could be a nuisance to neighboring properties. With no real central administration to which to turn, cities and the Länder were the first to take action.[36] Antismoke municipal ordinances were thus adopted in Breslau (1874), Nuremberg (1876), Brunswick (1883), Stuttgart (1884), Dresden (1887), Heidelberg (1890), and Munich and Freiburg (1891). Berlin remained under state control. Unlike in the United States, the civic movement in the German empire was loosely organized, although in Austria an "Association to Combat Smoke and Dust" did exist. Municipal regulations remained fragmented with little coherence. The few cities that did adopt ambitious solutions, such as Hamburg with its Association for Fuel Economics and the Reduction of Smoke, were anomalies. Overall, legislative application proved to be indecisive, despite a concerted effort by engineers. Even a national scientific journal (1910), *Rauch und Staub* (Smoke and Dust) was dedicated to the subject.[37] In particular regions, the Ruhr Basin for example, or in strategic sectors like chemistry, regulatory policy throughout the century seemed to lean toward the assumed protection of industry.[38] In other, less industrialized and less polluted European countries, similar trends unfolded. For example, after Italian unification in 1860, a piece of legislation on insalubrious establishments was adopted in 1888. It focused on the location of factories (where they could pollute without restriction), and was largely ineffective. The fight against pollution ultimately lay with local authorities, who were prompt to aid any activities likely to boost the country's industrialization and spur its economic growth.[39]

Outside of Europe and North America, industrialization was less advanced and there is painfully little accessible documentation, apart from some on European colonies. Japan is one focal point we can use to learn a little about regulations in Asia. The beginning of the Meiji Period saw the introduction of mercantilism and economic nationalism, which promised a "modern" industrialization whose goal was to ensure independence and power for Japan. Abolishing its preindustrial regime, Japan underwent an institutional metamorphosis that would shake up pollution regulations, similar to what happened in Europe and North America at the end of the eighteenth century.[40] At the majority of industrial sites, as with the tragic case of the Ashio mines, pollution was managed through the payment of compensation for damages. No other controls were instituted before 1896, at which point a law dealing with rivers was adopted to protect local residents. In 1897, a ministerial inquest provoked the "Pollution Prevention" decree, which launched the state's involvement in environmental matters. But this decree, which sought to bring parity between property owners and included technical provisions, was almost never applied. As a result, a new ministerial commission led by scientists was put in place in 1902. In its report, the link between pollution and industry was clearly established, but it suggested using the Western world's model to minimize negative repercussions on health, and stated that pollutions dispersed into the environment were acceptable. The solution to the problem was found in reconfiguring the waterways. Rather than closing the mines responsible for polluting the rivers—a regulation that preexisted the Meiji Period—the new quick fix was to reconstruct nature with the help of engineers, in step with industrialization. This, according to Robert Stolz, marked the beginning of state-controlled nature—indeed, the control of nature—in Japan.[41]

In India, as industrialization took hold in cities such as Bombay (cotton), Calcutta (jute), and New Delhi, the measures put in place echoed those in Europe: weak laws, prejudiced against older putrefying artisanal crafts in favor of the newer, high-capital industries that enjoyed greater access to power. Varying inspection methods were set up, notably in the fight against smoke. But all lacked sufficient means or enforcement. Bombay got as far as relocating all abattoirs to the outskirts of the city in 1867, but in Bengal a *Smoke Abatement Act* (1863) fell into disuse. In this same region in 1905, after the *Bengal Smoke Nuisances Act* had been passed, only 16 of the 10,000 cases of excessive smoke reported by inspectors were brought before the courts. In 1915, industrialists pressured to have the Act abolished, arguing predominantly for more open markets for business, employment, and technical improvement, as was the case in Europe.[42]

Can we then conclude from this overview that national regulatory systems for industrial pollutions existed in the nineteenth century? We must take into account that social, economic, and political contexts created very different situations. However, beyond the scope of judicial and institutional choices, certain common tendencies emerged. In the first place, state intervention occurred in all countries because pollution grew to be problematic at a national level and therefore could not be dealt with solely under municipal jurisdiction, as had been the hope. The fact that the sovereign public power intervened in these countries where industrialization was revered meant that pollutions from industry evaded criminal proceedings more often than not. In countries with common law, it had long since been possible to pursue a polluter through public nuisance proceedings—and this had had the potential of resulting in prison time. But with the preventive intervention methods of local and national administrations, recourse to criminal justice became gradually less accessible, leaving civil justice as the only acceptable means for compensation.[43] The change-over was even more apparent in countries with Roman law, in which the administration issued authorizations, where the state had decided that there was no risk to public health. In this case, an authorization could not be withdrawn and thus it was up to civil justice to decide on compensation to third parties. Naturally, the system had its flaws, but in general the administration worked to avoid criminal proceedings. For example, in France, a fishing law from 1829 gave criminal courts the right to pursue an industrialist for pollution in the case of fish mortality. This possibility was used very infrequently and, although a judgment by the Court of Cassation in 1859 confirmed the content, a decree in 1870 added the possibility for a settlement between disputing parties before criminal proceedings might ensue. These legal compromises and accommodations bore witness to the industrialist spirit of the central administration: industrialists could no longer be considered delinquent or criminal in the century of all-conquering industrialization.[44]

Where industrialists eluded penalties for environmental damage, compensation to injured third parties did increase. Apart from notarized, mutual, and private agreements—about which we know very little—industrial polluters managed to assuage the deleterious effects of their establishments by way of paying compensation to plaintiffs. Soda and aluminum manufacturers in the Maurienne province and producers of sulfuric acid in Bouches-du-Rhône, Gard, Alsace, Lorraine, and the Namur province in Belgium all managed to appease complaints with monetary payments, hardly onerous for them but substantial for their poorer plaintiffs. Similarly, zinc manufacturers in Liège, in Belgium, and in Viviez, in the Aveyron department in France, included financing for private contracts and legal fees linked to pollution in their company's budget. This generally accounted for 1 percent of their operational costs.[45] Apart from this negotiated "tax," another strategy industrialists liked to use involved buying up neighboring properties, which would create a larger nuisance buffer, remove prospective plaintiffs, and hopefully reduce or avoid compensation payments. Inasmuch as this constituted a remarkable adaptation to the new industrial landscape, in hindsight it affords the historian an opportunity to reflect on a real commodification of nature.[46]

It is hardly surprising that in the end the results were the same, regardless of whether pollutions were governed by a liberal system or an administrative one. Apart from the law, two things counted above all else: the industrialist project—and thus the law's ability to adapt to the economy—and the pace of innovation, both of which served to prioritize industrial interests and activities. Public action was often content to aid the process in adapting the law to meet the necessities of production. The dominant course of action was definitely that of sacrifice, targeting particular areas and populations in the name of public interest, as it related to industrialization. If the law seemed inadequately equipped to deal with nuisances, this might have been because other means had the upper hand in the battle against pollutions: technical progress, recycling, and dilution.

The Best Practical Means

Initially, industrialists did not view the intervention of public powers as a positive move. With systematic opposition and lobbying, they argued that as a result of restrictive regulations they were incurring extra costs that impacted their accounts. Added to labor blackmail was the worry about a national legislation that would negatively impact international competition, which was becoming ever more prevalent. In order to avoid any regulatory interference, industrialists employed a preemptive practice of technically improving their establishments to reduce pollution, a system already employed in Europe since the beginning of the nineteenth century, and which became widespread, reaching as far as Japan. Confidence in industry's capacity to regulate and solve its own problems became ever stronger; improving technical know-how was the perfect way to appease the critics.[47] Even if industrialists were not inclined to research new methods for pollution reduction of their own volition, they were encouraged to do so under pressure from local residents and public powers, or because they needed to make the byproducts of their industry more profitable. By mid-century, a frenzy of technical innovations was underway in response to regulations by authorities to reduce industrial smoke pollution. Countless innovations were legitimized in the name of public health and well-being, as every polluting sector made some attempt to adapt its manufacturing processes. Thousands of clean-up operation agreements and technical reviews, published around the world, took charge of sharing the details.[48]

It is worth examining two representative production methods typical of the time. The first example drew on traditional artisanal practices while the other exemplified the habits of more technologically advanced industries. Plant textile retting was tolerated less and less, but its prohibition was difficult in part because of its importance and also because its establishments were widely scattered. In the latter half of the century, in France, Belgium, Great Britain, and the United States, multiple new innovations mechanized the insalubrious retting process. Retting was moved into more controlled settings in factories, carried out in vats filled with artificially heated water, using mechanical, and then chemical methods. The so-called Schenck, or "American process," was tried out in Ireland in 1847 with the support of the Flax Society in Belfast, which had been created to encourage plant textile development after the Great Famine had ravaged the country. In 1850, the French Minister for Agriculture sent the chemist Payen to investigate the "salubrity of this new process." Prussian engineers soon followed.[49] Amongst the new innovations was "fairy electricity," celebrated continuously after 1880 for its capacity to render coal burning and gas lighting unnecessary. When electric lighting was developed after 1890, the principal argument in its favor was that consumers noted how much cleaner it left the air. Electricity was supposed to pave the way for hydroelectricity—"green coal" from rivers and "white coal" from glaciers—which unlike black coal, would decrease pollution. However, in spite of high hopes, nuisances continued to present themselves, as the majority of electricity was produced from coal-powered plants. Consequently, in 1890 New York City, the leading producer of electricity, was faced with a concerning level of atmospheric pollution. Electricity spread so quickly that rather than lowering levels of pollution, it actually added to the emission levels. Although some sites reduced their smoke levels thanks to electricity, this new power source simply moved the smoke pollution to where the generators were located.[50]

Rather than embarking on an impossibly long inventory, two principal problems which faced industrial societies in the nineteenth century might serve as examples: smoke and acidic and sulfuric vapors. In the first case, smokeless furnaces became the archetypal technical device designed to resolve atmospheric pollutions. Great Britain played a pioneering role in its transnational dissemination of knowledge and technical discussion; British experiences were closely watched by other states.[51] Even though smoke-reducing apparatus appeared concomitantly with the first steam engines, they only became more widely available after 1825. In swift succession, many other methods were adopted, improved, and abandoned. Various machines to condense or filter smoke—or even to "wash" it by passing it through jets of water—were conceived, but improving combustion remained the central concern. This involved circulating oxygen and experimenting with double furnaces. The second furnace was fed charcoal, which permitted an undeniable reduction in coal smoke.[52] Numerous models of mechanical loaders refined the process. With machines loading the coal, the doors to the furnaces were opened less frequently, resulting in more optimal and regular combustion. Among the more successful furnaces were those patented by British engineers Charles Wye Williams between 1838 and 1841, and Fairbairn in 1844. London's 1853 antismoke ordinance, and a similar one in Paris in 1854, stimulated further innovations. In 1855, the engineer Adolphe Mille (from the Ponts et Chaussées technical college) testified to the diversity of devices available on the market in London. He also observed the prevailing optimism the new technologies provoked: "If the smoke around us has not disappeared high above out a chimney, it has otherwise been *notably* reduced."[53]

This optimism, shared by industrialists and public powers, could not entirely conceal the failures encountered throughout the century. In 1894, a report presented in France to the Society for the Encouragement of

National Industry (SEIN) affirmed that the problem of eliminating smoke emissions from industrial furnaces "has not as yet a general or truly practical solution."[54] The multitude of promises and technical processes actually had a counterproductive effect in Germany and the United States. Presented with such a range of processes, their high costs, and their low practical efficiency, industrialists became skeptical about combating pollution at all. Renewed attacks on environmental associations saw their cause weakened and discredited.[55] Authorities were aware that the technology was imperfect, often fragile, costly, and restrictive. Further, even after installing the smoke-reducing technologies, industrialists were loath to use them to their full potential. As a result, municipal and national regulations engaged in a politics of pragmatism: they specified that industry should consume as much smoke as possible. This was the priority, but rather than rigid preventive measures and metrics, industry was left to demonstrate its own efforts. The guideline "consume or burn as far as possible all the smoke" and the request to use "the best practical means" were vague enough for manufacturers to accept and interpret as they saw fit. The belief in improved technology actually led to a surge in smoke-reducing equipment. Technical progress was however undeniable: by the end of the century, smokeless furnaces had advanced thanks to new apparatus.[56] Therefore, in order to understand the continuing increase in smoke levels in Western cities, it is important to consider the rebound effect. In 1865, the economist William Jevons explained that increases in energy production efficiency led paradoxically to more, not less, consumption.[57] The same reasoning could be applied to industrial smoke: within the workings of a single furnace, fumes were largely diminished; but as more factories were permitted or tolerated, global pollutions increased by dint of the aggregate number of pollution sources. Faced with this logic, municipal police and courts were powerless, because the industrialists' willingness to install and use modern processes generally managed to silence plaintiffs. Added to this was the difficulty of achieving a verifiable measurement of pollutants. It is shocking to discover the lack of rigorous criteria dealing with emissions surveillance, and how little progress had been made in this domain since the first parliamentary inquest in 1819. Attempts by the French engineer Maximilien Ringelmann in 1898 to scientifically define and quantify smoke density were circulated in Europe and the United States around 1910. These measurement grids—containing a series of differently shaded grey squares with which to compare against ambient smoke—were rudimentary and difficult to manipulate. Nevertheless, they provided the means for antismoke militants and local associations to acquire a counter-expertise in the smoke pollutions debate, by gauging and measuring smoke for themselves.[58]

The condensation of acidic gases followed the same logistical path. The production of caustic soda was particularly problematic. So, too, was the production of sulfuric acid and nitric acid, as well as all metallurgical production, such as refined copper from iron pyrites. Amongst tested condensation methods, the so called Gay-Lussac tower—named after the French chemist who first introduced it to the Saint-Gobain factory in Chauny, north of Paris, at the end of the 1830s—is certainly the best remembered. Although a series of trial-and-error improvements enhanced its performance, the basic principle hardly changed. It involved passing the acidic gas through towers filled with brick, carbon, and/or lime, into which a highly concentrated sulfuric acid was added to condense the gases. After 1860, the Gay-Lussac tower was generally used in industry in France and Belgium. In Great Britain, the chemist John Glover improved the process after 1859 and obtained a condensation close to 80 percent. The United States adopted the principle after 1880.[59] Jacques-Louis Kessler's towers, put into use in 1877 at his Clermont-Ferrand factory, were also popular around 1900 in France, Great Britain, Germany, the United States, and Russia.[60]

The quest for new innovations that would free industrialists from public regulations and court cases was particularly advanced in copper metallurgy. Beginning in the 1860s, the most efficient systems were developed by Moritz Gerstenhöfer, a German industrialist, even if they only succeeded in condensing 20 percent of the sulfur. Condensation with the aid of electric flux, invented in 1886, was not widely used before the First World War.[61] But it was the soda industry that posed the biggest problems; especially since until the middle of the nineteenth century hydrochloric acid use was so low that manufacturers were reluctant to condense it. It was already known that hydrochloric acid was condensed using water and lime. In 1836, the Englishman William Gossage designed the first truly effective absorption tower by passing the gas through long pipes and towers into a pool of water. Having become a copper refiner in Wales in the 1840s, and then in Widnes in the 1850s, where he linked copper refining (from imported copper pyrites) to the production of sulfuric acid and soda, his experiments gave him a yield of 90 percent condensed liquid.[62] However, Gossage towers did not find favor until requirements of the Alkali Act of 1863 put them into general use in factories in Great Britain. The next step to extract chlorine from the

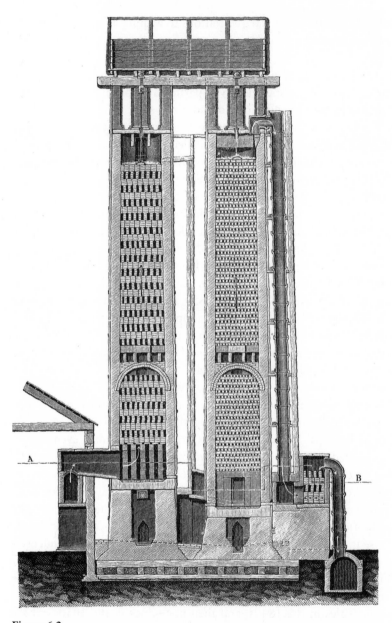

Figure 6.2
The Gossage tower in the Hutchinson sulfuric acid factory in Widnes (1880).[63] *Source:*
Georg Lunge, *A Theoretical and Practical Treatise on the Manufacture of Sulphuric Acid
and Alkali with the Collateral Branches* (London: John Van Veerst, 1880), 2:252.

condensed product was carried out after 1880, to the benefit of the textile and cellulose paper industries. By 1900, Gossage towers and their substitutes were used around the world, in particular in the United States, which became the world's leading producer of sulfuric acid.

These few examples permit some general conclusions. Far from reducing pollution, faith in technological advancements conversely led to its increase instead. Moreover, it established a technical fatalism which favored political inaction and hope for hypothetical progress in the future. Often a reduction in one form of pollution produced secondary effects, through either the rebound effect—as in the case of smoke—or by simply displacing the source of pollution. Thus, after 1860, with the success of removing hydrochloric acid from the air, rivers bore the brunt of the transferred pollution until after 1880, when the waste was disseminated and used in the paper and textile industries. Finally, these technical innovations became a part of discussions surrounding recycling, which industrialists latched onto in order to justify their activities. Since the chemical industry posed the most significant pollution hazards, the industry was tireless in inventing new uses and reuses for its byproducts. The perpetual race for innovation only complicated the nature and spread of pollution.

Recycling: Trend to Crisis

Instead of being spurned as threats, toxic substances and waste were embraced as assets to be valued, materials to be reinjected into production cycles and born again in new form and for new purpose. The middle of the nineteenth century saw recycling become more of a trendy endeavor, although the term "recycling" had yet to be coined. It was perceived by many as the best option to deal with the problem of unsolvable pollution, a promise of prosperity and abundance, a full-fledged way to manage industrial residue.[64] The idea of reusing waste was not a new one of course; the novelty, therefore, lay less in the systemization of this practice—sadly though, untreatable and unrecyclable waste was on the rise—and more in its use as a means to justify industrialists' activities that emitted highly toxic byproducts.

Recycling had long been concerned with waste for strictly agricultural use, especially manure and human excreta used as fertilizer for their nitrogen content.[65] Victor Hugo noted in a famous chapter from *Les Misérables,* that "all the human and animal manure that the world wastes, if it were put back into the land instead of being thrown into the sea, would suffice to feed the world."[66] The socialist philosopher Pierre Leroux, while in exile on the island of Jersey, developed his theory on "circulus": the idea that waste feeds production, in a circular vision of material flux.[67] One proposed solution to remove excess waste from cities was *sewage farms*. They were cost efficient and, mid-century, were an exemplary demonstration of the benefits to be reaped from recycling. Historians have studied at length this thrilling enterprise of revitalizing the countryside purged of its nutritive elements by hungry cities. In return for food, the countryside received the elements necessary for replenishing its soils in the form of urban wastes. This urban-agricultural cycle demonstrated cultural, spiritual, and political resilience; after industrialization, it also witnessed expression through technical and chemical efforts.[68] Following London's example, numerous cities sought to fertilize surrounding land with sewage carried through pipelines, sometimes over long distances. Not surprisingly, these sewage farms became cause for debate about the supposed virtues of this mode of fertilization in the hinterlands of Berlin, around Edinburgh, and Brussels, as well as on the plains of Gennevilliers downstream from Paris.[69]

Industrial waste was also partially reemployed in agriculture. Thus, byproducts from the production of rapeseed oil, flax oil, and other oleaginous seeds were turned into oilcakes to be used as fodder for livestock. This consequently gave rise to another international business in the nineteenth century. In 1863 on his experimental farm in Essex, John Joseph Mechi attempted to systematize the circulation of oleaginous seeds on a much larger scale.[70] Dregs from the steel industry were used as a more economical source of phosphorus—a good substitution for bones—which heretofore had been used in chemical operations to produce products rich in phosphorus. Essentially, phosphorus could be extracted from calcium carbonates that had absorbed phosphates during the steelmaking process. But the principal source of industrial waste introduced into agriculture in the nineteenth century was from gas lighting. The distillation of coal produced ammonium sulfate, which contained nitrogen, but industrialists were not sure what to do with it. During the 1870s, agriculture in Britain consumed 46,000 metric tons of this ammonium sulfate per annum. But its nitrogen content was inferior to that of guano or sodium nitrate mineral imported from Latin America. As a result, farmers were reluctant to make use of this bluish, damp, sticky, and foul-smelling substance, which also contained enough sulfuric acid to corrode farm machinery. However, the reformed 1881 Alkali Act encouraged the use of this troublesome product, which otherwise contaminated factory depots in the heart of cities as it sat unused. By 1913, some 400,000 metric tons of ammonium sulfate—the byproduct of gas factories—were being put to use; the majority was exported to Germany, where beet growers were less discerning. Even though the use of these industrial byproducts may have remained limited, their constant promotion enhanced the economic policy of industrialization and seemed to justify the promise to recycle industry's waste and toxic products.[71]

Recycling was mainly encouraged within industrial sectors, particularly within the chemical industry, where chemists, technical experts, and engineers attempted to find ways to reuse the waste in place of having to simply dispose of it. From 1862 to 1868, at the government's request, the engineer Charles de Freycinet—a member of the Corps des Mines, and future Minister of Public Works between 1877 and 1879—led an inquiry in England and Europe (notably Belgium and the Rhine Region in Prussia) to study industrial sanitation. The results were published in various reports that offered a very detailed array of remediation or depollution strategies and practices for the reuse of dangerous substances available at the time. Freycinet asserted that "whereas industrial residues may infect the soil when they are buried indiscriminately or dumped without measure, they can be rendered inoffensive and even useful when they are spread on land in a methodical and suitable manner."[72]

Certain major industrialists specialized in recycling contaminated matter; Payen and Frédéric Kuhlmann were prominent among them. The latter made his money by recuperating byproducts from lead chambers for use in paint production, and waste from chlorine soda plants to manufacture large quantities of barium chloride.[73] In the soda industry, where each metric ton of soda produced incurred two metric tons of sulfur waste containing 90 percent sulfuric acid, it was tempting to find a way to reuse these foul-smelling residues to recuperate the sulfur and acid content; the former otherwise had to be imported from Sicily at high cost. In 1837, Gossage enjoyed some modest success in treating dregs with carbonic acid to yield hydrogen sulfide, which when burned to ashes, furnished sulfur dioxide. The sulfur dioxide was subsequently reintroduced into the production of sulfuric acid. The Alkali Act of 1863 standardized this practice. The Mond process—named for Ludwig Mond—succeeded in recuperating 40 percent of sulfur from soda factories in Liverpool.[74]

As a guarantee for innovation and profit, recycling contributed to recharacterizing pollution as primarily a technical problem. Chemists involved with coking and the production of lighting gases searched for ways to enhance the many byproducts of the coal industry: tar, ammonia, bitumen, and other volatile compounds. This was for all intents and purposes the beginning of carbon-based chemistry. The extremely polluting nature of operations within this branch of chemistry was immediately evident, with even Parent-Duchâtelet himself admitting as much with respect to the distillation of tar at gas plants.[75] But in his *Treaty* in 1870, Freycinet actively promoted the recycling of carbon derivatives. At this time the quantities in circulation were still quite limited. For example, benzol—a toxic mixture of benzene, toluene, and xylene—was extracted from coal to make aniline, but it accounted for only 1 percent of available tar, itself a byproduct of coal; another 0.7 percent was used to produce naphtha. The majority of tar and other residues from the gas lighting industry—whatever the champions of recycling might have claimed— actually continued to be burned or buried.[76] However, at the end of the century, carbon-based chemistry underwent considerable developments, notably with progress in organic chemistry and later the development of recovery coke ovens in Germany. These new ovens recovered byproducts such as tar, benzene, anthracene, and phenols. Such symbiotic potential helped make sense of the regularity with which chemical plants were established near coal mines.

Benzene was a precursor to the synthesis of numerous organic compounds. In 1845, the British chemist Charles Mansfield isolated benzene in coal tar, and then in 1849 began producing benzene from coal on an industrial level. In 1869, in France, Bertholet was chemically deriving benzene from acetylene. This very toxic, carcinogenic compound was used as a solvent in glue, varnish, paint, and ink; it was also applied in dry cleaning and degreasing metals. In the middle of the century, Great Britain began producing aniline industrially for use in dyes and artificial colorants. In 1856, William Perkins discovered mauveine (a colorant derived from aniline) and founded in Greenford, a London suburb, history's first factory for the production of synthetic dyes. At the same time in Lyon, the Renard brothers launched production of fuchsin (from aniline and toluidine), which resulted in byproducts such as benzene and toluene. Germany would become the world leader in industrial recycling of coal for the dye industry. In 1869, chemists Carl Graebe and Carl Libermann, who worked for BASF (Badische Anilin und Soda Fabrik), took out a patent for the synthesis of alizirin (red in color), which was derived from anthracene extracted from coal tar. Vegetable madder, the previous source for the color red, was quickly abandoned. Likewise, the plant source for indigo blue was replaced by its synthetic equivalent, manufactured from benzene and aniline—and marketed by BASF in 1897. By 1914, Germany was producing 85 percent of the colorants used globally and became one of the most chemically contaminated places on the planet.[77]

If a large number of intermediate products or byproducts of the chemical industry stimulated the creation of new branches of activity, it is important to recall that more and more residues were thrown away or burned, and that the overall quantity of waste produced increased, despite the ongoing promotion of recycling. This led to the creation of garbage dumps and incinerators, two ultimate destinations for waste whose function and locations were emphatically debated in every country. Once again, Great Britain pioneered the way with the first garbage incinerator put to use in the 1870s. Popularity grew, and by 1893 fifty-five cities were equipped with incinerating devices.[78] Britain exported its incinerators around the world, thanks to its colonial empire (Calcutta, Madras, New Zealand). At the beginning of the twentieth century, the engineer Walter Francis Goodrich listed the places in the world where the British Destructor had been installed. His list was published in works that celebrated the wonder and superiority of British sanitation systems, which converted incinerator heat into electricity.[79] From 1905, New York and other American cities built incinerators.[80] However, incineration of urban waste was slow to spread to the European continent. The first commitment to incineration was found in Hamburg in 1892, following a cholera epidemic. In France, incinerators appeared very gradually, perhaps because of the continuing exchange between city and countryside.[81]

Dilution: The Ultimate Solution

Throughout the entire century and all over the world—taking into account the limits of administrative and judicial regulations and imperfect methods of recycling—the preferred method for dealing with pollutions was dilution. Dilution was both atmospheric and aquatic, but separating the sources of pollutions from populated areas remained a favored practice as soon as railways made that a viable possibility. In his 1870 *Treaty*, Freycinet perfectly summed up the guiding principles for authorities and technicians: "The isolation of factories ... is the optimal means to achieve our goals." The next best option was to construct tall chimneys to disperse toxic emissions high up into the air because "the damage caused must, to say the least, be diminished as a result of the chimneys' height." The third method was "that which we refer to as the natural outlets": dumping into fast-moving rivers and directly into the sea.[82]

Uncertainty on how to best evaluate air and water pollution meant that the enduring belief in the earth's ability to absorb harmful substances and to perpetually regenerate itself held strong. Arrhenius's warnings about carbon emissions gained little traction; in 1901, the German chemist Clemens Winkler, a member of the Royal Science Academy of Sweden and a professor in Hamburg, asserted that "the volume of coal consumed disappears without a trace into the vast ocean of air."[83] Tall chimneys remained costly to build and maintain; without pressure from their neighbors, industrialists were reluctant to invest in building them. Between 1830 and 1850, minimum height imposed on the chimneys for factories with steam engines was set between 20 and 30 meters, with individual countries setting their own standards. But already a race to build the tallest chimney had been launched by the chemical industry. In 1841–1842, two chemical factories built smoke stacks 124 meters and 133 meters tall in Liverpool and Glasgow. In Manchester, some exceeded 100 meters, and in 1857 Peter Spence, an alum manufacturer, proposed constructing a 200-meter chimney that would collect the entire city's smoke emissions and resolve the problem once and for all.[84] This project failed, however, due to the public financing it would have needed. "Copperopolis" (Swansea) devolved into an industrial oven: feeble protests from citizens did little to slow the expansion of the city's smokestack forest, between 20 and 50 meters tall. Conversely, in 1860 in the neighboring town of Llanelli, local protests led to the construction of a chimney almost 100 meters tall. Constant disagreements and disputes did not impede the quest for taller chimneys, and in 1889 the tallest one in Europe, at 140 meters, was built in Freiberg, Saxony. Once condensers were installed in chemical plants, mines and metallurgical foundries (for copper, silver, and lead) set new chimney height records. In 1908, the Boston & Montana Consolidated Copper and Silver Mining Company constructed a smokestack at Great Falls, Montana, that was 154.2 meters high. Of course, as an 1851 pamphlet from an Orléans industrialist wryly appreciated: "from the summit of this immense pedestal, the enemy escapes, hovers and falls on his prey, spreading even further due to the height of its point of departure."[85] By dispersing smoke out of the immediate vicinity of the plant, the tall chimneys had the merit of rendering their pollutions invisible. However, at the beginning of the twentieth century, detrimental effects on the Tharandt Forest, located 15–20 kilometers from Freiberg, were already visible.[86]

It might seem a familiar, almost repetitive refrain: the Emscher and the Rhine in Germany; the Mersey and the Tyne in Lancashire; and other rivers became the dumping grounds for some of the period's most

Table 6.1
Examples of tall chimneys in the 19th century (1835–1913)

Enterprise	Fabrication	Location	Date	Height (m)
Adam's Soap Works	Soap and soda	Birmingham, Great Britain	1835	95
Soda Ash	Soda	Liverpool, Great Britain	1841	123.75
Tennant & Co.	Chemical products	Glasgow, Great Britain	1841–1842	132.75
Marrel Frères	Forge, metallurgy	Rive-de-Gier, France	1867–1868	108
Schneider	Steel	Le Creusot, France	1869	85
Lead Mining Company	Lead	Mechernich, Germany	1884–1885	131
Halsbrückner Hütte	Metals foundry	Freiberg, Germany	1888–1889	140
Compañía de Peñoles	Lead and silver	Mexico City, Mexico	c. 1900	91.5
Boston & Montana Consolidated Copper and Silver Mining Company	Copper and silver	Great Falls, Montana, United States	1908	154.2
United Verde Copper Company	Copper	Jerome, Arizona, United States	1913	122

Source: James Douet, *Going Up in Smoke: The History of the Industrial Chimney* (London: Victorian Society, 1989); and Frédéric Pillet, "Les cheminées d'usine métalliques au XIXe siècle: Une question de caminologie particulière," *L'Archéologie industrielle en France*, no. 43 (2003): 5–15.

critical pollutions. Dumping wastes into rivers became a common disposal solution for the dominant pollutions of the nineteenth century, a situation in stark contrast with the attitudes of the previous century. Before the First World War, it was widely accepted that large rivers could dilute and evacuate pollutants; the same notion carried over to dangerous condensed gases that held no other industrial purpose. As Freycinet explained: "When the body of water is sufficient, not only does the insalubrious element become, by dilution, a little less offensive, but it might also be rendered completely nonexistent—in other words it might be chemically destroyed by the special action of the environment." This is why, for Freycinet, "the first care of a factory must be to look for an abundant stream."[87]

But what to do with solid waste? In Switzerland, industrialists in Basel discharged solid waste from their synthetic dyeing process into the Rhine from an opening on a bridge. As was the case in numerous such locations, the dumping usually took place at night.[88] In the middle of the century, the inhabitants of Chicago and Zurich felt the lakes that bordered their cities were large enough and deep enough to easily absorb harmful waste. This hope was disappointed at the beginning of the twentieth century, when locals noted the proliferation of algae and bacteria that had destroyed the entire aquatic ecosystem.[89] Although there is little quantitative documentation, and inasmuch as the process was largely invisible, discharges into the sea seem to have been widely practiced in the littoral industrial zones. From the eighteenth century, it was traditional for soap makers in Marseille to dispose of their waste on the coastline; the refuse became the foundation for the port's piers. This led prefects to regulate discharges into the Mediterranean, requiring that they be made further and further away from the shore. In 1830, the extent of waste dumped into the water was such that 100 meters from the coast, the sea's depth was only 4 meters. This hindered navigation, forcing authorities to order new, more removed areas for dumping. Successive port developments between 1844 and 1853 (in which 150,000 cubic meters of soap waste was used as filling substrates), inspired a period of *tout à la mer* ("everything into the sea"), which received considerable support from doctors and engineers. In 1895, the emissary of Cortiou facilitated these discharges, despite the warnings of the biologist and oceanologist Antoine-Fortuné Marion. A dredging to 20 meters revealed the brutal devastation suffered by the ecosystem:

> The machine only dredges up pottery fragments, debris of all kinds, in a black mud. The earth from the soap residues forms a crust in places which is quite hard, and which the dredger scours, finding underneath a greenish slime with a sulfurous odor. These are the polysulfides that have accumulated and that have destroyed almost everything.[90]

These discharges and overflows served as archetypal outcomes of Marseille's chemical industry from 1880 onward. One typical example: as soon as it was installed at L'Estaque in 1883, the Mining Company of Río Tinto chose to dispose of its soda waste by dumping it by the bargeful at sea.[91] The French Mediterranean was not an exception. Metallurgical plants created jetties by dumping their slag into the sea. In Wales, Yorkshire, and Middlesbrough, for example, wagons would dump their dregs over clifftops to the sea below; piles of waste—12 to 15 meters high— would creep out to sea for more than 200 meters. Rails were subsequently built on top of this new foundation, allowing for even more extended sea dumping grounds as they gradually grew longer. In 1870, at the mouth of the Tyne River in Lancashire, the Jarrow chemical factory chartered a large boat every day to dump the residues of its soda production 2 kilometers from the shore.[92] In New York, the oil industry quietly poured its refuse into the sea that adjoined its facilities. The water acquired an ink color and destroyed the flourishing oyster culture and fish resources. Such pollutions continued—and increased—up to World War I.[93]

Dilution had a spatial dimension to it, since the ability to physically remove pollutions away from inhabited areas has always been a sensible thing to do, if not the most efficient in the fight against pollution. According to Freycinet, the most toxic factories could operate safely 4 kilometers from the nearest town if the population could tolerate slight inconveniences. This process of spatial dilution played out on three different levels: urban, regional, and global. In the context of polyfunctional cities, in addition to the gradual eviction of dirty factories to their outskirts (as already noted) all studies show the spatial and social differentiation that took place as a result of pollution. Working-class neighborhoods dealt with the most nuisances, in essence because workers' housing grew up around factories. In America, however, a more insular approach was adopted. For example, in New York—which welcomed numerous petroleum refineries linked to chemical product plants at the end of the century—factories were gradually concentrated in districts like Newtown Creek and New Jersey (Standard Oil and Rockefeller among others), and on the East River estuary, through a combination of actions at municipal and judicial levels. Real estate and social and environmental factors also played a role. All of this resulted in a clear-cut social division: it fostered the creation of densely populated, working-class areas in industrial zones. Middle- and upper-class populations migrated to places like Manhattan, which remained unmolested by heavy industry. Any improvements in the fight against urban pollution generally only benefited the middle class or those who were more comfortably off, and were usually even more detrimental to the working-class neighborhoods to which insalubrious factories were relocated.[94] Such was the common pattern in early American struggles against air and water pollution.

Authorities volunteered no urban or infrastructural planning solutions
to pollution. Germany, however, revealed a glimpse of the emergence of
pollution featuring in planning. Darmstadt and Mannheim, old regional
capitals that industrialized over the course of the century away from the
big basins of heavy industry, sought balance between healthy living and
the demands of an industrial economy. After 1890, in an effort to promote public health, authorities adopted patterns of spatial differentiation
between districts left to industrial contamination on the outskirts, and
city centers devoted to residential life. This was achieved by encouraging and building railway networks to attract industry to its newly designated zones, as well as reinforcing control over steam engines within
the city limits, and encouraging gas lighting and chemical companies to
vacate the city center. The result was that—even in the absence of a truly
coercive plan—authorities managed to reconfigure pollution's geography for the benefit of the more privileged classes.[95] In 1870, in a similar move in Great Britain, the Metropolitan Board of Works in London
ordered highly polluting manufacturers to move their premises beyond
the boundaries of the metropolitan area. The result was to transfer nuisances to the suburban district of Blackheath.[96]

Pollution's spatial distribution can be observed on a smaller scale at
the regional level; thanks to the railways, it became increasingly easy to
establish factories near mines, and to sacrifice isolated zones for heavy
industry. This was the case, for example, in Middlesbrough, in England.
A small coastal village in Yorkshire with a population of 200 people in
1830, its exceptional growth (it supported 100,000 inhabitants in 1900)
was due exclusively to heavy industry—steel and metallurgy—that set up
there. This safeguarded the regional, middle-class capital city of York, as
well as the pluri-functional, industrial cities of Sheffield and Leeds. As
Britain's new steel capital, nicknamed *Ironopolis*, Middlesbrough relied
on its railway system and port. It became one of the most polluted steel
cities in the world, but locals complained little about living conditions,
in no small measure because most of them depended on the industry for
their livelihoods.[97] These new spatial configurations were made possible
by railway networks connecting mining districts and factories. Coal was
refined into coke on site at the mines, which kept this highly polluting
and toxic process in isolation away from cities. The coal was then easily
transported in its less polluting form into urban areas for consumption.
Of course, urban pollution reductions often meant a rise in pollution
levels elsewhere.[98] Soda factories also took advantage of the railway and
relocated farther away from city centers. As such, in France after 1870,

they were found in outlying areas such as Dieuze (Lorraine), Thann (Alsace), the hinterlands of Marseille, and even the Salindres commune in the Gard department. These zones, and others like them, fulfilled the principles for spatial isolation for industrialists. In 1854, the owner of the soda factory in Salindres started out by compensating his neighbors for nuisances before eventually buying them out to create a protected and isolated buffer for himself, amounting to 144 hectares (about 355 acres). Here he built a train station that connected his property to the Cévennes mines. In the 1870s, the Pechiney company took control of the business, and local life was thus defined by its patronage, its control of the local workforce, and its degradation of the environment.[99]

Lastly, pollution can be diluted on a global scale. In the latter half of the century the United States began constructing manufacturing sites on wide expanses of virgin soils that had recently been vacated of native peoples. From the gold rush in California in 1848 to the explosion of *Big Corporations* in the 1900s, which amalgamated the banking, railway, mining, and electric sectors (Morgan, Rockefeller, Guggenheim, and others), a veritable mining of the New World led to an exponential rise in the production of precious metals, copper, and lead—with the polluting foundries that accompanied them. Out of sight and out of mind: this massive exploitation occurred far from the eyes and lungs of middle-class Americans, in places like Montana, Michigan, the mountains of California, and the deserts of Utah, Nevada, and Arizona. Around 1900, Australia, South Africa, and Canada (notably Ontario) each took their turn in developing metallurgical mines and foundries, due in large part to the reduced cost of merchant shipping. Pollutions experienced their nascent globalization.[100] In some sense, colonialism at the end of the century—when looked at through a particular lens—could be said to have also played its part in the global spread of pollution. In India, for example, regulating pollution was first and foremost a response to racial segregation. In 1863, the air in Calcutta, connected to mining districts by rail, was just as polluted as in some European cities. The city issued an order to regulate industrial smoke in accordance with British metropolitan norms. In Calcutta, as in Delhi, policies to fight against pollution and for cleaner, healthier urban environments were inspired primarily to protect the districts where European settlers lived. Areas where indigenous people lived were described as dirty and polluted, proof of the backwardness of local populations.[101] In the French empire, regulating polluting industries (though there were far fewer than in India) also stemmed from segregation between colonists and indigenous populations. In Algiers, around 1850, the governor refused to allow the construction of a sulfuric acid factory because it was going to be too close to a military hospital for European soldiers.[102]

*　*　*　*

Throughout the nineteenth century, controlling pollution seemed like an impossible task. Despite undeniable achievements and the adoption of some laws after intense debates, controls, regulations, and solutions varied greatly between countries, time periods, the scale of the task at hand, and of course the particular industries in question. On a more general level, however, industrializing nation states welcomed polluting industries; to them they constituted a symbol of wealth, prosperity, and their rise in importance in the world. More than the recourse to common law—which was largely ineffective—or to the often fragile, defective, and easily circumventable laws that sought to impose controls, pollutions diminished in some places when industries began to relocate to more suitable areas, far from the madding crowd. Rail and maritime transport revolutions permitted the conquest of new territories, which made it possible for industry to find more isolated locations. At the same time, the century of progress held fast to the notion that technical advances would solve pollution problems. Promoted by regulatory authorities, it was a nice idea, but it accelerated and exacerbated the rebound effects and further dispersal of pollution. The institutional legacy of the period between 1830 and 1914 remained prominent throughout the century that followed; even as the twentieth century would have to reckon with an increase in contaminants—as well as new contaminants—as a result of wars, globalization, and continued industrial growth.

III

New and Massive Scales of Pollution:
The Toxic Age (1914–1973)

Contrary to what might be suggested by a quick comparison of the clearer skies over European cities in the 1970s and those drowned in the smog of the 1890s, the twentieth century did not witness an overall reduction in pollution. Rather, the nature of pollutions changed, but their growth was always rapidly increasing and correlated with the demographic explosion, the increase in the consumption and production of manufactured goods, and the remodeling of physical environments by synthetic chemistry. While the population quadrupled during the twentieth century, energy consumption grew ninefold and industrial production increased forty times over.[1] Despite cyclical and regional crises, economic growth accelerated in the industrialized world after the Great War, reaching its peak during the period known as the Glorious Thirty (1945–1975), with growth rates of 3 percent to 4 percent in England and the United States, 5 percent in France and Eastern European countries, and even 8 percent in Japan. Between the great industrial war of 1914 and the oil shocks of the 1970s, a massive acceleration in environmental changes led to increased pollutions and their spatial redistribution.

A new standard, endowed with all the virtues, stands out as a paragon of the desirable society: the gross domestic product (GDP) inspired visions of infinite growth and a society pacified by an abundance of material benefits on the horizon.[2] Its algorithm for determining the economic health of a nation became a kind of *lingua franca* for rulers, business leaders, unions, and historians alike. Developed in the United States to solve the serious crisis of the 1930s, the language and concept of GDP spread after 1945 to Europe and the world. Measuring the production of goods and services within a country during a given period, however, this indicator neglected heritage and nature. As a result, environmental damage and the depletion of natural resources was rendered economically invisible or irrelevant. Worse, far from supporting the fight against pollutions, the obsession with the rise in GDP tended to exacerbate environmental problems. Pollutions were the byproducts of manufacturing goods, which weighed positively on the GDP spreadsheet. Services provided to attend to accidents or medical treatment of environmental diseases were regarded similarly.

In this perpetual race for more, humanity entered surreptitiously into a new "toxic age" par excellence.[3] New chemical substances, in particular, caused expansive wastes that spread through air, soil, and water. They were often dangerous and sometimes resisted biodegradation in natural environments or organisms. Still, industrialists mostly disposed of their untreated waste in pits, dumps, rivers, or the sea. Such contaminated spots were still relatively isolated and modest in number in the early twentieth century; their number and size expanded rapidly after 1945. In 1980, estimates suggested that the United States alone was home to 50,000 hazardous waste dumps.[4] Moreover, these substances did not only pollute at the time of their production; rather, they became defined by their persistence through every aspect of their cycle: production, consumption, and disposal into the ultimate outlet, the oceans. Poisons that do not die: this created a kind of "universal poisoning," according to the journalist Fabrice Nicolino, "a new age of toxicity," according to historian William Cronon, "an age of poisons," according to biologist Rachel Carson.[5]

Like an uncontrollable oil spill, the development of pollutions took on a metastatic character that attacked societies in myriad ways. They aggravated the social, economic, and environmental imbalances between regions preserved for healthy living for those who could afford it and those corridors sacrificed to industrial production and the reception of its waste. At the end of the twentieth century, human activity extracted some 70 billion metric tons of materials annually to feed its rapacious appetite, but only 15 percent of the population—living mainly in the industrialized countries of the North—consumed half the mineral and fossil resources.[6] Despite this inequality with regard to environmental dangers and hazards (which remain an inequality linked to economic development), pollutions were unmistakably a global—globalized—problem. Although each particular industry tended to pollute less, thanks to the adoption of new, more efficient processes and because producers found means to reduce the quantity of their waste, demand and the scale of production meant that global pollutions, in aggregate, continued to increase. In fact, four phenomena counteracted the effects of technical improvement: first, industrialization's standardization throughout the world mechanically increased the overall volume of toxic discharges; second, the exhaustion of mining resources meant that the industry was forced to exploit veins that were less and less pure, which in turn demanded more aggressive extraction processes; third, mass consumption and population growth shifted pollution downstream to the act of consumption itself; finally, globalization and global economic integration accentuated the transformation of nature into a commodity: consumption of—or the perceived need for—new goods helped to soften opposition to pollutions, often rendering them less a feature of public discourse.[7]

Several processes played a decisive role in the advent of this new toxic age. First, two World Wars and the Cold War promoted a certain disinhibition of—or prioritization away from—the damage wrought by progress and the ravages of industrial expansion (chapter 7). The establishment of a fossil fuel energy regime and the ascendancy of oil also accentuated global pollutions (chapter 8). The lifestyles and birth of the "consumer society," which depended on industrial growth for its goods, multiplied and disseminated pollutants (chapter 9). Finally, the persistent weakness of legal regulations and intervention methods, which failed to slow the phenomenon, made it possible to understand the emergence of many new hazards in the 1970s (chapter 10).

7

Industrial Wars and Pollution

If wars before 1914—starting with revolutionary and imperial conflicts, the Crimean War (1853–1856) and the Civil War in the United States (1861–1865)—were terribly destructive with a variety of environmental effects, none reached the scale or intensity of warfare in the twentieth century.[1] A new era of total war brought with it new methods of widespread contamination. Napalm and the chemical defoliants used in Vietnam in the 1960s and 1970s were emblematic of the ecological ravages and pollution caused by twentieth-century conflicts. Gas warfare was launched during the First World War and, with the help of aeronautics, Spanish and French colonial powers dropped the first incendiary bombs on Moroccan Berber tribes during the Rif War (1921–1927). Napalm—a gasoline-based substance initially used for flamethrowers—was developed at Harvard University in 1942, before being used extensively in 1945 against Japanese cities, during the Greek Civil War (1944–1949), then in the Korean War (1951–1953) and in Vietnam. In addition to the innumerable human casualties, fauna, flora, soil, and water were permanently contaminated by its liberal use. The massive dropping of the famous herbicide nicknamed "Agent Orange" by the American Air Force during the Vietnam War to destroy the forests in which the Viet Cong soldiers were hidden, is considered one of the most important ecocides in history; it also reveals the close links between the civilian and military uses of this type of chemical.[2]

Total war—as a concept and a practice—began in 1914. Rather than just mobilizing soldiers to a battlefront, the entire population and economy of the belligerent countries were recruited into complete subservience to the conflict. War also became more and more industrial, enlisting chemical resources, and commandeering scientific and industrial innovation on an unprecedented scale to inflict maximum destruction.[3] As

Figure 7.1
Bombardment of Agent Orange from American planes during Operation Ranch Hand, 1962–
1971. Too often we forget how wars are major sources of pollution. The concept of "ecocide"
comes from the Vietnam War because of the massive bombardment of Agent Orange by the
US Army over the forests where Communist fighters hid. More than 3 million hectares (about
7.4 million acres) were contaminated by this Monsanto-manufactured herbicide.

La Nature (a French magazine aimed at the popularization of science)
observed in January 1916:

> If there is an industry that is equally valuable in both war and peace, it is the
> tars and benzols industry. In times of peace, it allows our elegant women to
> dress in bright colors and surround themselves with delicate perfumes; in war-
> time, it supplies our cars with a valuable substitute for exotic oils, and above
> all it serves as a basis for the manufacture of the most powerful explosives:
> melinite, trotyl, amatol, etc.[4]

Scarcely visible for a long time and considered secondary to the countless human casualties—an estimated 180 million deaths during the
wars of the twentieth century—environmental damage and destruction has recently emerged as an important research interest.[5] The

relationship between military action and environmental change is the subject of a growing number of works.[6] Indeed, war reconstitutes the interactions of—and between—nature and society, engages the world in the service of action, and intervenes in the redefinition of representations of nature, as in the construction of knowledge and ecological mobilizations.

We can ascertain that more than the pollution itself, wars were marked by the destruction and radical transformation of landscapes. The most spectacular damage to the natural environment was linked in the first place to policies of terror and scorched earth. The destruction of the Yellow River's dykes by Chinese nationalists in 1938—a desperate action to halt the Japanese advance—undoubtedly had a much greater impact on ecosystems than pollution. Widespread deforestation of vast areas is arguably the most important environmental impact of twentieth-century warfare.[7] Similarly, upstream from the war process, military-industrial complexes undertook the construction of massive infrastructures and developed scientific programs that shaped environments. But as a matrix of new technoscientific trajectories, as a space in which exceptional mobilization and unprecedented lack of restraint dictated social activities, wars marked the sea-change of scale in pollution's history during the twentieth century. In war, environmental protection and public health were relegated to the background as the theater of combat—and its imperatives—took center stage. Conflicts were the inspiration for technological developments that would otherwise have been inconceivable: from destructive chemistry to nuclear energy, to say nothing of the considerable acceleration in the design of new materials and logistics, a whole range of new, highly polluting products emerged.

Wartime Mobilizations and the End of Restraint

In the "Age of Extremes," military interventions transformed landscapes and the environment. In certain circumstances, it could be said that military actions led to the creation of sanctuaries conducive to the special protection of ecosystems, away from industrial influence.[8] Indeed, war could reduce local pressures exerted on certain natural environments. This was the paradoxical case for the 250-kilometer long demilitarized zone created in 1953 between North and South Korea; which has seen the development of a rich biodiversity. To a lesser extent, the more than

3,000-kilometer zone of the former Iron Curtain which divided East and West Germany manifested a similar condition. In times of war, restrictions and rationing policies reduced the footprint of some particularly polluting activities. Some countries experienced shortages of basic consumer goods; for example, Italy and France suffered a decrease of 40 percent and 50 percent, respectively, in their GDP between 1938 and 1945.[9] In France during these years, fishing for salmon in the Seine became possible again for the first time since they had disappeared in the 1880s resulting from pollution of the river; the same phenomenon was noted in Finland.[10]

Shortages sometimes forced a return to older production techniques that required less energy consumption and generated less pollution. Energy conservation was the rationale behind the introduction of daylight savings—adopted in 1916 by Germany and the Austro-Hungarian Empire, followed by France, Great Britain, and the United States. After entering the war in April 1917, the United States tried—unsuccessfully—to limit the dense fumes escaping from the factories, which appeared to be wasteful and a sign of antipatriotic inefficiency. During the Second World War, the American government encouraged the use of bicycles as a patriotic duty, while Britain promoted recycling.[11] In other cases, rationing forced agriculture to dispense with chemical fertilizers or to develop low-energy farming techniques. In Japan, for example, during World War II, the ammonia shortage, the use of which was monopolized by the industrial sector in order to prosecute the war, pushed farmers to practice composting in order to enrich their soils with organic matter.[12]

Though such examples of constrained wartime management of resource shortages exist and abound, they should not obscure how much—on a more macroeconomic level—the belligerent nations engaged in a total mobilization of their productive apparatus. On average, nations' GDPs grew by almost 35 percent during the Second World War, and in the United States by 214 percent between 1938 and 1944.[13] During the wars, the decline in pollution, if and where it occurred, was still limited and circumscribed, usually followed by a phase of intense economic—and concomitant pollution—growth after the end of hostilities. Although the situations varied greatly according to landscape, the economic structures of the countries, and the location of the battlegrounds,[14] war favored a recomposition of markets, nourished the "insatiable appetite" of the great powers for natural resources—wood, sugar, oil, or uranium—and encouraged their intense exploitation.[15]

Wartime priorities also redefined ways of thinking about and exploiting natural environments. War allowed for unprecedented logics of predation, lifting controls and inhibitions, silencing opponents, and imposing a modernizing framework favorable to the unprecedented growth of polluting emissions. At the heart of this industrial mobilization was the production of shells, machine guns, airplanes, tanks, and trucks, but major projects also featured: the manufacture of large-scale equipment as well as transport and energy infrastructures were critical to the war effort.[16] For example, after 1941, the US Army ordered 500 million shoes, and the auto industry was required to produce military vehicles. Until 1945, American industry alone made two-thirds of allied military equipment: 2 million trucks, 300,000 aircraft, 193,000 pieces of artillery, 86,000 tanks. Such production intensification was inevitably accompanied by numerous environmental impacts; its continuation required a relaxation of stiff environmental regulations and some tolerance of legislative infringements. War and production pressures dictated that polluting discharges were poured directly into watercourses or hastily dumped.[17] In this context of industrial reconversion to meet military demand, ecological considerations and worries were subsumed by the urgencies of the time. Administrative and legal controls on pollution were eased and protest movements lost considerable influence.

The Great War was the first highly polluting energy and chemical conflict. It propelled oil production from 40 million metric tons in 1910 to 100 million metric tons in 1921, while Royal Dutch Shell's dividends multiplied fourfold between 1914 and 1919. Mechanized war made oil such an essential requirement that in December 1917 the head of the French government, Georges Clemenceau, urged US President Woodrow Wilson to send him 100,000 metric tons of oil quickly: "If the Allies do not wish to lose the war, then, at the moment of the great German offensive, they must not let France lack the petrol which is as vital as blood in the coming battles."[18] Demand for oil forced the major powers to increase their influence over the resources of the Middle East. During the conflict, the patriotic "sacred union" in France and the defense of the nation imposed in most belligerent countries a consensus around the needs of production and fostered an intense techno-industrial modernization. The conflict convinced many participants—such as the French socialist Albert Thomas, union organizer of arms production workers and the first director of the International Labor Office after the war—that the growth of production and the use of new technical processes were necessary.[19] In

1917–1918, the United States broke records for the quantities of coal burned. In so doing, it reversed previous efforts to reduce industrial fumes; air quality declined dramatically in many cities, including Pittsburgh, St. Louis, Cincinnati, and Chicago. The press reported that smoke clouds had never been so thick. In New York, an anthracite coal shortage forced manufacturers to use dirtier, lower-quality fuels. In addition, war and its patriotic propaganda silenced protest movements and antismoke activists. For example, private sector engineers came to dominate and take control of the International Association for the Prevention of Smoke.[20] In Germany, too, the necessities of war production pushed the administration to put a moratorium on legislation that limited industrial fumes: in 1915, the Prussian Minister of War invited local authorities to suspend environmental quality regulations.[21]

The suspension of certain environmental regulations seemed a universal rule that applied as much to the Second World War as the First. In the United States, the antipollution regulations adopted in St. Louis and Pittsburgh before the American entry into the war were quickly suspended after 1941; they were not reinstated until the postwar period. In Nazi Germany, the economy was geared to victory on September 3, 1939 and, from 1943—a year marked by a succession of defeats in Stalingrad and in North Africa—the whole nation and its human and ecological resources were mobilized. Admittedly, the Nazi administration was torn over the question of air pollution and its control. The subject did not directly interest Hitler, but it did breed debate, nevertheless. Some lawyers wanted to promote a policy of reducing fumes in the name of the good of the nation and the Germanic race, but authorities finally decided and ordered the limitation of the legal constraints that governed polluting emissions. However, complaints against nuisances and pollution never completely disappeared during the war.[22] After 1945, the rivalry between East and West further polarized tensions, and the specter of the Cold War significantly slowed down any concerted return to the regulation of pollutions. If it featured at all in the great international conventions of the postwar period, pollution was only ever at the margins, until the 1972 United Nations Conference on Humans and the Environment, held in Stockholm. Israel's case marked another rationale for alleviating environmental regulations. Given the tensions and contexts surrounding its creation in 1948 and the many conflicts with its neighbors—in the first place, the Six-Day War in 1967 or the Yom Kippur War in 1973—the Israel Defense Forces acquired a special status within the State of Israel, and an immunity from adhering to the country's environmental regulations.[23]

As with conflicts, war preparedness and the militarization of societies in the 1930s—or during the Cold War—legitimized major projects, removed the political constraints that could curb them, and set aside prudence for the benefit of a total mobilization of industrial capacity and resource appropriation.[24] Armed conflict notwithstanding, military maintenance added to overall pollution levels, accounting for 6 percent to 10 percent of air pollution, for example, since 1945. Daily military operations, such as research, development, testing, and storage—as well as training, maintenance, or deployment of troops and arsenals—created vast amounts of waste and polluting emissions. For example, the US military alone generated 750,000 metric tons of toxic waste per year, more than the five largest chemical companies in the country combined. That waste needed to be processed or stored in the 8,500 military enclaves in the United States and installations in allied countries, such as Canada, Germany, Great Britain, Italy, Panama, the Philippines, South Korea, and Turkey.[25] The expansion of the number and influence of military bases stemmed from the increased warlike tensions of the twentieth century, which the Cold War only exacerbated: at the end of the 1980s, military encampments and facilities covered 1 percent of the surface of the globe. Many housed some of the most polluted places on earth. At that time, the United States had 375 major and 3,000 smaller military bases scattered around the world. These bases were responsible for environmental transformations that gave rise to much opposition, particularly in Great Britain and France. In France, the fight against the extension of the Larzac base in the 1970s also precipitated the emergence of an environmental consciousness that goes beyond purely military concerns.[26]

The American military presence around the world, which provided the basis for the post-1945 international order, created multiple localized pollution incidents. Generally, military secrecy ensured that such infractions were ignored, excused, or covered up. In Panama, for example, chemical weapons experiments—conducted over several decades at large US military installations—provoked many nuisances and local complaints. After the bombing of Pearl Harbor in 1941, the United States Armed Forces purchased most of the island of Vieques, near Puerto Rico, as a strategic naval base. After the war, the military turned the island into a munitions depot and base for testing and maneuvers. For sixty years, the small island has been under colonial control, with a veritable deluge of bombs causing rates of cancer or cirrhosis well above the average observable in the Caribbean.[27] In Europe itself, the presence of US troops and North Atlantic Treaty Organization (NATO) forces to stem the threat of communism weighed heavily on some ecosystems. In Sardinia, for example, the Italian Government authorized in 1956 the establishment of shooting ranges and various military installations which, for several decades, heavily contaminated the island.[28]

Energy consumption related to maintaining the military presence of the great powers has increased steadily along with the relative gigantism of military equipment. After the Second World War, military technologies became ever more energy hungry: a US Army Abrams tank consumed 400 liters of fuel per 100 kilometers, a B-52 bomber—commissioned in the 1950s—burned 12,000 liters of kerosene per hour, an F-15 used 7,000 liters, more than a car over several years. Between 1945 and 1975, CO_2 emissions linked to military training and the production of weapons in the United States alone represented around 10 percent of all emissions.[29] On the other side of the Iron Curtain, the desperate quest for strategic mineral resources for the war effort and the rapid construction of a vast chemical and metallurgical industry to meet military demands transformed entire regions of the USSR into militarized camps and "industrial deserts."[30] After the Second World War, during which the Ural regions—rich in mineral resources—were mobilized to feed the war effort, mining and industrial cities flourished. This caused massive increases in pollution, especially in the north of the Soviet Union. In the same regions, mushroom towns were born around mines and military bases, typified by gulags. Vorkuta, located in the heart of a vast mining basin beyond the Arctic Circle, constituted an archetypal example. Connected to the rest of the country by railway from 1941, and hosting a major military base for bombers during the Cold War, the Vorkuta region offers a paradigmatic example of the pollution of water and air by heavy metals; the same phenomena were observed in the cities built in Siberia to exploit the mineral resources that fed military needs.[31] Though the scale may have varied, other countries' practices followed the same pattern. In Japan, for example, during the interwar period, air and water pollution worsened as the war effort intensified: local authorities reported numerous contaminations by the mining and quarrying industries and chemical plants. The fisheries administration within the Japanese Ministry of Agriculture identified more than a thousand cases of water pollution, and most of the country's major cities were victims of toxic emissions from industry.[32]

A Matrix for New Polluting Trajectories

By mobilizing human and material resources, removing restraints and marginalizing concerns about pollution, war set the stage for the great acceleration of environmental changes in the latter half of the twentieth century. The new armed conflicts were both total wars and "world wars," prosecuted on all continents. In effect, the two world wars reshaped categories and representations of space and time, and undermined previous balances of power—particularly colonial—while impelling new scientific and industrial dynamics on a spectacular scale. By removing all inhibitions, the world wars initiated the global triumph of brutal and destructive technologies.[33] They also prepared the ground for new pollutions by shaping the technical and legal framework of the mass consumer society, stimulating a military-industrial complex under state control, and boosting motorization, aeronautics, and new, especially polluting sectors such as pesticides and aluminum.

Wars in the twentieth century ushered in the establishment of "Big Science," which made a new world of research increasingly dependent on partnerships between the state, scientists, and industry, all in the service of pharaonic projects that had profound and lasting impacts on the environment.[34] This "heavy science" stemmed from developments during the First World War and interwar period, which set decisive milestones; in 1916 the National Research Council was created in the United States to place academic research at the service of military objectives.[35] But especially after 1940, science settled at the heart of political issues around major research programs (particularly nuclear), and scientists acquired a central position of power. While in 1939 fundamental research was still dominated by small teams with limited budgets, scientists found themselves after 1945 at the heart of military programs whose polluting capacity had increased tenfold by the alliances it established with industry.[36] After chemistry, physical science—particularly nuclear and high energy physics—maintained ever closer relations with military research. By the Cold War, the great powers were investing massively to improve their knowledge of the physical world and matter and to develop new, more destructive weapons; satellite imagery, oceanography, seismology, meteorology, and climatology were recruited for military control of the planet.[37] Some universities, such as the Massachusetts Institute of Technology (MIT) in the United States, were at the heart of the nascent "military-industrial complex." Its director stated at the end of World War II that "[MIT's] military value was equivalent to that of a squadron or an army."[38] Although CERN's European project was more independent of the military because of the unique diplomatic issues it presented, most major postwar research organizations, such as the French Atomic Energy Commission (CEA), were closely beholden to or derived from military interests.

Without the war, motorization across European and North American societies—like the development of aeronautics and new chemical substances—would have been much slower to develop, and some new and particularly toxic products would probably never have emerged or spread with such ease. If the rise of the automotive industry received a boost during the Great War—Renault and Citroën in France, for example—aeronautics was the key sector that greatly benefited from twentieth-century conflicts. While 5,000 aircraft were built worldwide in 1914, their number surpassed 200,000 in 1918; the French army had fewer than 350 aircraft in August 1914, but by 1918 it had built 41,500, including nearly 10,000 for the Allies. The increase was not only quantitative: in 1918, planes flew twice as fast and five times higher than before the war, and their energy consumption increased.[39] In 1918, the Royal Air Force was created in Britain, and in the 1920s aerial warfare was studied and theorized. New aerial technology meant that from now on civilian areas would be part of the destruction of war: aerial bombardment became an essential element of total war.[40] But the advent and ascent of aeronautics helps in understanding the fundamental impetus of the Second World War in a very energy-intensive mode of transport that became commonplace for civilians in the second half of the twentieth century. In 1945, the United States alone produced 20 million metric tons of aviation fuel (compared to 2 million metric tons for Great Britain).[41] The age of jet engines also emerged during the Second World War: simple experimental prototypes in the 1930s, they became operational during the conflict. They were subsequently employed during the Korean War, before being adapted for civilian use.[42]

War as a matrix of new production models also affected a wide range of technologies, of which aluminum and chemical substances were probably the most representative. At the beginning of the century, aluminum was still a precious metal produced in small quantities. However, it became a fundamental metal in aviation, ideal for its lightness and physical characteristics. In the 1930s, fascist Italy and Nazi Germany increased their demands on the aluminum industry. The authoritarian regimes also repressed protests—such as those from the villagers of Mori, in northern

Italy along the Adige River—affected by fluorite pollution from aluminum smelters.[43] The Second World War tripled global production capacity. To meet the war effort and the needs of the military, American aluminum production rose from 130,000 metric tons in 1939 to 1 million metric tons in 1945: three-quarters of world production. Germany produced 200,000 metric tons in 1944. In order to profit from the vast amount of industrial equipment used during the war, it became necessary to find postwar civilian markets for aluminum, all the while without restricting its continuing military uses. The metal was touted as a modern material with almost miraculous properties; it was promoted for the manufacture of cooking utensils, packaging (the famous soda cans), or storefronts. Aluminum became one of the veritable symbols of mass consumption after the war.[44] In 1950, 1.5 million metric tons were produced globally, a figure inconceivable without the decisive impetus of the Second World War.

The two world wars also introduced a large number of new chemical substances, often already known but for which opportunities and markets had been lacking. Their proliferation contributed to the "universal poisoning" by chemistry.[45] One example: during World War I, the Haber-Bosch (1909) ammonia synthesis process for fixing nitrogen in air was improved because nitrate was critical in the production of explosives. Responding to the freezing of German imports and growing military needs, the American chemical industry expanded on an unprecedented scale. The war profoundly transformed American chemistry, notably under the influence of the American government's Chemical Warfare Service. In November 1917, Charles E. Roth, secretary of the American Chemistry Society, asserted that the country had "accomplished in only two years what had taken Germany nearly forty years."[46] The Second World War presented an industrial echo of the first on an even larger scale: while the large American companies in the chemical sector were still confined to local markets before 1940, they dominated international markets after 1945.[47] In a few decades, tens of thousands of substances were put on the market and infiltrated into and entrenched themselves in people's daily lives. Prior evaluation of their hazards and real administrative control over their production or distribution were largely dispensed with. Functional chemical oversight in the United States did not emerge until the Toxic Substances Control Act was passed in 1976; by that time, 62,000 chemical products were in circulation.[48]

Much of the proliferation of chemical pollutants was due to the need to recycle war products. After the hostilities of the First World War, the chemical gases used in the fighting were converted for civilian use, primarily as sanitizing materials. Chlorinated gases were reused for disinfection in medicine and in public places. Arsenic gases controlled pests, especially weevils. Despite intense debates, polemics, and resistance, the insecticides from combat gases spread widely in American agriculture in the 1920s and 1930s, a use supported by para-institutional propaganda and the prestige of the powerful chemical industry that sought to optimize the equipment it had developed for wartime production.[49]

The famous case of dichlorodiphenyltrichloroethane (DDT) offers another remarkable example. This substance, synthesized by an Austrian chemist in the 1870s, had remained a simple laboratory feat before the Swiss Paul Hermann Müller (1899–1965) discovered in 1939 its utility as an insecticide, first against moths and then against Colorado potato beetles (he received the Nobel Prize for Medicine in 1948 for this discovery). In May 1943, after the Food and Drug Administration's studies attested to its safety, DDT was manufactured on a large scale to supply the US military and eliminate disease-bearing insects from theaters of operations, particularly following the typhus epidemic in Naples in October 1943. Nearly 1.3 million civilians were treated with Neocide powder, a substance containing DDT. In 1944, fourteen US companies produced DDT in large quantities; it was used extensively in the Pacific War to fight mosquitoes—often portrayed in advertising copy as the treacherous Japanese (1943–1945).[50]

DDT was perceived and sold as a miracle product that had contributed to the Allied victory and the protection of soldiers. After 1945, manufacturers had to create numerous new markets for its overproduction; postwar agricultural pollution was largely the result of the disposal of stocks inherited from the war. Historian Edmund Russell has shown the incessant intersections and linkages between certain forms of "wars between humans" and "wars against nature" through chemistry, particularly in the form of pest eradication programs. War deeply shaped systems of norms governing the relationship with the environment—couched in a rhetoric of increased violence—which in turn further radicalized warlike action. Through the twentieth century, the brutalization of human societies and nature went hand in hand; the exceptional, justified in wartime, became normalized in times of peace. It is therefore hardly possible to understand the reinforcement of pollution's entrenchment after 1945 without reflecting on these large-scale military clashes.[51]

Under the Bombs: The Gases

Assessing the relationship between war and pollution in the twentieth century also involves diving into the heart of the battle, so as to scrutinize its impacts on air, water, and soil. Depictions of nature were profoundly redefined by experiences derived from war. This was especially the case during the unprecedented violence of the Great War. Peaceful country landscapes were evoked as a means of symbolizing hope and providing a serene atmosphere to calm anxiety and fear.[52] In contrast, the battle landscapes of desolation were nevertheless mainly represented in iconography and testimonies. Soldiers, educated or of more modest origin, painted the ecological ravages. Visions of death and of silenced spring, a longstanding symbol of rebirth and profusion, scarred war's witnesses:

> Visions of grass without grass, buds without buds, and flowers without flowers. Season of poisoned bulbs. Seasons of total war. Seasons of lovers without love. Lunar landscapes. Visions of dead stars. Seasons of unnecessary and bloody offensives. Seasons of sterile and desperate mutinies. Seasons of strange sources; seasons of flooded tombs. Seasons of gases and poisons. Nature hungry and carnivorous. Rusty locks.[53]

Trench warfare, along a front line stretching from its western extreme on the North Sea to the Swiss border, left behind the detritus of war: "polemological" landscapes strewn with sterile soils, riddled with metals and unexploded shells buried underground, unsuitable for agriculture. Of the 3.3 million hectares (about 8.2 million acres) of agricultural land affected by the fighting in France—studied with precision because they entered into the postwar calculation of war reparations—tens of thousands could not be cultivated after the war. Reforestation projects in the 1930s served as rehabilitation of spent lands.[54] The artillery's impact on the land corresponded to 40,000 years of natural erosion.[55] The ravages were also ecological: "red zones" were massively polluted by substances of all kinds. The risk of contamination of groundwater from the decomposition of corpses was a problem that was quickly realized—the French Academy of Sciences identified the issue as early as December 1915.[56] During the long siege of Verdun in 1916, experiments were conducted with chlorine to disinfect the water and counteract the risks of epidemics. The term "verdunization" stuck. But the war also posed longer-term problems of persistent pollution. Atmospheric fallout from weapons fired and explosions lead to high concentrations of pollutants in some areas. Very high levels of heavy metals such as lead (from shells) or mercury (used as primers) durably contaminated local areas where they were left in abundance. For example, surrounding the so called "*Place-à-gaz*," located in a forest near Verdun, the French army dealt with 200,000 chemical shells in the 1920s. Surveys and measurements carried out a century later in 2014 indicated intense and lingering quantities of arsenic (17 percent of the weight of the soil), but also of cadmium, lead, and mercury. If this site is still closed to the public by prefectural decree because of pollution, all the areas near the front line were affected by such residual pollution related to the First World War.[57]

In his famous novel *Under Fire* (1916), Henri Barbusse described the world of trenches as a "polar desert whose horizons fume," with "barren fields," invoking in his story recycled imagery of previous industrial pollutions. He compared the "dark," "nauseating" atmosphere of the trenches to "the smoke and smell of factories," as if the industrial war had brought to an end the environmental costs of the industrialization of the previous century: "Gigantic plumes of faint fire mingle with huge tassels of steam, tufts that throw out straight filaments, smoky feathers that expand as they fall—quite white or greenish-gray, black or copper with gleams of gold, or as if blotched with ink."[58] The "asphyxiating gases" mentioned by Barbusse as a chemical threat were, however, in 1915 only a whisper of the great danger they would become with rapidly increased use.[59]

The Great Chemical War began in earnest on April 22, 1915, when a wave of chlorine from German lines broke into an Ypres trench on the Belgian front. Fifteen thousand French soldiers were quickly incapacitated; 5,000 of them died. The significance of the new threat incited the Allies to redirect their chemical industry into a producer of offensive weaponry. At the peak of production in 1918, English factories produced 210 metric tons of chlorine a week for the war effort. But chlorine was quickly supplanted by phosgene, which had already caused 6,000 deaths on the Russian front on May 31, 1915. Research into toxic gas innovation and production—and, ideally, methods of firing them as projectiles—escalated. In addition to asphyxiating gases (including phosgene, diphosgene, and chloropicrin), the new research yielded an arsenal of incapacitating gases, based on bromide or cyanide, and subsequently sternutatories, based on arsenic, or compounds of chlorine and arsenic. Research in the Allied camp was concentrated in Great Britain, under the direction of the great chemist William Pope, who promoted the industrialization of laboratory tests on triphenylarsine dichloride (TD) and diphenylarsine chloride, and even chloropicrin. In Germany, chemical research was conducted under the direction of Fritz Haber. The introduction of *ypérite* shells (named after Ypres, which again served as practical laboratory for the military use of chemical gases—these were the now famous "mustard gas" or dichlorinated ethyl sulfide) on July 12, 1917, marks a much more important milestone: the new poison attacked any part of the body, causing extensive burns. Insidious and persistent, mustard gas required that soldiers keep mask and waterproof protective clothing readily at hand. As soon as it appeared, mustard gas became the main combat gas, quickly adopted by Germany and then by the other belligerents. Responsible for most of the casualties from gas, it was however produced in industrial quantities only from the summer of 1918.[60]

After the end of hostilities, "chemical warfare" raised many concerns. In Europe, pacifists and nationalists shared the same fear with respect to combat gases.[61] Speeches and writings from the interwar period warned against the risk of contamination by new military technical means; these debates shared many points in common with those surrounding industrial pollution. Physicist Paul Langevin, Vice President of the League of Human Rights, claimed that 100 metric tons of gas were enough to cover Paris with a toxic layer 20 meters deep. These strong concerns contributed to the imposition of regulations, such as the Washington Treaty (1922)—which was supposed to prohibit the use of asphyxiating gases but was never put into effect—and the Geneva Protocol (1925) on the Prohibition of Chemical and Bacteriological Weapons. In 1931, an international commission for the legal protection of civilians against the dangers of air warfare was also created.

The march to war during the 1930s nullified these tentative attempts at regulation. The bombing of Guernica in Spain on April 26, 1937, by the Nazis and the Italian fascists, which is one of the first raids in the history of military aviation on civilians and a defenseless city, signaled the destructive intentions of the Second World War that followed. In the Pacific, Japan and the Allies became ensconced in a terribly destructive conflict, while imperial powers reinforced colonial control over these territories. The ecological damage produced by the war on these often-fragile island territories was particularly important.[62] Direct pollution linked to the fighting is unfortunately poorly understood. From January to June 1942, for example, during the Battle of the Atlantic, German submarines sank dozens of tankers, which dumped their cargoes of crude oil into the ocean, but it is difficult to assess their range. American bombing of German cities was equally damaging, inflicting real environmental punishment and turning swaths of Germany into a desert. For example, on July 27, 1943, the Allies dumped nearly 10,000 metric tons of phosphorus bombs on Hamburg, while the bombing of Dresden on February 13, 1945, killed 25,000 civilians and totally destroyed the city. In all, 2 million metric tons of bombs were dropped on the Third Reich in May 1945 alone.[63]

The Asian wars after 1945, starting with that of Korea (1950–1953), extended the same destructive logic. The British introduced the military use of defoliants against the Malay Communists (1948–1957).[64] Since the Vietnam War, which spanned twenty years (1955–1975), historians have begun to speak of "ecocide," so devastated were landscapes that they became unfit for agriculture because of the extent of pollution.[65] During this conflict, the destruction of the physical environment became a primary military objective. The American infantry advanced with powerful bulldozers (Rome plows) that cleared forests and crops. A new 6-ton (metric) bomb, the specially developed "daisy cutter," was designed to instantly create landing zones in the middle of the forest for helicopters. Noting the failure of incendiary bombs and napalm to destroy the dense tropical Vietnamese rainforest, the US military finally pulverized the landscape with defoliants from the herbicide industry: more than 70 million liters of herbicides were dumped between 1961 and 1971, including 44 million metric tons of "Agent Orange," the notorious herbicide from Monsanto. Some 2.6 million hectares (about 6.4 million acres) were sprayed with this product, which subsequently seeped into soils and water, permanently contaminating vegetation and living creatures. In total, half of Vietnam's mangroves and 40 percent of the country's arable land were contaminated. The country lost 23 percent of its forest area. The effects on the populations remain still forty years after the end of the conflict.[66]

More generally, after the fighting, the impact of military pollutions was immense; much of this history remains unexamined. Many sites were contaminated for generations, banned from the public, secretly monitored by the military authorities. In other cases, egregious postwar spills, especially at sea, at sites that were forgotten or no longer monitored, produced disastrous ecological consequences many decades later. In France, thousands of metric tons of munitions have been immersed in Lake Avrillé (Maine-et-Loire) or buried in the Jardel Chasm (Doubs). In Belgium, part of the stockpile lies off Zeebrugge.[67] Contaminated black zones became even more ominous after World War II and the Cold War created new cemeteries of weapons, ammunition, and toxic products. Indeed, after World War II, nuclear waste was added to more conventional weapons stockpiles.[68]

Nuclear Contaminations

The atomic bomb serves as the singular and most spectacular symbol of twentieth-century warfare. Its enduring legacy, nuclear contamination, may rival the bomb's destructive significance. While most of the theoretical knowledge concerning protons, uranium fission, and the resulting "chain reaction" were available before 1939, wartime exigencies galvanized the switch to practical applications, and primarily military operations.[69] In 1939, US President Franklin Roosevelt ordered applied research on this new technology, and in 1942 the "Manhattan Project" was launched to move to the industrial stage of production. This led, in spring 1945, to the development of two bombs with plutonium and another with uranium: three bombs that were detonated, first in the desert of New Mexico (July 16) and subsequently over the Japanese cities of Hiroshima (August 6) and Nagasaki (August 9). The two assaults on Japan remain the only nuclear bombings as an act of direct aggression. The atomic bombs killed roughly 200,000 people at once. Measuring the impact of the radioactive fallout that persisted long after the war is a difficult proposition.

Since 1945, questions surrounding health and environmental impacts from nuclear weapons have been constantly debated. After the destruction of Hiroshima and Nagasaki in 1945, scientists found that deep-sea fish had survived. Similarly, an October 1947 survey of Hiroshima's fauna found that vertebrates and insects had returned to the city, while local flora also showed a strong resilience and was robustly reclaiming the ruins of the city itself.[70] But nuclear pollution was not limited to this solitary, explosive event. The race for arms and power resulted in a rapid proliferation of nuclear weapons. Increased weapons testing and its resulting nuclear waste caused much more damage than the bombs of 1945.

In addition to the Cold War's two superpowers, by 1975 the United Kingdom, France, and China had managed to build the atomic bomb. These five nuclear powers were the permanent members of the UN Security Council. Their atmospheric nuclear tests carried out in 1971 released energy equivalent to the explosion of more than 500 megatons of TNT.[71] In total, more than 2,000 explosions have been officially recognized by the major powers, more than half of them by the United States. Although the sites chosen for these trials were located in isolated or even desert areas, the extent of contamination is persistent. Between 1946 and 1958, the US military conducted thirty-three atmospheric nuclear tests on Bikini Atoll, previously emptied of its inhabitants. In March 1954, the detonation of a prototype of the H-bomb (or thermonuclear bomb) weighing 11 metric tons produced a crater 2 kilometers in diameter and permanently irradiated an area of more than 100 kilometers in diameter. In the 1970s, the contamination of fauna and flora was such that the local population was still prohibited from returning to the island.[72]

On the other side of the Iron Curtain, the USSR produced 45,000 nuclear warheads and officially conducted 715 nuclear tests between 1949 and 1991. But if we add accidents, the sabotage of submarines or the civilian use of military nuclear warheads to open canals, the egregious dumping of waste in the Arctic Ocean, and countless leaks, the Soviet military nuclear regime has helped to pollute the world on a scale hitherto unknown. Between 1949 and 1989, the Soviet Union detonated 468 atomic bombs (125 in the atmosphere and 343 underground) at a single site known as the Polygon, at Semipalatinsk in Kazakhstan. Many studies have since revealed the extent of radioactive pollution. The Western Siberian testing grounds remain one of the world's most radioactive places.[73] Although there is no comparison of scale with the magnitude of pollution the United States and the USSR have produced, other countries have contributed to the radioactive contamination of the environment with their own nuclear weapons programs. After 1960, for example, France carried out 210 nuclear explosions in the Sahara and Polynesia, with numerous health consequences for the local communities and populations.[74]

Pollutions related to the development and maintenance of the nuclear arsenal are also important, although they were hidden for a long time. The United States and the Soviet Union built tens of thousands of nuclear warheads in the utmost secrecy. In the United States, Hanford's manufacturing facilities near the Columbia River in Washington State, which were inaugurated during the Second World War, dumped thousands of liters of radioactive waste into the river over the course of fifty years. This polluted the Columbia and the region's groundwater over the long term.[75] In 1957, the French Atomic Energy Commission installed in a rural area in the north of the Côte-d'Or department the only French metropolitan site for the production and maintenance of nuclear charges for atomic bombs. Despite reassurances about the facility's safety, regular tritium discharges have slowly and discreetly polluted the surrounding environment.

Fifty years of nuclear weapons production and rivalry between major powers has resulted in the production of tens of millions of cubic meters of enduring, hazardous waste. This radioactive waste has either been temporarily stored in facilities, or simply released into the environment. Nuclear wastes also contaminate the oceans, as stringently pointed out in 1960 by Commandant Jacques-Yves Cousteau in France or, before him, Rachel Carson in *The Sea Around Us* (1951). In 1956, a study by the American National Academy of Sciences on the biological effects of atomic radiation concluded that the ocean could be safely used to store waste. Consistent with the traditions of nineteenth-century expertise—whose first priority was always technological advancement—such reports championed the environment's regenerative capacity: the oceans were deemed sufficiently large and wide to dilute and disperse all radioactive byproducts.[76] From 1947 to 1959, the United States dumped 50,000 barrels of nuclear waste into the Pacific and Atlantic Oceans; the British submerged radioactive containers in the English Channel.[77]

Starting in 1963—the year of the greatest Russian and American atmospheric nuclear explosions—international regulations were introduced to control nuclear weapons testing. The Partial Test Ban Treaty, signed on August 5, 1963, by the United States, the Soviet Union, and the United Kingdom, limited and regulated these atmospheric tests. Testing would continue, but the detonations would occur underground to limit the dispersion of radioactive fallout; the overall ecological impact of this new restriction, however, must not be underestimated. The magnitude of the pollution caused by these explosions is fraught with a large number of uncertainties, because they are difficult to assess as the extent of contamination varies according to meteorology, winds, and dilutions. For example, the explosion of a nuclear bomb in the upper atmosphere does not necessarily cause significant ground contamination, even though rains can bring radioactive dust or ash back to earth. Available data are poor and the effects of low doses of radiation remain controversial. In 2005, a report by the Centers for Disease Control and Prevention concluded that it was impossible to make any reliable estimate, while pointing to the possibility of 11,000 deaths from radiation-induced cancers in the United States since 1950.[78]

Despite these uncertainties, the atomic bomb plays an unequivocally important role in the evolution of ecological thought and the emergence of new representations of the environment. The destructive potential of military techniques gave rise to philosophies of fear and catastrophe that served as a matrix for the development of political ecology.[79] In the 1970s, the historian Donald Worster considered July 16, 1945—the date of the first nuclear explosion—the beginning of the "Age of

Ecology," where humans could no longer deny the effects of their actions on the earth.[80] For whistleblowers and environmental activists, war was often used as a metaphor for describing global environmental risks. In *Our Plundered Planet* (1948), the American biologist Fairfield Osborn warned against "this other world-wide war … that of man against nature." The same year, in response to the threat of a third world war, ecologist William Vogt also proclaimed the prospect of the apocalypse in his book *Road to Survival*.[81] In the United States, pacifist movements against the Vietnam War and opposition to nuclear weapons provided fertile ground for the emergence of new forms of environmental activism in the 1970s, as illustrated by the career of biologist Barry Commoner. Mobilized in the US Navy during the Second World War, Commoner became a recognized specialist in plant physiology before embarking on the fight against nuclear weapons testing in the late 1950s, from which his environmental activism was born.[82] The nuclear issue also led to the development of new international laws of the sea concerning the management of ocean pollution. In the wake of the United Nations Conference on the Human Environment (1972), the Convention on the Prevention of Marine Pollution by Dumping of Wastes and Other Matter, also known as the "London Convention," prohibited dumping nuclear waste in the sea. It was ratified by fifteen countries and entered into force in 1975. And in July 1974, the major powers signed the Treaty on the Limitation of Underground Nuclear Weapon Tests, which supplemented the 1963 text by banning the testing of nuclear weapons with a power greater than 150 kilotons.[83]

* * * *

In the twentieth century, war and new military actions played a central role in the evolution of the relationship between nature and society. By initiating new industrial trajectories and the dissemination of new toxic products—and in the establishment of wartime priorities that exempted industry from environmental responsibility, which was in turn conducive to the aggravation of pollution—war transformed environments in myriad ways. The magnitude of twentieth-century wars provoked unprecedented destruction, of which pollutions were only an often-neglected aspect. But these wars should not be analyzed as aberrant phases; rather, they radicalized polluting practices that existed in times of peace and that, in the emergency of conflict, found new horizons in which to unfold. The chemical, automotive, and aeronautic industries were the main beneficiaries of deadly conflicts. Without them, the enormous resources of men, raw materials, and energy would not have been gathered together. Groups like BASF, Monsanto, or DuPont de Nemours would not have become the industrial empires that flooded the planet with their products. These moments of exacerbated pathological power heralded new energy and industrial trajectories—the consumption and transport revolution—while consolidating the logics of Fordism and generalized mobility at the risk of their pollutions.

8

A High Energy-Consuming World

Between the First World War and the oil crisis of the 1970s, a new and particularly polluting energy model became entrenched around the world.[1] The decline in national energy intensities the world over—the ratio between energy consumption and energy production—was counterbalanced by an overall increase in global energy production.[2] To wit, total energy use jumped from the equivalent of 2.3 billion metric tons of coal in 1948 to 4.2 billion metric tons in 1960 and 6.8 billion metric tons in 1970, a threefold increase in just twenty years. According to John McNeill, the world consumed more energy over the course of the twentieth century than during the entire history of humanity prior to it.[3]

While coal was the leading source of energy in the nineteenth century, it was usurped by petroleum in the twentieth century. The annual growth in petroleum production averaged 5 percent per year, with eighty-four times more petroleum being produced at the end of the century than at the beginning. It had long been used in its natural solid state (bitumen and naphtha). During the nineteenth century, the distillation process was perfected to yield a fluid rich in carbon, which was used to produce kerosene to light cities, and oil for sea—and later land—propulsion. Drilling began in the United States, Central America, and the Middle East. Over the course of the century, more and more countries became hydrocarbon producers: from only a handful in 1914 to over a hundred in the 1970s. The United States led the way, and its supremacy was bolstered during the Second World War. In 1945, American production accounted for 65.4 percent of petroleum consumed worldwide, significantly ahead of Venezuela (13 percent) or the Eastern Bloc (6.7 percent). After 1950, petroleum production exceeded that of coal, increasing from 770 million metric tons to 2,334 million metric tons between 1955 and 1970. By 1960, American production only amounted to 34 percent of global production as the Middle East and Africa had emerged as major players in the industry.[4]

Figure 8.1
Protest against automobile pollution in United States during the 1970s. (Photograph: Charles Gatewood.) In the twentieth century, oil and gasoline engines added to the already wide range of pollutions across the environment. In countries with a high standard of living, numerous demonstrations denounced the ravages of the petroleum civilization. Here, young protesters parade in front of a motor show in the United States and make the link between the increase in population, automobile production, and pollution.

As the century unfolded, societies around the world became ever more dependent on petroleum. Although gross consumption of traditional energy sources like wood continued to grow, the relative part they played at this point was in drastic freefall (less than 10 percent in 1990). But the so-called Glorious Thirty period—that rapid, three-decade economic expansion after World War II—came at a high cost, with the use of fossil fuels causing many setbacks in terms of pollution. Every stage of fossil fuel production generated nuisances. From extraction, to transport, to refinement, to final consumption: fossil fuels polluted air, soil, and water. After the 1970s—marked by "the oil crisis" and higher ecological stakes—changing forms of energy underwent numerous investigations and critiques to examine their ecological impact.[5] The new global energy model for the twentieth century identified pollutions and pinpointed their locations. Yes, coal smoke was still a serious issue in many regions of the world, but new challenges needed to be addressed in the face of the expansion of the petroleum industry and its energy derivatives—gas, and expansive hydroelectric and nuclear infrastructure.

Coal's Deadly Fogs

While coal still represented 50 percent of the world's energy consumption in 1914, the twentieth century witnessed the relative decline of its previous supremacy. Coal dropped to a 40 percent share of global energy consumption by 1946, and to only 25 percent by the end of the century. Its collapse in North America was even more precipitous: from 70 percent down to 20 percent. However, the total global consumption of coal by volume more than doubled from 1,220 million metric tons to 2,500 million metric tons between 1913 and 1980. This increase in absolute consumption helps explains why—despite the change in the structure of the energy mix—smoke, dust, soot, and dirt from coal continued to sully the air in and around cities and industrial sites, to the point where the

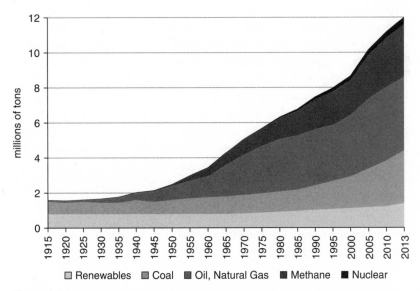

Figure 8.2
Global energy consumption since 1915. *Source:* Yves Mathieu, *Atlas Mondial des énergies. Ressources, consommation et scénarios d'avenir* (Paris: Armand Colin, 2014), 12.

term *smog* (1905) was used frequently to describe the mists and fogs that blackened the air with deadly fumes.

The location of coal seams also cut across geopolitical issues. Largely concentrated in the United States and Western Europe before 1914, coal mines and consumer markets expanded to other locations in the years between the two world wars. Most notably, these new territories were located in the Soviet bloc, around the large Kuzbass coal deposits in southern Siberia, and the Donbass region on the border between Ukraine and Russia.[6] Despite the failures of the "Great Leap Forward" (1958–1960), coal production in Maoist China continued around a few larger, highly mechanized mines and tens of thousands of smaller mines that employed 20 million miners at the end of the 1950s. In 1976, China became the third largest coal producer in the world.[7] In comparison, India was only a small producer, while coal production was in crisis in the United States (deposits in the Appalachians and Wyoming) and Western Europe.

Extraction processes were transformed and outputs improved in the larger mines, thanks to mechanization, their increasing size, and the rationalization of labor.[8] The miner was simultaneously a heroic icon of the proletariat—a hero of emancipation, combining strength and will while confronting the chronic dangers of the subterranean world, as embodied in the USSR by Alexey Stakhanov—and a polarizing figure, due to strong connections to labor organization and the trade union movement. If—in relation to the tonnage of coal extracted and the number of workers— mine shaft collapses and accidents diminished throughout the century, the number of poisonings by dust increased. Indeed, in addition to the deskilling that it caused in the profession, the mechanization of labor alleviated certain tasks but led to an increase in dust in the mines, caused by drilling machines and rock cutters operated by technicians. Silicosis, a pulmonary illness brought on after inhaling silica dust, was the archetypal mining malady of the twentieth century. It was first diagnosed in Great Britain, where a well-organized workers' movement had the ear of a more sympathetic government, compared to in the United States, where silicosis was not detected until much later.[9] Though exact numbers are hard to come by, it is estimated that in French coal mines alone about 100,000 fatalities were the result of silicosis between 1945 and 1987. Unfortunately, lowering the risk of silicosis never really became a "national cause" in France because employers strategized to employ immigrant and temporary workers as a solution. Also, certain doctors were complicit in efforts to overlook the illness, trade unions and the Communist Party were passive, accurate records were not kept, and there was a general sense of obedience to the paternalistic system in place at the time.[10] In Belgium, pneumoconiosis (a combination of pulmonary illnesses) was not recognized as a malady that warranted compensation until 1964. Such delays in awareness also featured in Italy.[11] In 1954, filmmaker Georges Franju documented the prevalence of dangerous particles in these workplaces in his documentary evocatively titled *Les Poussières* [Industrial Dusts].[12]

Industrial dusts from mines not only affected miners, but also the health of those living around the mines, as was shown in the United States by historians Gerald Markowitz and David Rosner. Coking plants were usually constructed close to coal mines and were recognizable by their thick plumes of smoke. Their production of purified coal also linked them to the carbon chemistry industry, a particularly polluting sector. Moreover, coal deposits and mining towns were affected by soil erosion, groundwater pollution, and destruction of the surrounding countryside, all of which sparked numerous conflicts over land use.[13] In China, health and environmental damage was linked to a dangerously high number of cave-ins, as well as atmospheric pollution, without provoking any major protests.[14] The filmmaker Wang Bing filmed the crisis of Shanxi province's vast coal and metallurgical industrial complex in northeast China, highlighting its ecological devastation and the helplessness of former workers. Developed in 1933 under Japanese occupation to produce military equipment, the mining region was transformed into a huge state industrial zone after 1949. It employed one million workers at the height of exploitation during the 1970s. In typical form for the grandiose projects of the Cultural Revolution, it left the region devastated; the land and its people still bear testimony to this today.[15]

But the main feature that characterized the metamorphosis in coal-based industries of the twentieth century was the change in production scale. In his novel *I Married a Communist,* set in McCarthyite America in the 1950s, the American author Philip Roth describes the narrator's surprise (despite his familiarity with industrial landscapes) when on a train ride he passed factories in Gary, Indiana (near Chicago), where coal and steel industries coexist under the control of one industrial giant, US Steel:

> Coming as I did from industrial north Jersey, I confronted a not unfamiliar landscape.... We had the big factories and the tiny job shops, we had the grime, we had the smells.... We had the black smoke rising from high stacks, a lot of smoke coming up everywhere.... We had the dirt and we had the stink, but what we didn't have ... [was] the open-hearth furnaces that lit up the sky when the mills were pouring steel, a red sky that on clear nights I

could see, from as far away as my dormitory window, way down in Gary.... Concentrated here was the power of the Midwest. What they had here was a steelmaking operation, miles and miles of it stretching along the lake through two states and vaster than any other in the world, coke furnaces and oxygen furnaces transforming iron ore into steel, overhead ladles carrying tons of molten steel, hot metal pouring like lava into molds, and amid all this flash and dust and danger and noise ... sucking in vapors that could ruin them, men at labor around the clock, men at work that was never finished. While I ... took in what looked to me to be mightily up-to-date, modern, the very emblem of the industrial twentieth century ... no fact of my life seemed more serious than that.[16]

A change in the scale of production also occurred in the USSR, where coal-based heavy industry became the showcase of the new regime's exploits and the extent of its power. Certain areas were particularly affected, such as the Ural-Kuznetsk combine that connected the iron mines at Magnitogorsk to the coking coal mines of the Kemerovo-Novosibirsk-Altai triangle, situated 2,000 kilometers away. The rapid industrialization of Magnitogorsk—following the model of the US steel factories at Gary—was a priority in Stalin's five-year plan in 1930. Linking these immense coal and iron ore reserves created the perfect site for a steel industry to thrive and rival any in the United States. Safety measures and air pollution controls were not high priorities; bronchitis, asthma, and other respiratory illnesses increased in what was to become one of the most polluted cities on the planet.[17]

Sometimes, a stifling, coal-dust-laden fog would rapidly descend. In December 1930, a "mysterious fog" enveloped the Meuse Valley in Belgium for several days. Many thousands of people complained of breathing problems; roughly sixty people died. In the search for the cause, in line with the expertise and decision-making processes of the nineteenth century, the state health services at first wanted to reassure the public that what had happened was the result of a purely meteorological event, or an aggressive flu epidemic. Ultimately, however, blame came to rest on industry. The valley was one of the zinc manufacturing centers of the world, producing more than 200,000 metric tons per year in the late 1920s, which amounted to 15 percent of global production or the equivalent of one third of Europe's overall output. Transforming mineral into metal required copious amounts of coal: four or five times the amount of zinc finally produced. The results of a ten-month inquest by a committee of experts concluded that "the noxiousness of the fog" was due to "sulfide particles resulting from the burning of coal," notably sulfur dioxide and sulfuric acid. Yet, despite these conclusions, the catastrophe, presented

Figure 8.3
Fog in the Meuse Valley, 1930. *Source:* Albert Humblet collection.

as inevitable, did not highlight the responsibility of the zinc industry, a significant contributor to the Belgian economy, and the second largest producer in the world after the United States. Instead the blame shifted to weather inversion, the valley's topography, and the fact that certain people were more susceptible to the adverse climate conditions that had manifested. Coal and coal-related industries were not to blame.[18]

In the face of disasters, absolution of industry was a common theme; the Donora catastrophe in the United States served as another example. Donora was an industrial zinc city in a valley near Pittsburgh where, in October 1948, a significant portion of the local population was poisoned by a smog linked to a weather episode that trapped carbonaceous particles in the air. As was the case in the Meuse Valley, a few dozen people perished, and reports from the inquiry into the incident incriminated natural conditions more readily than economic activity—proof that regulators and experts, notably in the Industrial Health and Safety Division (a branch of the Public Health Service), had vested interests in industrial production and its major companies, particularly US Steel.[19]

The death rate associated with coal did not only affect mines and industrial sites. Although mostly used in the form of coke, coal remained the most widely used form of fuel in homes, workshops, and urban factories. In spite of the advancements in smoke filtration systems, sulfur and nitrogen oxides, fine particles, and greenhouse gases were still being released into the air. It is true however that, on the coal front, the situation had changed since the nineteenth century. Coke had become widespread and the coking process took place mainly at coal mines while electricity was gradually replacing gas lighting and steam engines, much to the benefit of cleaner city air, and reducing the number of factories necessary for the distillation of coal. Of course, thermal power generation plants were often established in close proximity to cities, and they used a lot of coal as a fuel source. But at the same time, energy converters were constantly being improved with technological advances, so that while in 1955 it took on average 0.7 TCE (tons of coal equivalent) to produce 1,000 kilowatt hours, by 1967 only 0.458 TCE were required. Despite these structural changes, and before the spread of electrification and natural gas and fuel oil were in general use, large European cities were regularly engulfed in fog caused by the burning of coal. In Paris, for example, the height of coal consumption was between 1930 and 1950, where it accounted for 80 percent of the city's total energy use.[20] The city's coal dependency provoked widespread concerns in the press. In 1926, in an article on carbon dioxide, the French Communist newspaper *L'Humanité* lamented that "we do not know how to burn coal," "city air [is] vitiated by fumes and discarded debris": one "factory's chimneys in Paris emit 1,500 kilos of ash an hour."[21]

In order to better evaluate these emissions, measuring instruments became ever more precise, but slowly. The Ringelmann scale for measuring the apparent density of smoke "black, thick, and extended" was in use for quite some time. During the 1930s, devices were invented in Great Britain to measure levels of sulfur dioxide. Meteorological observations were boosted by new measuring apparatus and measuring stations—which became standard—although visual and olfactory perception remained in regular use until the 1950s. At that point, academic research led to the creation of more sophisticated sensors. Different types of pollutants could be identified and the instruments' calibrations became increasingly exact. Above all, not only sources of pollution were measured, but so too were levels of pollution in cities. The air became an object of permanent, technical surveillance.[22]

Unfortunately, advancements in surveillance equipment did not impede the increase in urban air pollution. After the Second World War, in Great Britain—the country with the most measuring devices—the mortality rate from bronchitis was 62 in every 100,000, compared with

4.5 in France. The majority of these deaths were caused by coal smoke. Great Britain was also the place where the biggest public health catastrophe struck after the war. On December 1, 1952, in response to the great demand for heat, the British government stopped rationing *nutty slack,* a poor-quality coal that produced a lot of smoke. A few days later, London was completely hemmed in by smog. The mortality rate for this period is the subject of heated debate, with historians estimating the death count as anywhere between 4,000 and 12,000. Such catastrophes stemmed from the fact that the fight against pollution still confronted— just as it had done in the nineteenth century—a laissez-faire regime, a blind trust in industry to install the necessary filters, and a general allergy to regulatory oversight. New urban pollutions—from growing numbers of automobiles—also introduced new challenges.[23] If, in the years that followed, air pollution from coal decreased in European and North American cities, this was the result of a displacement of nuisances and the adoption of a new energy system founded on petroleum and natural gas, as well as hydraulic and nuclear energies.[24]

The Curse of Black Gold

The rise of petroleum in the twentieth century transformed industrial zones profoundly and added new sources of pollution to carbon emissions. If hydrocarbons were more or less toxic depending on the type of product used and the particular environment contaminated, pollutions varied just as much depending on whether petroleum was being extracted, transformed, or burned.[25] Still in its nascent state outside of the United States before the Great War, petroleum's role in global energy consumption grew from 3.4 percent in 1915 to 50 percent in 1973, surpassing coal around 1965. It completely overhauled knowledge, global geopolitics, and consumer practices.[26] According to Timothy Mitchell, who examined the sociopolitical stakes of this liquid fuel, oil would play its part in the vast project of social control and the domestication of the labor movement, which was particularly contentious in the coalfields.[27] It would also influence the evolution of the portrayal of the working class, as the figure of the coal miner was gradually supplanted by that of the qualified worker in the petrochemical industry.[28]

While coal pollution remained for the most part limited to areas around mines, ovens, and steam engines, petroleum pollution spread over much vaster global reaches because its uses were far more numerous and it was far easier and cheaper to transport.[29] Pollution began with extraction, primary transformation, and transport. Ecological damage from oil began in the United States, where production jumped from 60 million barrels per year in 1900 to 600 million barrels by the early 1920s. New oil states emerged: Texas (where the first oil deposit was drilled in 1901) and California. Born in the California oil boom, Los Angeles' population increased from 500,000 in 1920 to 1.2 million in 1930. Wells and derricks sprang up even in the heart of the city, between Long Beach and the city center, and up Signal Hill, which was nicknamed "Porcupine Hill" because of the number of derricks poking out of it. During peak production in 1923, tourist associations promoting seaside vacations began to denounce the erosion and pollution of the shorelines.[30] In his novel *Oil!* (1927), the muckraking journalist Upton Sinclair gave insight into the obsession with profit, how the working class was being exploited, and how nature was being destroyed by drilling.[31] In the decades that followed, oil reserves diminished, forcing drilling to move farther from cities, inland to deserts and then out to sea in the 1960s. Each migration spread the pollutions that accompanied these activities still further afield.[32] The first derricks were equipped with modern pumps—the iconic *nodding donkeys*—in Texas in 1925, further increasing access to underground resources and the speed of extraction. Along the coast of certain southern states, ports that received oil and oil fields were rapidly sullied by hydrocarbons, much to the chagrin—and financial cost—of fishermen and residents. Damage was so bad that the US Bureau of Fisheries wrote a report titled *Danger to Fisheries of Oil and Tar Pollution* (1921). Even insurance companies were alarmed by the magnitude and rapid growth of the problem.[33]

Houston, a former capital of the cotton trade, became a petroleum hub, more thoroughly shaped by the oil industry than Pennsylvania, Southern California, Dallas, or Los Angeles. With its dense concentration of refineries, petrochemical plants, transport infrastructures, and storage facilities, Houston was both "the energy capital of America" and the site of the worst petroleum pollution in the United States in the 1930s.[34] Complaints were limited and also curtailed by a local political life controlled by business and conservative elites, who prevented any regulation of the extraction process.[35] In 1938, North America was producing 60 percent of the world's oil and its companies, Standard Oil (John Rockefeller's historic company), Exxon Mobil, and Chevron became multinationals and the largest of the mining giants.

After the fall of the Iberian empires, South America became an exclusive domain for American interests, and the second largest prospecting terrain for petroleum companies.[36] In Mexico, the second largest producer in the world after the Great War, the small port of Tampico, north of Veracruz and located close to numerous waterways and tropical forests, was by 1921 a city of 100,000 inhabitants. Fifty-eight different oil companies were established there, along with sixteen refineries, and 400 kilometers of pipeline. This unbridled infrastructure was the source of numerous emissions, leaks, and spills, all of which had a destructive impact on the local environment.[37] In 1938, after a drop in domestic oil production, Mexican president Cárdenas signed an order that expropriated nearly all of the foreign oil companies operating in Mexico, in turn leaving Venezuela to become the second largest producer of oil in the world. Lake Maracaibo—the most important freshwater reserve on the continent—was gravely contaminated by the petroleum industry as wells were drilled using archaic techniques. In addition, refineries dumped huge quantities of petroleum residues, which sterilized wide swaths of land and put an end to agriculture in the area.[38]

In the interwar period, the oil industry also developed in Western Europe, mainly in the Netherlands (Royal Dutch Shell) and Great Britain (British Petroleum), which relied heavily on their colonial empires in Asia and the Middle East. Although not quite as developed, the oil boom in France gives an idea of the extent of its pollutions. The Compagnie Française des Pétroles (CFP) was only created by the state in 1924. Although most of the petroleum and its derivatives were imported in the early 1920s, refineries in France produced 400,000 metric tons of petroleum products in 1931 and more than 8 million metric tons by 1938.[39] At that point, fifteen refineries, located mainly in the lower Seine valley and on the banks of Étang de Berre lagoon near Marseille refined petroleum from Romania and the Middle East. The state established new prospecting companies to stimulate the sector: Régie Autonome des Pétroles (1939), Société Nationale des Pétroles d'Aquitaine (1941) and the Bureau de Recherches de Pétrole (1945). They became one group in 1966, which eventually became known as Elf-Aquitaine. This new entity benefited from the opportunity to exploit oil in Algeria. Around the same time, the CFP became a subsidiary of Total in 1965, the same year that oil surpassed coal in global energy consumption.[40] A new generation of refineries was established at Feyzin, south of Lyon, as well as around Étang de Berre, to handle the oil imports from Algeria.

With such high stakes at play worldwide, risks and pollutions were given little attention in spite of complaints and recurring protests by locals. In 1936, the mayor of Royan, in southwest France, complained about pollution from two petroleum refineries in the Gironde estuary. The companies were exonerated from all responsibility by the prefect, who issued a decree prohibiting discharges of hydrocarbons from ships' bilges into sea or river water.[41] Around that time, the newspaper *L'Humanité* reported that 1,000 fishers on Étang de Berre had gone on strike "because of pollution of the lake water, resulting from ever increasing discharges from oil refineries."[42] Since the First World War, the area had become a jewel in the crown of the chemical and petroleum industries, in large part due to a deliberate desire of the Chamber of Commerce in Marseille to make it the dirty, industrial suburb of the Phocaean city.[43] After 1945, the rise in importance of hydrocarbons drove the petrochemical industrialization of Étang de Berre forward; the development of the neighboring harbor Fos-sur-Mer, situated on 20,000 hectares (about 49,000 acres), soon followed. This development had the support of both central and regional authorities. Arbitrating between incommensurable industrial, farming, and fishing interests, the state was swift in throwing its lot behind black gold: in 1957, fishing was prohibited in the lake because of the contaminated waters. In fact, this coastal region that had previously been a place to fish and walk and for birds to find shelter, was now gravely harmed by polluting hydrocarbons, which the banning of fishing could evidently do nothing about.[44] On the contrary, as a priority of the Fifth and Sixth Plans (1966–1975)—designed to help Marseille compete with the larger ports in northern Europe, such as Rotterdam—further industrialization was encouraged and abetted. In 1973, Marseille witnessed the highest level of oil traffic ever; close to 95 million metric tons passed through the city, while daily emissions of sulfur dioxide into the air were estimated to be between 800 and 900 metric tons at the beginning of 1975.[45] On the initiative of local elected officials—often communists or fishers—grouped under the umbrella of the CGT [General Confederation of Labor], complaints were drafted. In response, the prefect created the Secrétariat Permanent pour les Problèmes de Pollution Industrielle (SPPPI), charged with supervising industrial waste, coordinating construction permits, and setting standards amongst industrialists.[46]

In the East, despite the early exploitation of oil fields at Baku, Soviet authorities continued to focus on coal, and it was not until the 1950s that they embraced petroleum. The shift was remarkable: in 1960, oil accounted for 30 percent of energy consumption in the Soviet Union. At that point the country was producing 700 million metric tons of oil, and there was a frenzy of refinery and pipeline construction. Reserves in the Caucasus region of Azerbaijan were surpassed by those in the Volga-Ural region.[47] The economic stakes were such that the issue of ecological devastation was systematically suppressed even though damage was apparent, especially in the Baku region. The Caspian Sea's vast basin, into which the Volga flowed, also fell victim to petroleum pollution; rivers and the inland sea rapidly deteriorated. The sturgeon population diminished drastically, compromising caviar production.[48]

In Africa, oil exploitation did not begin in earnest until the 1950s, coinciding with a wave of decolonization. The French Equatorial Africa Oil Company (SPAEF), a branch of the French empire, was established in 1949 but the first oil reserves were not found until 1956 in Gabon and the Algerian desert at Hassi Messaoud.[49] In Gabon, refineries quickly established themselves in Port-Gentil and the region south of the town. Crude oil production increased from 1.4 million metric tons in 1966 to more than 11 million metric tons in 1976. Most was produced offshore; 85 percent of it exported abroad.[50] In Nigeria, after the civil war that led to independence in 1960, the petroleum industry grew rapidly around three large deposits in the Niger River delta. Production of crude oil increased from 396 million barrels in 1970 to more than 823 million barrels in 1974. But numerous spills seriously contaminated the delta, damaging fishing areas and agricultural lands at the expense of local populations, among them the indigenous Ogoni people who subsisted on local resources. Their protests were suppressed by authorities, whose only real concern was to protect the interests of large companies.[51] As in many regions where the petroleum industry operated—notably the Middle East and the Persian Gulf—pollution remained invisible for quite some time, because of its relative isolation in a desert setting removed from larger towns and cities. This was especially the case in Iran, Iraq, and Saudi Arabia, all of which became major players in the petroleum industry after the Second World War. Because of geopolitical tensions, these countries had difficulty extricating themselves from British and American influence. In 1960, however, their struggle led to the foundation of OPEC (Organization of the Petroleum Exporting Countries).

The oil energy system was based on an increasingly globalized economy in which transport played a primary role in connecting production sites to their consumer markets. In 1966, petroleum in and of itself accounted for 53 percent of the total value of global trade. Transporting these great quantities of petroleum required immense infrastructure. Much of that infrastructure proved fragile and became the source of spills and accidents that resulted in long-term contaminations of oceans and soil. Between 1953 and 1967, gas pipelines grew to 2.4 times the length they had previously been to cover a total of 1.9 million kilometers. At the same time, the average size of oil tankers drastically increased as the age of supertankers dawned: a leap from an average size of 10,000 metric tons around 1950 to 200,000 metric tons by 1961. The closure of the Suez Canal after the Six Day War in 1967 further accelerated the race for bigger and bigger tankers. Also in 1967, the *Torrey Canyon* oil tanker was shipwrecked, causing an environmental disaster, the first in a chronicle of big oil spills. Some 40,000 metric tons of crude oil were spilled into the English Channel, with pollution spreading 350 kilometers from the coast, killing almost 100,000 birds. Ten years later, the *Amoco Cadiz* catastrophe resulted in more than 200,000 metric tons of crude oil contaminating 400 kilometers of the coastline in Brittany. In 1979, the explosion of an offshore oilfield in the Gulf of Mexico poured 600,000 metric tons of crude oil into the sea. These spectacular oil spills overshadowed many more not covered by the media. In fact, innumerable spills that were less visible contributed overall to the destruction of environments around petroleum sites. At least 1,500 maritime spillages greater than 500 metric tons of oil were recorded between 1951 and 1999. Add to these the routine discharges of oil at sea, by definition uncontrollable, but numerous nonetheless.[52]

The Smell of Gasoline

In the final pages of *A Simple Life,* a novel by Émile Guillaumin about country life in France in the nineteenth century, the narrator, in the twilight of his life, reminisces about how the world around him has changed over the course of his life. His vantage point was 1902. With the sudden loud entrance of an automobile intruding on the peaceful pastoral surroundings, he intuits that a change is coming to civilization.

> Now there are carriages that have no need of horses.... I would suddenly hear a noise, shrill and disagreeable, which got louder and louder, and the motor car, carrying men fantastically dressed in caps and oilcloth jackets and wearing goggles like stonebreakers, would pass rapidly, raising a cloud of dust and leaving behind it a disgusting smell of petrol.[53]

In 1899, the journalist and satirist Léon-Charles Bienvenu, known by his penname Touchatout, ranted against the automobile: it "threatens, poisons, and deafens everyone with its puff ! ... puff! ... bowling us over with its petrol fumes, and its unconventional appearance." From the outset the sudden entrance of the automobile was frowned upon, with drivers drawing disapproval. Those smelly and noisy machines were considered destructive and dangerous. The dust that they kicked up and their speed were yet other symptoms of the "oil madness" that character-ized them, and over several years, a struggle for space and class seized fossil fuel transportation.[54]

Those in favor of the motor car—typically the elite who could afford to buy such a luxury item—argued that the "car" driven by an internal combustion engine (invented in 1893) would be better for the environ-ment, because horses would no longer be needed and their waste would therefore no longer sully the roadways.[55] Industrialists, for their part, managed to focus attention on accidents by defining the problem in terms of risk behavior, thereby putting an educational morality on the political agenda, while ruling out the possibility of questioning the act of driving in and of itself.[56] The outcome was an historically important success and a founding feature of modern civilization in the twentieth century: the social acceptance of a major risk, the automobile. Every year since World War II, car traffic has killed a few hundred thousand people worldwide (1.3 million people die every year in the twenty-first century), peaking in France and the United States in 1972 with 16,000 and 54,400 killed, respectively.

The automobile's environmental consequences were considerable. Without even taking into consideration its symbolism, social uses, or components, but focusing only on the issue of energy, the motor car opened up a new Pandora's box for fossil fuel consumption, mainly oil. Both the internal and external combustion engines, developed in the 1860s and adapted for individual locomotion, were introduced into the first automobiles at the end of the nineteenth century. An engine is a sensitive piece of machinery. It was necessary to refine the oil to elimi-nate the impurities, then to distil it, which meant separating the different components to obtain—in a schematic fashion—fuel for automobiles, kerosene for airplanes, diesel for trucks and tractors, heating oil, and heavy oil (essentially bitumen) for industry and maritime navigation. These refining and distilling operations were extremely polluting in themselves, which explains why petrochemical plants were often situ-ated far from populated areas (in France at Feyzin, south of Lyon, Étang de Berre, and Fos-sur-Mer), which did nothing to slow down global pol-lution, of course.

Producing diverse types of fuel progressed exponentially throughout the century, in keeping with the growing number of cars that needed fuel. The United States was one of the first countries to produce a high quantity of cars. In 1913, it boasted one car for every 77 people. In France there was one for every 318 people; in Italy only one for every 2,000 people. In Africa and Asia this luxury item was still very rare. Motorization spread through Europe and the United States during the interwar period: a significant jump from 1 million motor vehicles worldwide in 1910 to 50 million in 1930, to 100 million in 1955, and 500 million by 1985. In 1950, five million cars were manufactured. The rapid increase in production meant that another one million were added each year.[57] A symbol of Western individualism, the car took longer to become commonplace in the USSR: in 1970, only 1.5 million individual vehicles were in circulation (compared with 13 million in France at the same time). It should be noted, however, that Russian cars had a reputation for being especially polluting.[58]

Accompanying the increase in fuel consumption, road traffic became the primary cause of pollution in the 1960s. Apart from carbon dioxide and methane, two greenhouse gases, automobiles also emitted gases toxic to humans and ecosystems: carbon monoxide, ozone, nitrogen oxide, sulfur dioxide, fine particles, VOCs (volatile organic compounds), and lead (added to gasoline in 1921 to improve combustion). By the end of the century, lead additives had been banned in numerous countries, but gasoline still contained ordinarily more than 150 other chemical substances added to hydrocarbons to improve the octane value, increase combustion efficiency, and protect the motor and exhaust components. Diesel, typically limited to trucks and tractors prior to the late 1970s, emitted a complex mix of gases including nitrogen oxide, carbon monoxide, formaldehyde, acetaldehyde, benzene, and polycyclic aromatic hydrocarbons (with and without nitrates). Diesel motors spewed up to one hundred times more fine particles than gasoline engines. Being only 0.01 of a micrometer to 1 micrometer in size, these particles were small enough to lodge in the lung tissue, and those smaller than 0.1 micrometers (ultrafine particles) were able to invade the lungs through the vessel wall to enter the bloodstream to reach other systems in the organism, such as the cardiovascular system.[59]

But air pollution from automobiles was only slowly identified as a public health risk. In 1910, the German manufacturer Wilhelm Maybach (of carburetor fame) wrote an essay titled *Ueber Rauchbelästigung von Automobilen* [On the Problem of Automobile Fumes]. In 1926, the *Journal of the American Medical Association* carried out a now famous and often cited study that revealed the high level of carbon monoxide in police officers' blood in Philadelphia. For scientists and engineers, however, fumes and public health problems were first and foremost a technical issue, with a few proposing perfuming the gaseous exhaust fumes to stifle this inconvenience. Public authorities were hesitant to attempt to regulate this rapidly growing sector, which provided an increasing number of jobs. German regulations did try to control gaseous exhaust fumes through the creation of a special police patrol in Berlin during the interwar period, but it was unsuccessful.[60]

Concern over automobile pollution grew during the 1960s and 1970s. Despite the importance of the automobile industry for the region's economy, the United Automobile Workers (UAW) union in Detroit rallied around improving working conditions and controlling fume emissions. In 1969, it helped to create the "League Against Pollution," whose goal was to improve the environment for local inhabitants.[61] Difficulties in regulating automobile pollution and the relationships that civil societies had with this problem were also shaped by the influence of large industrial lobbies and national contexts. In Greece, the absence of car manufacturers led to more ambitious public policies than in France, where the number of automobiles was much larger and car manufacturers more politically powerful.[62]

A turning point in awareness took place in Los Angeles as a result of chronic postwar smog. Ironically, the City of Angels had originally been founded by immigrants seeking pure air to heal tuberculosis, but because of its topography and weather conditions, its location and low atmosphere tended to trap fumes. The city, which had a population of barely 100,000 in 1914, had become a megalopolis of six million people by 1960. Few cities were more designed with the automobile so explicitly in mind. The railway tracks and tramways had been dismantled in the 1920s when the decision was made to create a car-friendly city. In 1950, close to three million motor vehicles were in circulation in Los Angeles, contributing levels of pollution akin to those coming from industry (1,160 metric tons of toxic gases per day compared with 1,280 from industry). Smog was an issue almost all year round, affecting public health and tree growth within an 80 kilometer radius around the city.[63] Indeed, every city that had based its development around the automobile experienced myriad problems; those like Los Angeles that did not have the benefit of wind to eliminate pollution were particularly susceptible. In Athens and Mexico City the number of vehicles increased, with the latter seeing a rise from 100,000 in 1950 to two million in 1980. In Bombay, the 400,000 vehicles in circulation at the beginning of the 1980s were responsible for roughly 90 percent of air pollution.[64]

Influential city planners such as Jane Jacobs and Lewis Mumford, vociferously condemned the nuisances derived from "auto-centric" urban development, while other protest groups launched boycotts and demonstration campaigns to put pressure on automobile manufacturers. In this context, technological improvement continued to be put forward as a viable solution. In 1966, the catalytic converter—which converts certain exhaust fumes into less toxic matter—was built into engines in California. Car manufacturers worldwide did eventually comply with new standards and measures put in place to reduce pollution at the beginning of the 1970s.[65] This was necessary for them to remain competitive in foreign markets, in particular the North American market. In fact, competition is what drove the continued improvement of filtering fine particles and the energy efficiency of engines, which led to an overall reduction in pollutions over the remainder of the century. Or, rather, polluting substances were replaced by others, the deleterious effects of which would not be revealed until later. Regularly, the industry promised the "clean car." In August 1968, the American newspaper *Nation's Business* announced, perhaps a little prematurely, that "the day is close at hand when the automobile will no longer be considered a major source of pollution."[66] Three years later, while pollution from cars continued to rise, the propaganda magazine *Pétrole-Progrès* published by Esso (SO, Standard Oil) announced the next triumph of the electric car while asserting that in theory hydrocarbon combustion "should only produce harmless products."[67] But the gains were in fact largely offset by the increased distances traveled, the soaring number of cars on the road, and the power of the engines in them. This explains why—similar to the case with smoky furnaces in the nineteenth century—the undeniable drop in toxic releases per unit actually added to, through addition and progression, the global volume of pollutions emitted.[68]

The Unbearable Lightness of Natural Gas

Faced with the environmental risks associated with other fossil fuel energy sources, "natural" gas seemed to be a welcome alternative in the 1960s, because it produced fewer toxic fumes. As a source of energy, natural gas was already well known—it is a fossil fuel naturally present in gaseous form under pressure in the porous rocks of the subsoil, composed of a mixture of hydrocarbons, mainly methane but also propane, butane, ethane, and pentane—but its implementation was complex. Prior to 1945, gas consumption was limited and concentrated, because of the absence of infrastructure to transport it, even though the development of leakproof seals dated back to the end of the nineteenth century and the first major gas pipeline network was constructed in the United States in the 1920s.[69]

In 1950, the United States was producing 90 percent of the world's supply of natural gas and the number of homes using it doubled between 1945 and 1955, particularly in new residential suburbs in California.[70] Between 1950 and 1973 in Western Europe, consumption increased 800 percent, and global consumption climbed from the equivalent of 266 million metric tons of petroleum in 1949 to 1.2 billion metric tons in 1973.[71]

Natural gas gradually replaced the former gas supply known as "city" gas, which had been derived from the distillation of coal and the cracking of petroleum products. The latter was a particularly toxic process, because it emitted carbon monoxide. Natural gas, however, released no dust, only a little sulfur dioxide (SO), nitrogen dioxide (NO_2), and less carbon dioxide (CO_2) than other fossil fuels. Its methane content—a significant greenhouse gas—was only a marginal issue before 1980. Old factories and their extremely harmful (and dangerous) manufacturing processes in the heart of cities gave way to the more discrete infrastructure of natural gas.[72] Its high calorific power, double that of coal gas, and low toxicity propelled natural gas over the next few decades to become an essential fuel in the effort to overcome carbon pollution, while also meeting growing energy demands around the world. Fluid, efficient, and easy to use domestically and industrially, natural gas has continued to be presented and promoted as a "clean energy" source, which has contributed to its success.

The emergence of new marketing techniques such as liquefied gas in 1964, which enabled natural gas to be transported in liquid form (at -161°C), reducing its volume by 600 times, galvanized the worldwide distribution of this form of energy far from the site of its natural deposits. This mode of transport by gas carrier completed a network of pipelines that were difficult and costly to build. Japan, anxious to diversify its energy supply, became one of the biggest importers of liquid gas from Alaska. Japanese authorities saw this as an opportunity to reduce air pollution in larger coastal cities, which up until that point had been powered by coal and oil.[73] In the 1960s, new deposits of gas were tapped in the Volga, the Caucasus, the central Asian republics, Algeria, Iran, Libya, and Indonesia. In Europe, significant gas fields were discovered in Norway, the Netherlands, and the United Kingdom. In all of these countries, the gas sector was considered strategic, and controlled by the state; big national public companies such as Gaz de France, British Gas, ENI (Italy), Distrigas (Belgium), and Gasunie (Netherlands) maintained national monopolies on imports, transport, and distribution.[74] However, as deposits were depleted, Europe became more and more dependent on gas imports from abroad.

Although it was far less polluting than coal or oil, natural gas was still a fossil fuel that emitted greenhouse gases such as nitrogen oxides, sulfur oxides, and reactive hydrocarbons at every stage of the exploitation process. Indeed, once extracted, in order to be commercially viable, it had to be treated and purified through a series of transformation steps from extraction at the source, to storage, compression, transport, and distribution. Byproducts—such as helium or corrosive compounds like sulfur—required disposal. Moreover, in regions around gas fields, soil and water became heavily polluted. Prior to 1960, considerable quantities of residual gases were simply burned off, as companies were not sure what else to do with them. In Iran, for example (the world's second largest gas reserve after Russia), billions of cubic meters of methane from refineries or natural gas were simply left to burn. The National Iranian Gas Company continued to simply burn off methane until 1966, when it constructed a network of pipelines to export the gas abroad.[75]

The environmental impact of exploiting natural gas was still not completely understood. On a local scale, it provoked new cycles of pollution, as was the case at Lacq, one of France's biggest natural gas deposits in the southwest of the country. The discovery of natural gas in 1951 was met with great enthusiasm during the postwar reconstruction phase. Lacq was hailed as the "French Texas"; the press celebrated the "prodigious force of nature tamed by French genius."[76] The political and economic elite sang the praises of this industry, which they imagined would bring wealth, employment, and modernity to a region perceived as isolated and backwards. Within a few years, however, the province of Béarn, traditionally a small agricultural corner of France, was overwhelmed by the construction of factories, wells, and pipes that branched out over a ten-kilometer radius. The brutality of the process and the first environmental breaches led to complaints and conflicts, but these were muffled in a general sense of euphoria. However, some protests by local residents against nuisances produced by the infrastructure for natural gas extraction were relayed by certain unions like the French Confederation of Christian Workers (CFTC) in 1961:

> The residents of Arrans, a neighboring village of Lacq, are apprehensive about living so close to industrial plants, which they fear might negatively impact their health. The gas and sulfur fumes make the crops in the fields wither and die, the leaves on the grape vines turn yellow long before harvest time, and farmers, who have lived here in peace for generations, have come face to face with the gas mask and fear.[77]

Local complaints were in response to sulfur odors. Fishers complained that the fish they caught had acquired "the taste of petroleum." Farmers' land was bought from them; new pipeline construction damaged many of their fields. In September 1961, a union committee was formed to study the agricultural problems of the Lacq commune and protest the toxic fallout of sulfur dioxide and fluorine. In the face of economic optimism, a growing countercurrent denounced these nuisances for what they really were.[78] In 1959, Bernard Charbonneau, who taught and lived a few kilometers from the site, also spoke out about the "Lacq myths" and how his beloved Béarn had been transformed into a "garbage dump." A few years later, François Mauriac lamented the irreparable damage caused by the discovery of hydrocarbons: "There is no help for nature, violated day and night—on all the roads and from the sky above, or the assaulted earth—and condemned to chemical overproduction, gutted by prospectors there where we thought it best protected ..."[79]

Fluid Electricity

From the end of the nineteenth century, electricity became one of the fastest growing forms of energy. It could fulfill myriad uses, without foul-smelling, toxic smoke, or vapors associated with other fuels needed for steam engines, combustion engines, and lighting. In the 1920s, advertising focused on how quiet, clean, and safe electric energy was for its consumers and where they live, as opposed to coal and its grime.[80] However, generating electricity required primary fuels, which meant that rather than eliminating pollution, electricity simply relocated pollutions upstream to sites of production. The electric current and the fantasy of its cleanliness contributed to the naturalization of energy pollution by hiding it from the consumer. Sources of pollution were moved to sites away from populated areas, thanks to sophisticated distribution networks that linked the fuel source to consumers.[81]

Initially scattered between a number of small production sites—in Great Britain, for example, half of the providers were municipal around 1900—large coal power plants were built further from city centers and therefore required a means by which to transport electricity to consumers. A network of high-voltage wires was developed for this purpose over the course of the century. The first line to carry 100,000 volts was erected in California in 1908, followed by 200,000-volt lines in the 1930s. The USSR constructed a 500,000-volt line in 1961. It did not take long for electricity to be supplied to cities in the West but supply to rural areas was slower.[82] The USSR lagged behind and its electricity supply remained well behind that of the West. However, Soviet engineers, confident in the power of technology, built many more power plants under Khrushchev.[83] At the beginning of the 1970s, all industrialized countries had adequate electricity supply and networks; Africa and most of Asia did not.

To respond to the extraordinary development of electricity networks, it was necessary to mobilize many sources of primary energy, harnessed and converted at large power stations. Electricity was subsequently transported by means of metal cables, made primarily of copper, which was extracted from vast mines scattered around the world with each one producing its own litany of environmental nuisances.[84] Electricity provides an effective illustration of one of the central features of modern technology: it makes invisible the resources and waste required for its operation and removes the waste, pollution, and risks that accompany its production. Thus, to the consumer, electricity is made to seem a clean, nonpolluting commodity. Thermal processes used to generate electricity by burning wood, coal, gas, or oil in a turbine that drove a generating alternator remained dominant in the twentieth century. Ever-expanding electricity networks and the growing demand for energy increased the consumption of fossil fuels. Therefore, while individual cities reduced their smoke levels during the century, global pollutions continued to grow, as the case with coal in Germany demonstrated.[85]

Another of electricity's laudable virtues was the fact that it could be produced without the use of combustible fuels, thanks to running water, wind, and the fission of atoms, all of which could power turbines. Hydroelectricity was first developed at the end of the nineteenth century, initially near cities. Later, with the construction of more extensive and sophisticated networks, larger hydroelectric power stations—working in conjunction with dams and reservoirs—were built further afield. The first giant dams were built on the Colorado River in the United States and on the Volga River in the USSR in the 1930s. China was one of the best-equipped countries, but the majority of its dams were small and used primarily for irrigation up until the end of the 1970s.[86] Over the course of the twentieth century, one large dam—over 15 meters high— was opened every day somewhere in the world. Their number swelled from 5,000 in 1950 to 30,000 by 1975.[87] These large dams involved heavy investments and were backed by enough clout to overcome the strong resistance to them. They embodied state power, their capacity to "transform nature," and to tame the elements in the name of national prestige and independence. All of the world's top political leaders used this tactic to reinforce their power: Roosevelt during the 1930s crisis; Stalin in his attempt to bend the rivers to the will of the state and its planners; Colonel Nasser in Egypt, anxious to modernize his country to assure political independence; and Nehru, in an independent India after 1947, in order to build modern "temples" that were supposed to encourage progress.[88]

This direct link between dams and politics explains why many of these projects were unprofitable and posed ecological problems. The artificial flooding of valleys upstream from dams involved firstly displacing huge populations—20 million people in India between 1947 and 1992, 50,000 Nubians between 1960 and 1971 during the construction of the Aswan Dam in Egypt, and a total of between 60 million to 80 million people worldwide since the 1930s.[89] Even if actual pollution while building a dam is only a partial aspect of the process, it must be noted that a dam permanently alters the environment around it. Dams accelerated the salinization of soils, diminished alluvial deposits—increasing the need for pesticides in the valley—increased waste as a result of evaporation, modified local climates and landscapes, and contributed to the contamination of their surroundings and the loss of fish and fishing grounds. Moreover, new energy supplied by the dam attracted new industry. After the construction of the Aswan Dam in Egypt, agricultural and industrial pollutions drastically increased due to the use of chemical fertilizers to compensate for lost intakes of silt, and the newly established industry close to the power plant.[90] In France, the Rhône was the heart of a vast national energy independence program; it was completely transformed by hydroelectric equipment. The Compagnie Nationale du Rhône (CNR), one of the prominent electricity companies in France, was established in 1933; and its first major achievement, the Génissiat Dam, began generating power in 1948. However, the joint constitution of the chemistry corridor south of Lyon, the physical alteration of the water's course, the modification of its flow, and barriers that curtailed the movement of fish and sediment, all added to local pollution, which was denounced by various environmental movements in the 1960s and 1970s. One such group was the Association in Defense of Nature and the Fight Against Pollution in the Rhône Valley, which opposed and tried to impede large-scale projects by the CNR. This association, founded by Camille Vallin, senator for the French Communist Party and mayor of the Givors commune (near Lyon), shed light on the ambivalence surrounding hydroelectricity. Hydroelectricity was a paradox: celebrated globally as a source of power but denounced locally as a source of nuisance.[91]

In the same spirit of technological power and energy independence, the civil nuclear power industry developed in a similar guise, promising unlimited and clean energy. Nuclear energy appeared after 1945 as a means of relaunching modernity out of the postwar ruins.[92] As with all thermal power plants, nuclear fission created a water vapor, which turned a turbine connected to an alternator. The first experimental nuclear power station was created at the Idaho National Laboratory in the United States in 1951. It was quickly followed by commercial nuclear plants outside the United Stats: at Calder Hall in England in 1955, and Marcoule in the Rhône Valley in France in 1956. The push for nuclear energy accelerated after 1960; global nuclear power surpassed around 1 gigawatt (GW) in 1960, and 100 GW by 1980.

Because of reassuring speeches and rhetoric, a framework of modernization, and other tools employed to render this exceptionally dangerous trajectory socially acceptable, the question of pollution emerged quite late in critical discourse and actions against the civil atom. Certainly, radioactive contamination was known in the science world: at the first International Congress of Radiology in 1925, the need to reduce and limit exposure to radiation was imperative and the International X-Ray and Radium Protection Committee was created in 1928. It provided details on acceptable limits of radiation dosage in 1934, the same year Marie Curie died from aplastic anemia, brought about by her exposure to radiation. After the Second World War, the International Commission on Radiological Protection (1950) was responsible for specifying acceptable thresholds and studying the relationship between doses and their effects on health, flora, and fauna.[93] As well, as part of the production of nuclear energy, radiation exposure also took place before and after the materials were used to create fission: in the uranium mines—far from production and consumption sites—and while managing and stockpiling nuclear waste. In addition, a "thermal" impact followed the release of hot water into local waterways, which could be harmful to flora and fauna. In the production phase, every nuclear power plant continually emitted radioactive and chemical effluent in liquid and gaseous forms in controlled quantities and within the framework of regulatory authorizations for waste deemed acceptable by public authorities. But these limits were sometimes surpassed if there were accidents or incidents such as the Three Mile Island accident in the United States in 1979.[94]

Health and environmental risks were only assessed after nuclear power plants were already in place.[95] In France, diverse groups voiced fears over health risks for the workers in nuclear plants. Once Marcoule was operational, a French uranium miner remembered that "the extraction of that mineral was not without its element of danger for the miners … one could not stop the rays that emerged from the mineral, nor the radon (a gas also coming from the mineral) from escaping or accomplishing their sinister work: death." At a congress of the French Confederation of Christian Workers (CFTC) in 1957, a worker's delegate warned against the effects of radiation as well as the danger associated with the plant's unusable waste being "absorbed into earth and water."[96] Despite the fact that nuclear pollution was rendered invisible due to a lack of sensory stimulus and the fact that the mines were far removed from consumers— and that radioactivity released during the mining process was difficult to distinguish from natural levels in the mines—movements in opposition to nuclear pollution formed after 1968 and enjoyed considerable publicity. In Cherbourg, various groups came together to form a committee against atomic pollution at the main waste treatment center in Europe, which began operation in 1966 in La Hague, at the end of the Cotentin Peninsula.[97] In the 1970s, Friends of the Earth investigated the uranium mines in the Limousin and Morvan regions in France, and warned against "extraction [which] is the cause of significant 'normal' (as opposed to accidental) radioactive pollution, in particular for the miners."[98] It was precisely at this time that the extraction of fuel and its local nuisances were increasingly pushed abroad. The Atomic Energy Commission (CEA) had its geologists locate prospective mining resources in Africa, Madagascar, Gabon, and Niger, where sizable deposits of uranium had been found in 1969 at Arlit in northern-central Niger. Thus, environmental damage was confined to postcolonial peripheries, where fewer complaints were made, even if such damage and pollution was a significant source of environmental injustice.[99]

*   *   *   *

In the sixty years that separated the Great War and the oil crisis, the world functioned with an intensive and highly polluting energy model while dealing with a population explosion, the need for a development policy, and an ever-increasing hike in demand. New energy converters were developed, but these simply added to the list of energy possibilities (and their pollutants), rather than replacing former energy sources. The energy-consuming world of the twentieth century was a world predicated on accumulation, which also extended to the range of pollutants it generated. Thus, while natural gas, diesel, and nuclear power produced new toxic effluents, the amount of coal consumed had never been higher, in spite of its apparent decline. The process also reflected a deep global imbalance. Northern countries, mainly industrialized and capitalist, who engaged in these new riskier and more polluting trajectories, essentially pillaged resources from countries in the global South. In 1974, the United States, which accounted for only 6 percent of the world's population, consumed 44 percent of the world's coal, 33 percent of its oil, and 63 percent of its natural gas.[100] Next to these industrialized countries, who became ever more dependent on hydrocarbons, stood the other rural and agricultural regions in Africa, India, and China, where human and animal labor provided sufficient mechanical energy. In 1971, of the 15.7 million tractors in use in the world, only 7 percent of them were used in countries in the Southern hemisphere. In these countries, draft animals and human strength remained the principal forces of labor and transporting goods, contributing little to the growing pollutions in the Northern hemisphere, with its higher standards of living and societies infatuated with mass consumption.[101]

9

Mass Consumption, Mass Contamination

If the production of energy is generally conspicuous, noisy, and smelly, the act of consuming energy seems much more benign. During the twentieth century, it became the norm for societies to symbolically and physically keep at some remove the nuisances, inconveniences, and risks generated by the increasing profusion of material goods. However, each product hid subtle pollutions incorporated into the heart of the production process, from the mine to final consumption. During the twentieth century, between the extraction of materials and their use in the form of manufactured goods, the operations and stages of production became more complex. Goods incorporated more and more composite products, many of which—for the average consumer—were almost magical in their derivation, but also profoundly affected the environment and human health.

Surreptitious and chronic, the pollutions of the consumer society that flourished in the postwar period also provoked rapid and serious problems. Thus, in the early 1950s in Minamata Bay, in southwest Japan, cats were seen dancing strangely and frenetically before suddenly dying. Shortly after, fishermen were afflicted with a comparable mysterious evil. It gradually became apparent that these unexplained deaths were due to releases from the Chisso petrochemical plant, which specialized in the manufacture of acetaldehyde—an organic chemical derived from ethylene, used in particular for the production of dyes and the synthesis of rubber, in high demand by the automotive industry. The plant was installed on the bay in 1907, and in 1932 it adopted mercury oxide as a catalyst for the synthesis of acetaldehyde. Mercurial compounds (400 metric tons between 1932 and 1966) were added to the many other heavy metals released into the bay, which contaminated marine fauna, accumulated in the food chain, and poisoned the local populations.[1] Once mercury poisoning was revealed as the source of the mysterious ailment, the scandal was immense: 900 directly attributable deaths were recorded between

VOLUME XL.   NEW YORK, NOVEMBER 20, 1902.   NUMBER 1047.

Entered at the New York Post Office as Second-Class Mail Matter.
Copyright, 1901, by LIFE PUBLISHING COMPANY.

WHO OWNS IT, ANYWAY?

Figure 9.1
"Who owns it, anyway?" Drawing from *Life* (November 20, 1902). Symbol of mass consumption, from its inception the automobile provoked popular concern because of its noise, speed, and pollution. In this satirical drawing published in the American magazine *Life* in 1902, the artist depicts the feeling of loss of control that accompanied the appearance of the new fetish in the American Way of Life. In fact, it is difficult to identify twentieth-century pollutions not exacerbated by its advent. Materials that needed to be mass-produced (including rubber, plastics, and metals) resulted in blackened sites in many parts of the world.

1949 and 1965 and more than 10,000 patients were compensated. But these figures represent a minimum that obscured long-term neurological disorders and the disease's—which subsequently became known as Minimata disease—lasting effects over several generations.

Based on the growth of carbon-based chemistry and petrochemistry, synthetic organic chemistry became more complex during the century and pushed the boundaries of technological advancements. With synthetic rubber and plastics, both derived from fossil fuels, and other carbon-based derivatives, the range of materials used to produce consumer goods—in addition to wood, glass, vegetable matter, and metals—increased sharply in the twentieth century. Further, the global population explosion (1.8 billion inhabitants in 1920, 4 billion in 1975), combined with rising standards of living, led to a dizzying increase in the demand for materials that ushered in the visible manifestations of the world's entry into the Anthropocene, or "Great Acceleration," that some date back to the postwar period.[2]

In developed countries, the consumer society's attraction deeply reshaped the needs of individuals. In so doing, it also altered their ecological footprint. That is to say, consumer societies exerted new and greater pressures on natural resources and ecosystems. So-called mass consumption, linked to the strategies of industrial capitalism in search of new markets to absorb the increasing production capacities allowed by the turn to Taylorism, became a global phenomenon, even if it took on highly variable forms. On the demand side, the emergence of advertising, consumer credit, and the revolution in distribution chains amplified the potential for the dissemination of products and their impact on the environment. Be it transportation, food, housing, or recreation, the new synthetic lifestyles created waste and pollutions in increasing quantities. Contaminated life was embodied in a consumer society of which some iconic products—plastics, automobiles, aluminum objects, and pesticides—shed light on the change of scale in twentieth-century pollution.

The American Way of Life

The twentieth century witnessed the flourishing of consumer society. Admittedly, "the system of objects" and the elite's and bourgeoisie's consumer habits were already well established in Western cultures, and had been for several centuries.[3] But, in the United States first, from the 1930s, then in Europe and other prosperous areas of the postwar world, the consumerist way of life grew and the commercial tools to develop it became more defined and standardized. To respond to the economic crisis following the stock market crash of 1929, the US government adopted a policy of stimulating consumption, an important part of the New Deal. Around the same time, the British economist John Maynard Keynes wrote *The General Theory of Employment, Interest and Money* (1936), which became the dominant theoretical work for demand-driven economic policies. Keynesian economic theory and American economic policy were both engaging with the rise in purchasing power.[4] Especially after 1945, the rise of the welfare state provided the framework for many developed countries, which permitted access to this consumer society to unprecedented swaths of their populations. The entry into the consumerist frenzy took place early in the United States, where purchasing power increased by 60 percent between 1939 and 1944. After 1945, consumption was widely affirmed not only as the precondition for economic prosperity, but also for democracy and peace. In the land of the "American way of life," a veritable "consumer republic" was reshaping the idea of happiness.[5] But already, the problems of pollutions transcended the ideological confrontations between the capitalist West and the Soviet bloc: American university studies drew parallels between the two worlds, and insisted that the communist world's environmental contaminations were no less severe, even if their ideologies were still supposedly resistant to liberal ideas of consumption.[6]

This desire for a growing number of objects and services was deliberately cultivated by states seeking peaceful social relations and by companies seeking profits, through the use of advertising, access to credit, and the valuation of macroeconomic indicators such as GDP.[7] While savings, sobriety, and repair became obsolete values—even dangerous for the progress of nations—the frenetic consumption of products with limited lifetimes asserted itself as an imperative. Consuming became a patriotic duty and a fundamental freedom that profoundly reshaped political economy. During the Cold War, the promotion of the North American consumerist model in the world, including through the Marshall Plan, also aimed to stem the lure of communism by the enactment of a "popular capitalism."[8] Capitalism in the developed world promised a hopeful future and provided the stability of social compromise, which was supposed to end misery and shortages. The resulting prosperity and consumerism defined the celebrated "Glorious Thirty," an expression coined in 1979 in an essay by the noted French economist Jean Fourastié. In 1956, Boris Vian could sing his "Complainte du Progrès," while Georges Perec triumphed with his 1965 novel *Les Choses*. Both served as testament to the irruption of objects into the affective lives and innermost experiences of individuals. But the triumph of mass consumption cannot be reduced to a global process of Americanization. Between the two world wars, in Europe as in Asia, there existed various national traditions mixing the call for austerity with the quest for material abundance. The entry into consumerism took variable paths, and ambivalence often dominated— between attraction for wealth and individualism on the one hand, and respect for frugality and community on the other. This phenomenon was observed in Nazi Germany as in postwar Japan.[9]

The triumph of mass consumption also coincided with the growing urbanization of societies. While on the eve of the Great War only 20 percent of the world's population was urban, nearly 40 percent was concentrated in cities in the 1970s, despite significant contrasts. In the developed West, the share of urban dwellers already exceeded 70 percent; in China and Africa, urban populations made up less than 25 percent of the total. Urbanization was linked to car travel, which in turn disrupted existing marketing practices. The centralization of commodification led to the creation of supermarkets, then hypermarkets (or superstores: more than 2,500 square meters of floor space), a model born in the United States in the 1930s which then spread to Europe. The first French hypermarket opened in the suburbs of Paris in 1963. In 1970, however, supermarkets and hypermarkets still accounted for only 6.2 percent of retail sales in France.[10]

Mass consumption—socially very differentiated, but universally buoyed by the feeling of emancipation by objects—was a major source of pollution, because new consumerist lifestyles imposed strains on energy, materials, and toxic products. Take the example of leather, a very old material, but whose high global consumption reconfigured production. After the discovery of a new potential chrome tanning process based on a chemical treatment by Friedrich Knapp of Germany in 1858, industrial applications began in 1884 under the impetus of August Schultz, another German chemist established in New York, who worked for an importer of artificial dyes. Major changes appeared at the beginning of the twentieth century: after millennia of vegetable tanning, the entire industry converted to chrome tanning in fifty years. At the end of the twentieth century, 85 percent of world leather production was manufactured using this method. Chrome tanning was perfectly adapted for the consumer society, since the speed of manufacture was improved by a factor of thirty (a day instead of a month). In addition, the leather was sufficiently inert, allowing it to be transported over long distances.[11] However, chrome tanning is particularly toxic. Though the chromium(III) salt that is used does not present a health or environmental hazard, it can oxidize during the production process or the incineration of leathers. The resulting (hexavalent) chromium(VI) is a true poison. It was commonly released into the environment by many tanneries and was also present in some leathers. In particular, it became a scourge in India, the main country of leather production after 1945, for both workers and residents.[12]

The postwar consumerist revolution produced a lot of waste but, paradoxically, it promoted a cult of cleanliness and personal hygiene, activated by advertising propaganda aimed at women. Electric home appliance equipment was synonymous with liberation and progress. By the 1930s, 50 percent of American homes were already equipped with vacuum cleaners and washing machines; only 8.4 percent of French households had the same appliances in 1954. Twenty years later, however, that percentage had increased to 66.4 percent. The cult of cleanliness insisted that these appliances were necessities: modern living demanded tools to clean homes, kill bacteria, and remove substances dangerous to health. Indeed, along with other new home appliances, they became symbols of modernity.[13] The postwar period was also the heyday of industrial design: objects were redesigned to hide their inner workings while suggesting efficiency and cleanliness.[14] As for detergent, it became iconic, while hygiene and beauty products enjoyed massive growth.

Ironically enough, a gigantic increase in urban waste stemmed from more disposable products and the obsession with packaging, which were invented in the name of hygiene.[15] The relationship with objects changed dramatically. Ephemeral consumption—even programmed obsolescence—as theorized in the 1930s by the promoter Bernard London in the United States in order to stimulate the production cycle, became a common and preferred approach. In his 1932 novel, *Brave New World*, Aldous Huxley already anticipated this new disposable consumerist imaginary through the slogan "ending is better than mending," or "the more stitches, the less riches."[16] The case of nylon stockings is a good example. As a synthetic product invented by the chemistry company DuPont de Nemours in the 1940s, nylon was hugely successful. But its strength caused a stagnation of sales: the original formula for nylon was subsequently revised, voluntarily decreasing the strength of the fiber to weaken stockings and force consumers to buy more. The light bulb was another symbol of planned obsolescence, whose lifetime was deliberately shortened from 2,500 hours to 1,000 hours by the Phoebus cartel of American light bulb manufacturers, created in 1924. Even before computing and electronics, consumer society embraced the logic of disposable and replaceable goods and their spiral of waste. As early as 1925, the American economist Stuart Chase described this phenomenon in his book *The Tragedy of Waste*, and ten years later the historian Lewis Mumford denounced its mechanisms. During the "Glorious Thirty," planned obsolescence spread in many sectors, encouraged by designers such as Brook Stevens. But it was equally criticized by economists such as John Kenneth Galbraith (*The Affluent Society*, 1958) and journalists like Vance Packard (*The Waste Makers*, 1960).[17]

The decoupling of sites of production—subjected to direct pollution—from sanitized sites of consumption rendered almost imperceptible to the public the increase in emissions of toxic substances in the environment. Yet, the extent of the waste produced by this mass and disposable consumption quickly emerged as a problem. Even if the installation of new reprocessing plants and sewer systems—more expensive and more technically advanced, to be sure—relocated nuisances and made it possible to clean up some cities, particularly in developed countries, the spiral of waste was incessant. Though new waste management methods mitigated some of the problem, urban society's rapacious appetite for goods demanded ever more artificial products. The quantity of waste produced ballooned. Where efficient infrastructure existed, garbage was collected and dumped elsewhere, just like in New York City, which in 1948 opened a gigantic landfill on Staten Island. It received thousands of metric tons of waste every day and became the largest landfill in the world. Management methods varied in different countries: Japan seems to have deployed considerable ingenuity in recycling urban waste; repurposing it, in some cases, as building materials. On the other hand, in many megacities in the global South, waste management became a source of increasing contamination. For example, in the 1950s Mexico City produced 3,000 metric tons of garbage per day.[18] The thrall of the consumerist movement left little room for environmental dissent. In the 1960s in France, some trade unionists, such as Fredo Krumnow of the French Democratic Confederation of Labor (CFDT), denounced "the waste civilization."[19] In the United States, the first Earth Day in April 1970 drew attention to this problem; the model of frenetic consumption was contested by activists who challenged the nuisances provoked by the "American way of life."[20] Yet these oppositions failed to radically transform the bounds of the debate. For the most part, they remained outside of mainstream discourse and marginalized by the addictive attraction of access to new goods.

By the end of this period of rapid growth, humanity's ecological footprint had exceeded the planet's carrying capacity. That strain on the earth's biocapacity was calculated retrospectively: it was an indicator born from the Rio Conference in 1992 and has since been calculated by the Global Footprint Network. It measures the amount of space needed to produce the goods and services consumed, regenerate resources, and absorb waste, including the carbon dioxide derived from these activities. With regard to the environmental footprint, it assesses the impacts of a product type or organization.[21] Another approach (virtual water) focuses on the consumption of water needed to produce each good or each food; it shows that demands on water increased proportionately more than aggregate consumption, because consumption's structure changes with the addition of new products that use greater amounts of water.[22] As noted by Jacques Theys in the early 1980s, each growth point seems to become more toxic; traditional industries, based on natural materials, declined in favor of composite and synthetic products that were subject to multiple operations and resulted in many intermediate wastes.[23] As a result, quantities of very bulky but low-polluting waste—such as ash or inert organic matter—stagnated or decreased, while the most dangerous forms of waste—be it nuclear waste or the chemical industry's production of long-lasting plastics or hydrocarbons—tended to grow. For example, a study of Wellington's ecological footprint in New Zealand between the 1950s and the early 2000s showed that food production quantities had changed little; meanwhile, transportation, habitat, and consumption of manufactured goods and services all rose sharply. It should also be noted that these ecological footprints, which are often calculated on average, depend on consumption practices, so they are extremely diverse in the world according to social groups and regions. The decades of strong growth in the middle of the century deepened inequalities and disparities in the amount of waste between North and South, between richer and poorer populations.[24]

The Plastification of the World

In this consumer society, carbon and petrochemistry profoundly transformed the material foundations of life. Already initiated at the end of the nineteenth century, the synthesis of new materials expanded during the interwar period. Though oil offered some competition, coal was still king. Industrialists and engineers continued to marvel at the mineral's potential to create a vast array of products. In 1930, during the Liège World Exhibition, a radiating diagram synthesized the omniscient power of coal's byproducts.[25]

Figure 9.2
L. S. Lowry, *Industrial Landscape*. (Tate Museum, London, 1955.) This oil on canvas is typical of urban scenes painted by the British artist L. S. Lowry (1887–1976) during his career. Despite being an imaginary composition, the painting is punctuated by real elements that give the idea of an urban environment dominated by factories and chimneys spewing their smoke. This was typical of the imagination of the previous decades. The foreground acts as an invitation to enter this terrifying world.

The quantity of synthetic chemicals released into the environment increased in order to meet the demand for new materials for clothing, housing, construction, transportation, recreation, and agricultural production. According to DuPont's famous 1935 slogan, "Better things for better living ... through chemistry," chemical expansion was designed to preserve nature by decreasing demands on biomass and offering substitutes for rare or expensive natural materials such as silk, ivory, or rubber. Despite uncertain estimates, about 10 million chemical compounds were synthesized during the twentieth century, of which 150,000 acquired commercial applications.[26]

But after 1945, oil reigned supreme in the design of new materials, spurred by a full range of plastics, with properties considered miraculous.[27] Roland Barthes analyzed the new plastics craze in his 1957 classic *Mythologies*:

at the far reaches of these transformations, man measures his power ... , the very itinerary of plastic gives him the euphoria of a prestigious free-wheeling through Nature.... The hierarchy of substances is abolished: a single one replaces them all: the whole world *can* be plasticized, and even life itself.[28]

After the first polymer synthesis derived from hydrocarbons in 1910 (Bakelite, the first plastic, was produced on an industrial scale until the 1950s), the interwar period experienced an explosion of research and patents on plastics and synthetic rubbers, in Hitler's Germany as in the United States and the USSR. This resulted in the universal spread of polyvinyl chloride (PVC), industrialized production of which its German creator IG Farben launched as early as the 1930s. Its cost effectiveness and properties, such as being light and waterproof, ensured its wide distribution. PVC production began in 1933 in the United States (by Union Carbide). Annual production multiplied 120 times during the Second World War, to reach 160,000 metric tons in 1952, then 2.7 million metric tons in 1973. The two world wars stimulated investment: during the Second World War, DuPont marketed the nylon stockings which would subsequently invade the world. The first commercial patent for plexiglass, which was transparent, smooth, resistant, and lightweight, was filed in 1948. Chemists from BASF and Dow Chemical developed polystyrene manufacturing, which was used for packaging, coffee cups, noise barriers along roadways, and housing insulation. Enumerating the long list of these plastic products is an almost interminable venture: in 1935 polyethylene (plastic bags) appeared; in 1937 polyurethanes (paints, varnishes); in 1938, polytetrafluoroethylene, which in 1945 became known by the commercial name of Teflon and was soon used as a frying pan coating and then for countless other uses; in 1954 polypropylene (bumpers and car dashboards). In the music industry, naturally derived shellacs were replaced by chlorinated vinyl polymers to make records. After 1945, the American record industry invented the microgroove and the now famous vinyl record became a worldwide success.[29]

Viscose—or "artificial silk" or "rayon"—patented at the end of the nineteenth century, is one of those plastic products that grew in popularity after 1918. Derived from cellulose treated with various chemical substances, its manufacture required the use of toxic materials and the emission of large quantities of pollutants. In 1930, 2,000 kilograms of soda, 1,500 kilograms of sulfuric acid, and 550 kilograms of carbon disulfide (CS_2)—produced from coal—were needed to make 1,000 kilograms of viscose. The adoption of viscose manufacturing initially experienced strong resistance from the silk industry surrounding Lyon. In 1920, it represented less than 5 percent of overall production. But afterwards the bulk of the city's principal silk producers embraced this material—in particular, the Gillet company, which built an industrial empire at the juncture of textiles and chemistry.[30] Other viscose plants also opened in the United Kingdom, near Coventry, and in Italy, Japan, and the United States. Viscose production remained controversial for a long time: the hazards of carbon disulfide to workers were documented early on in the medical literature and social surveys. A report from the 1922 Labor Inspectorate in France and Alice Hamilton's investigations in the United States described cases of poisoning in these factories. In England, local residents also complained about air pollution and carbon disulfide emissions, but the authorities dismissed the dangers; industrialists provided financial compensation to quell dissent. In the 1930s, manufacturing conditions were even worse in Fascist Italy and Nazi Germany. Workers and residents were victims of very toxic waste, but the location of the factories and the exploitation of a quasi-servile workforce prevented any curbing of polluting production practices.[31]

After 1945, taking advantage of cheap energy and unprecedented demand for materials considered modern meant rewarding the standardization of processes and the triumph of mass production: plastics became virtual necessities. In West Germany, for example, the consumption of plastic rose from 1.9 kilograms per capita in 1950 to 15 kilograms in 1960.[32] But these new substances were also responsible for very different types of pollution, both in their manufacture in large petrochemical plants using hydrocarbons, and at their final point of consumption where a considerable amount were then abandoned in soils, rivers, and oceans.[33] Petrochemical complexes, located near oil and natural gas deposits, first developed in the United States and Europe before expanding into Third World countries beginning in the 1960s.[34] The expansion of petrochemical corridors—significantly called "Cancer Alleys" in the United States—precipitated expansion of pollutions, especially surrounding vinyl chloride manufacture. Many cities and communities had their water and air contaminated by the plastics industry. This was especially the case along the Lower Mississippi River.[35]

In France, the first large petrochemical plant was opened at L'Estaque near Marseille in 1949–1950. Ten years later, another fifty had already spread around oil refineries (Lower Seine and Étang de Berre, south of Lyon) and places of natural gas production like Lacq. As in the United States, the plants produced many plastic products but also fertilizers. They were also responsible for significant environmental contamination and abnormal cancer rates. South of Lyon, for example, the imperatives of industrialization led to the acceptance of these nuisances, which were relocated to peri-urban and working-class communities. In 1964, the opening of the vast modern refinery of Feyzin inaugurated a "chemical corridor" in Saint-Fons; two years later a tragic explosion at this refinery fostered debate on industrial risks and their pollutions.[36]

The entire history of the twentieth-century petrochemical industry is predicated on chemists' efforts to imitate and improve upon nature's materials while being confronted with the excesses of their own practices. The new "age of chemistry" aroused the doubts of researchers such as Pierre Baranger, holder of the Chair of Chemistry at the École Polytechnique from 1949. In a synthesis on the subject published in 1956, he observed that plastics are only "pale imitations of the admirable natural substances," their success being especially due "to the deep and unconscious tendency of man to do without nature and to adore himself through his own creations in the center of an ersatz pantheon." Baranger also denounced the frenetic and unthinking squandering of fossil raw materials to obtain that pantheon of new creations.[37] At the turn of the 1960s and 1970s, concerns about plastics were redefined as a public health issue and a source of major pollution. Modern society had to reckon with the fact that plastics were very difficult to recycle and that their degradation—which also contaminated the environment—was particularly slow, several hundred years in general. In addition, the microparticles derived from it were ingested by a large fringe of the marine biotope. Among hundreds of letters sent to the German Interior Minister in the course of an investigation into environmental issues, seventy blamed plastics and their waste as the main source of pollution, while a petition that collected hundreds of signatures in May 1971 protested

> against ever-increasing pollution of the environment and the deterioration of the living conditions of humans, animals, and plants due to automobile emissions, industrial waste, and the multiple non-degradable wastes of our affluent society (packaging or plastic bottles).[38]

In 1970, the American architect and designer Paul Mayen also asserted that industry should stop producing products "with such high pollution potential," and especially to strive toward developing plastics "with high reconversion potential."[39] Finally, although the issue of waste was generally muted, with few studies before the 1980s on environmental contamination by plastics, two American biologists, Edward Carpenter and Kenneth Smith, raised the alarm by revealing the presence of high concentrations of synthetic polymer particles in the sea off Florida in 1972. The following year the oceanographer Elizabeth Venrick reported the same finding north of Hawaii.[40]

It is impossible to be wholly comprehensive in recounting the number, variety, and quantity of the new petrochemical synthetic products that were assimilated with the plastics heretofore discussed. But it is necessary to acknowledge the scale of their significance. As an illustrative example, it is useful to focus on polychlorinated biphenyls (PCBs), synthesized at the beginning of the century and used for its dielectric properties and thermal conduction as electrical insulation in transformers, but also in capacitors, hydraulic fluids, paints, adhesives, and other applications. Their growth was due mainly to the Monsanto Company, which became the main producer after 1929, before production spread to Europe after 1945. Although Monsanto was aware of their toxicity

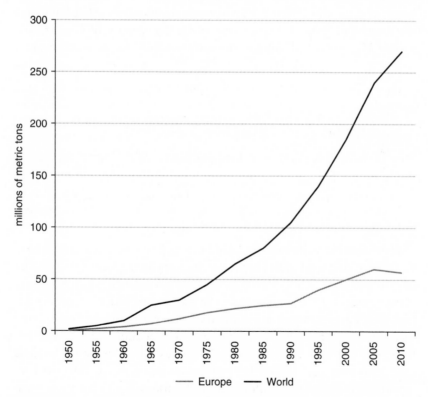

Figure 9.3
Global plastic production, 1950–2010. *Source:* Plastics Europe, *Plastics, the Facts, 2013: An Analysis of European Latest Plastics Production, Demand and Waste Data,* 10.

as early as 1937—the company's main production site in Anniston, Alabama became highly contaminated—PCBs were used extensively until the 1980s and caused widespread pollution.[41] As a persistent organic pollutant, it was indeed a hypertoxic threat to human health and the environment. Before 1975, the Hudson River near New York City, and the area around Bloomington, Indiana, paid a heavy environmental cost for PCB production, but that was ultimately the case at all production sites, before their release and dissemination affected the entire planet to varying degrees.[42]

Automania: Supreme Fetish

Chemistry's plastics and new synthetic substances infiltrated all everyday objects, first and foremost the totem of consumer society: the automobile. It is difficult today to grasp the full effect of the automotive revolution and the shock that it caused its contemporaries. After lamenting the sound and smell of these machines, the old narrator of *The Life of a Simple Man* remarked sadly in the early years of the twentieth century that

> these motor-cars really are the instruments of the devil, invading the roads, upsetting us and doing us mischief…. A man from a bygone age, a grandfather shaking his head, it is not for me to pass judgment on that subject. The young people will get used to these vehicles of progress.[43]

Fifty years later, in his *Mythologies*, Barthes noted the cult of the car:

> I think that cars today are almost the exact equivalent of the great Gothic cathedrals: I mean the supreme creation of an era, conceived with passion by unknown artists, and consumed in image if not in usage by a whole population which appropriates them as a purely magical object.[44]

Symbol of individualism, status indicator, chasm of fuels and materials, cult object, consumerist fetish: the car dethroned the railroad in the order of fast mobilities. It also served as the medium through which humanity slipped toward a chronic acceptance of risk and pollution. A socially differentiated acceptance, it is worth stressing: while large engines offered their owners more safety, while emitting more polluting substances, traffic was relegated to the fringes of urban areas.[45] In addition to the toxic dust from fuels, cars rapidly became the source of very serious environmental degradation and multiple contaminations due to chemical additives in fuel, tire manufacture and disposal, the production of composite materials, and the gigantic infrastructures necessary for their mobility.

For the automotive system to become fully entrenched, it was first necessary to create a favorable environment. For starters, in the 1920s and 1930s, it chased trams from all cities that had them. Behind the scenes, the oil and automotive lobbies pushed for the supremacy of the private car in major cities all over the world. France became a pioneer in this shift toward the automobile. The network of tramways in Paris—at the time, the largest in the world—was removed in the 1930s. The inner-city line removal was followed by the closing of lines in its suburbs. Trams were replaced by Renault buses. In the decades that followed, all the other provincial cities carried out the same process.[46] In the United States, where tramway equipment was extensive, large automobile and oil consortia also exerted a major influence. As early as 1922, the head of General Motors set up a working group whose mission was to design a strategy to replace electric streetcars with buses and then cars. In 1935, Congress passed a law prohibiting energy companies from owning streetcar companies. This act encouraged their buyout by General Motors, who subsequently allowed their infrastructure to deteriorate. Two years later, the 150 kilometers of Manhattan tramway tracks were removed. General Motors was also, in one way or another, involved in the dismantling of more than 100 streetcar networks in 45 cities, including New York, Philadelphia, Saint Louis, and Los Angeles. In April 1949, a federal jury found General Motors guilty of conspiring with Standard Oil and Firestone Tire, but the damage had been done. The tram networks were not coming back: the car had taken their place.[47]

Buoyed by a powerful imaginary, the car was part of a huge technical ecosystem that shaped new desires and new landscapes. Automotive infrastructure became a second nature that inspired the invention of a world adapted to its measure, with the expansion of highways and the new urban forms that accompanied them.[48] The highway was initially a concept of fascist regimes. The first section was inaugurated in Mussolini's Italy between Milan and Varese (80 km) in 1924. Then, shortly after the rise of the Nazi party in 1933, the first major German motorway projects were launched. These *Autobahnen* provided jobs, economic recovery, and strengthened the country's centralism. They also facilitated the rapid movement of troops in anticipation of the war. Major instruments of Hitler's propaganda, they demonstrated the capacity of the new man to tame nature.[49] On the same principle, these gigantic infrastructures were built in the United States during the Cold War. Under President Dwight Eisenhower in the 1950s, the construction of the interstate highway system cost more than $100 billion. In France, the Gaullist regime launched a similar movement by adopting in 1960 an ambitious master plan for the development of a road network that provided for the construction of nearly 2,000 kilometers of motorways before 1975.[50] In cities, the priority given to car traffic burst a tenuous urban equilibrium; cars promoted ever more sprawling suburban enclaves. Los Angeles became the emblematic case: the city grew from 1.6 million to 6 million inhabitants between 1930 and 1960, devouring space and requiring incredible quantities of energy as it spread spectacularly and tentacularly. The new urban-suburban landscapes had a considerable environmental impact. Tons of asphalt, from petrochemicals, were poured to build roads so that cars could carry citizens between urban work and ever more distant suburban homes.[51]

Systematized Taylorism in the factory plant—built on automation and efficiency—meant that the automobile also revolutionized workspaces and production methods. In the 1920s, Henry Ford developed the famous Model T with the first assembly lines.[52] To control the production process, in 1928 Ford opened the largest factory in the world in Detroit (Michigan), on the banks of the Red River. As early as the 1930s, 80,000 employees produced 10,000 vehicles a day.[53] Unlike other car manufacturers who spread the nuisances of their activity among many dispersed factories and suppliers, this integration strategy lead to focused environmental impacts at the same site, making pollution particularly visible and palpable. In anticipation of this problem, the plant was equipped with methods of treating solid, gaseous, and liquid waste. Ford's waste management system followed the logic inherited from the nineteenth century: unwanted waste was diluted and released into the environment, along the Red River. The main criterion guiding the treatment of toxic and polluting waste was financial profitability.[54] After 1930, the world's leading automakers followed this pattern of concentration in large factories and the trend became stronger after the war.

Cars are assembled from so many diverse materials—many of them modified or substituted in and out over time—that it is impossible to list them. While iron and steel made up nearly 75 percent of the weight of the first Ford Model T, the range of metals diversified over the following century to such an extent that one might assert that hardly any metal was not used in automotive manufacturing. In the first environmental history of automania in the twentieth century, Tom McCarthy drew a very complete picture of the controversies and many pollutions associated with the automobile at each stage of its life cycle.[55] Heavy metals, in particular, became increasingly critical. In postwar Japan, the cadmium spilled by Mitsubishi manufacturers in the Jinzu River caused a terrible bone disease, while the increase in sulfur dioxide emissions in the heart of cities exacerbated respiratory ailments.[56] For their part, plastics gradually became primordial materials. As early as 1936, at the Berlin Motor Show, the chemical company IG Farben introduced a new kind of synthetic rubber for tires, manufactured at its Saxony plants.[57] The overconcentration on the atmospheric pollution of fuel emissions means that studies on the environmental impact of cars too often forget the considerable quantity of materials required for their production, the various solvents needed for their maintenance, and the waste that results; as well as the design of more and more composite and synthetic products: for dashboards, brakes, bumpers, bodywork, and so on.[58]

It is possible to get an idea of the breadth of these pollutions by drawing on a representative example of one of these composite and synthetic accessories—tires. Initially, tires were made of rubber cultivated from *Hevea braziliensis*, the tropical rubber tree. The first car to run on pneumatic tires was designed by Édouard Michelin in 1895. Its production was itself a source of major pollution, given the vulcanization process designed in the nineteenth century.[59] Ten times more watertight than rubber, butyl, a synthetic elastomer, was adopted in the early twentieth century for inner tubes and membranes of tires, which incorporated chlorine to promote chemical bonds with the rubber. Production industrialized in the 1930s: DuPont invented neoprene, based on polychloroprene, the first real synthetic rubber. Neoprene was mythologized in popular culture after the actress Claudine Auger, a former Miss France, wore a revealing synthetic wetsuit in 1965 in the fourth James Bond film, *Thunderball*. After the war, tire materials diversified: sulfur, oils, and the very toxic carbon black were added, along with various braided metals (zinc, cadmium), to improve their resistance.[60] In addition to the pollution generated in composite rubber's production, it is also necessary to add those of the daily abrasive wear with the road, which released a volume of fine particles probably as important as the fuels. These microscopic residues massively contaminated oceans.[61] Finally, tires came to constitute the archetype of waste that befuddled waste management. In the early 1970s, piles of millions of tires in landfills posed insoluble problems. They were difficult to recycle, their quantity was enormous, and fumes from their incineration were very harmful. Incineration was initially quite popular, but it was gradually regulated. Official initiatives were subsequently taken to throw tires into the sea. In 1972, two million used tires were dumped a mile off the Florida coast near Fort Lauderdale. The rationale for dumping tires at sea was grounded in environmental interests: the scheme aimed to create artificial reefs for marine life. Out of sight; out of mind. This invisible clearance of cumbersome toxic waste was copied in France, where 90,000 cubic meters of used tires were immersed in the Mediterranean Sea. Similarly, Portugal, Spain, and Italy followed suit. Japan laid claim to world champion status in this regard, dumping 20 million cubic meters of tires at sea. The cost of operations, as is often the case for pollution, was passed along to following generations. Unsurprisingly, the environmental damage of the enterprise was considerable. Future generations were forced to extract the tires from the sea at considerable expense.[62]

Many chemical additives were also incorporated into the operation of the automobile. Predictably—as with most automotive innovations—their origin was the United States. Starting in 1922, leaded gasoline constituted the ultimate automotive plague. General Motors added tetraethyl lead to improve engine performance. Although workers were poisoned during its manufacture and regulators prohibited its use for a year in 1925, time enough to confirm its transgressions, powerful lobbying interests eventually impelled lead's inclusion in gasoline. Tetraethyl lead's benefits to the automotive industry and to the economy were sacrosanct. The Kettering Lab, founded in 1930, was responsible for evaluating the product. The work, under the direction of the toxicologist Robert Kehoe, was funded by General Motors, DuPont, and the Ethyl Gasoline Corporation. Kehoe claimed tetraethyl lead's hazard was innocuous. His argument—that the burden of proof rested on determining the severity of a hazard rather than proving its safety, that "demonstrable economic benefits should always outweigh unproven risks"—delayed action on lead for several decades. Lead's toxicity was a real concern, and against Kehoe's principle, several independent laboratories demonstrated it over the following years. Eventually concerns prevailed; in the early 1970s, gradual bans on leaded gasoline were introduced. But for half a century, the spread of lead in the environment resulted in one of the most significant cases of widespread pollution, found as far away as Arctic ice samples. Lead poisoning also heavily affected people living near road infrastructure.[63]

At the turn of the 1970s, the automobile increasingly became associated with the concept of pollution. Its omnipresence raised mainstream questions and doubts, even strong criticism. For the first time, televised reports were devoted to this question: on October 6, 1973, for example, on the occasion of the inauguration of the Paris Motor Show, a fifteen-minute report was broadcast on French television, in which the Deputy Director of Pollution and Nuisance Prevention at the Ministry of Quality of Life (1973–1976) discussed technical solutions to reduce nuisances, and the engineer Gabriel Boulaton, head of the Department of Engineering of the Geneva Institute Batelle, considered the automobile a pure "anachronism" in contemporary city life.[64] The following year, the comic book *Les Mange-Bitume* (The Bitumen Eaters) presented a vast cautionary futuristic fable about the worst of all possible worlds. In this highly exaggerated mechanized future, people essentially live in their cars and consumerist obligations overlap with totalitarian social control. Humans mutate into drivers and leaders conceive of a society where traffic jams become a way of life.[65] That year, 39 million vehicles were produced worldwide (compared with 4 million in 1946). But, contrary to what the emergence of these criticisms might suggest, and despite the oil shocks of the 1970s, automotive appetites continued to grow unabated. Production exceeded 50 million at the end of the century.[66]

Light Metals, Heavy Impact

The mass production of manufactured goods—like the car, the weight of which was constantly growing—required huge quantities of metals. In addition to steel compounds, the exploitation of copper, lead, zinc, or a "light" metal, aluminum, experienced significant growth over the twentieth century, considerably exacerbating preexisting nuisances and pollutions.

For example, world copper mining grew from 400,000 metric tons in 1900 to 10 million metric tons a century later. As with most nonferrous metals, that production was extracted from increasingly poor deposits. In 1800, English ore contained more than 9 percent copper, and 6 percent in 1880. But the veins were depleting; American ore contained 3 percent copper in 1880, 2 percent in 1930, and 1 percent in 1975.[67] In comparison, demand was going up. This decline meant that in order to extract the pure metal, more aggressive means had to be found and even larger quantities of ore needed to be mined per unit produced. In ever larger mines, the productive geography changed. Copper deposits in the Belgian Congo and Zambia, and especially Chile, a copper giant since the 1920s, played a growing part. In 1970, Chile led the world in copper production, thanks to the Anaconda Copper Company, located in the Chuquicamata mine in the Atacama Desert since 1922. Production reached 130,000 tons in 1929. The United States retained an important role. For example, the giant open-pit mine at Bingham, Utah, opened in 1906 by Río Tinto, became the world's first copper refining complex in 1912. During the following decades, large areas around it were contaminated with copper, arsenic, and lead; the environment was devastated by sulfurous gases.[68] The same was true in Montana, around the Washoe mine, operated by the Anaconda Copper Company (see figure 10.1).[69] These giant mines initiated a novel kind of "mass destruction," which also occurred around smaller mines. For example, a dramatic collapse of aquatic life and a general contamination of the environment occurred in just a few years of operation (1964–1966) at the relatively small (13 hectares, or about 32 acres) Mt. Washington copper mine on the Tsolum River on Vancouver Island.[70]

The ore's decline in metal content necessitated new refining processes in the early twentieth century. The Manhès-David (1880) and Peirce-Smith (1908) processes, which used a converter to oxidize undesirable chemical elements with air, were used throughout the century. In addition to these pyrometallurgical processes, a hydrometallurgical method was developed, which dissolved sulfides by watering the ore with sulfuric acid, cyanide, or ammonia. As a general rule, despite the use of filters, the sulfurous fumes, also loaded with zinc oxides, arsenic, and lead, were significant. The converter remained the main problem, because it was difficult to seal and its intermittent operation disrupted flue gas treatment installations.[71] As a result, the copper industrial complexes were distinguished by chimneys that beat height records. In Washoe, a 178-meter chimney was built in 1918. It emitted more than 100,000 cubic meters of smoke per minute. The smoke was collected and traveled in pipes, sometimes for several kilometers, in which the particles were supposed to be deposited (see figure 10.1).[72] For the most part, regulations were almost nonexistent before the 1980s, because of the relative isolation of these mines, especially in emerging countries.[73]

Because mining zinc and nickel required similar techniques to copper, soil and water contamination and the emission of toxic dusts, loaded with sulfur and arsenic, lead, and other heavy metals, were also common features of their extraction process. In addition to prominent disasters in the Meuse Valley in Belgium (1930) and in Donora, Pennsylvania (1948), where waves of toxic smog from the smelters killed numerous local inhabitants, zinc refining caused chronic pollution wherever it was practiced. The primary pollutant was cadmium. The Broken Hill Mine in the remote outback of New South Wales, Australia, became the capital of zinc extraction in the twentieth century.[74] While nickel was hardly exploited before 1900, it enjoyed increased industrial use in fine and resistant alloys. The twentieth century's arms race, in particular, fueled its development. New Caledonia has long been one of the world's major centers and reserves of nickel production. It was joined in the 1920s by

Canada, Russia, and South Africa—and after World War II, by Australia and the Philippines.[75] In the gold and silver mines that dotted the African continent, cyanide residues—which replaced mercury as the refining technology of choice for precious metals after 1887—devastated landscapes for many generations.[76]

But if there was an iconic metal of the twentieth century, it was aluminum. Its production experienced the greatest increases over the course of the century, and it better typified the advent of consumer society and its new contaminations. Aluminum was originally a luxury metal before it became one of the most consumed metals in the world, beginning in the 1960s. It was widely touted as a miraculous product with seemingly infinite commercial possibilities, lauded in advertising catalogs and exhibitions, promoted by contests and in specialized magazines (in France, the periodical *Revue de l'Aluminium* was created in 1924 and ran until 1983).[77] Aluminum provided another example of stimulated wartime production initiatives seeking postwar consumer outlets.[78] It became the symbol of mass consumption of the postwar period. Gradually, it was also used as a food additive.[79]

The harmful effects of this new everyday material were generally ignored by a powerful industrial sector that was concentrated and organized in cartels; and by the state, which supported its development. The dominant Bayer process extracted alumina from bauxite with soda, but it produced toxic waste that contained caustic soda and heavy metals (lead, mercury, chromium), called "red mud" because of its color that came from high iron oxide content. Once extracted, the alumina was converted into aluminum in large plants through extremely energy-intensive electrolytic processes.[80] This conversion stage created significant air pollution; before the 1970s, 30 to 60 kilograms of fluorite were emitted for every metric ton of aluminum produced.[81]

In the Alpine valleys, the legacy of pollution inherited from the turn of the twentieth century persisted, despite continued contention and very strong litigation. Near the big factory of Saint-Jean-de-Maurienne, opened in 1904, a report of the Hygiene Council (1908) noted deforestation, the death of bees, and workers' health risks as problems. Farmers organized themselves into unions and received renegotiated benefits each year according to the amount of livestock affected due to the pollutions. Similarly, in Switzerland, peasants close to the Martigny plant blamed emissions of fluorine for the decline in their vegetable crops.[82] In Italy, the Italian Aluminum Company (SIDA: Società Italiana dell'Alluminio), founded in 1927 under the consolidating context of the Mussolini fascist regime, established a vast factory in Mori, a small village in the province of Trento. Shortly after, local silkworms began to die, tens of thousands of people became victims of fluorite fumes, and pollutants spread over 4,000 hectares (about 9,900 acres). In 1932, a committee of locals confronted industrialists, who adopted a wait-and-see mentality, and the state, who simply paid compensation for damages to locals. Unimpressed, these inhabitants rose up in 1933. The factory was temporarily closed. Engineers traveled to France to study filtering techniques, while children from the surrounding villages were sent to homes beyond the range of the pollutions, and the company engaged in a massive propaganda campaign that denied links between plant activity and local health problems. When the factory reopened in 1935, the context had changed: the rise of international tensions and the march to war pushed the Mussolini regime to strongly repress the disputes.[83] The United States reinforced its advantageous territorial and resource capacities in the interwar period. Spurred by the economic incentives provided by the New Deal and the war that followed, construction of major hydroelectric dams emboldened Alcoa (Aluminum Company of America, previously the Pittsburgh Reduction Company) and its industrial production.[84] In Quebec, Alcoa (whose international holdings after 1928 were housed under a new company: Alcan, Aluminum Company of Canada) built a giant plant in 1925, the largest in the world. The plant required the construction of its own city, Arvida, named after the president of the company, Arthur Vining Davis. Urban planning was designed to stabilize the workforce entirely devoted to production, which explains why complaints about pollution were limited. In 1945, the plant produced nearly 500,000 metric tons of aluminum.[85]

The French aluminum example shows both uninterrupted growth after 1945 and the pollution problems that accompanied it. Production increased sevenfold between 1956 and 1972, from 160,000 to 1.14 million metric tons, often on sites away from the old industrial ones, near bauxite deposits and near access to hydroelectricity. In 1950, 85 percent of production was carried out far from major steel and coal mining centers.[86] Aluminum's importance to the economy generated an agreement between the state, the industry, and the unions to limit opposition to the contaminations that became evident at the refineries, in surrounding vegetation, and among the alarming number of workers who contracted cancer. Companies also implemented paternalistic policies to ensure social peace and order in the factories.[87] In the 1960s, the main bauxite processing complex in France, operated by Pechiney in Gardanne near Marseille, discharged a metric ton and a half of caustic residues for every metric ton of alumina produced. Huge amounts of red mud, stored in reservoirs, threatened to contaminate groundwater and soils. A pipeline—about 40 kilometers long—was built in 1966 to offload the sludge into the Mediterranean Sea. As ever an historical constant when waste is dumped at sea, fishermen and local authorities anxious to attract tourists protested.[88] In the Maurienne Valley, fluorine air pollution from the aluminum smelting plants created tensions, especially because of the valley's depth, which trapped contaminants. The topography made the pollutants even more visible and intolerable, especially when the valley was also developing a tourism industry.[89] In the 1950s and 1960s as production increased to meet ongoing demand, the plant adopted an electrolytic process using so-called Söderberg tanks, which expelled fluorine in high amounts without taking effective measures to capture it before it escaped into the environment. Contamination thresholds of 40 ppm (parts per million) were greatly exceeded in some areas. In 1966, the National Forestry Office identified 10,000 hectares (about 24,700 acres) of affected forests; the contaminated area was 40 kilometers long. The

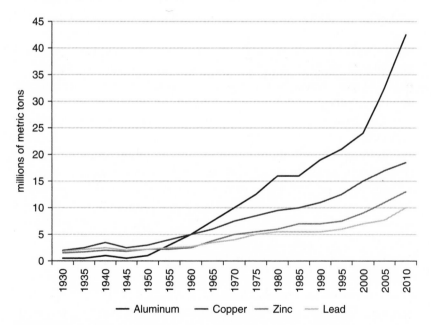

Figure 9.4
Global metal production in millions of metric tons (raw aluminum; unalloyed, refined copper; zinc; lead) *Source:* derived from l'*Annuaire statistique mondial des minerais et métaux* [SIM et BRGM] as reproduced on the Société chimique de France website.

aluminum company, Pechiney, was forced to pay more than 4 million francs in annual compensation for a production of 99,000 metric tons.[90] Widely covered by the major media and the emerging environmental press, the pollutions of the Maurienne—renamed "Valley of Death"— played a prominent part in the boom in environmental coverage by the media in the early 1970s. Fights backed by local environmental associations, and some unions such as the French Democratic Confederation of Labor (CFDT), emerged. An antipollution committee even appeared in the valley.[91] Only in the context of stern opposition did manufacturers choose to invest in modernizing their facilities and significantly reducing emissions.[92] Similar processes—oppositions forcing changes in production—with similar chronologies were common all over Europe, such as at the Martigny plant in Switzerland, for example.[93]

"Green Revolution," Gray Soils

Like most economic sectors, agriculture became increasingly dependent on chemistry. Whereas before 1914 few chemical inputs were deemed useful, chemistry transformed Western agriculture after 1945, especially with fertilizers and biocides designed to fight against undesirable weeds and pests. In the 1970s, agricultural chemistry began to expand throughout the rest of the world. At the heart of this diffusion was a beautiful utopian idea: feeding a rapidly expanding world and eliminating famines and shortages by introducing food abundance. "Green revolutions" heralded this agricultural modernization, which constituted a transition from peasantry to agroindustry. The terminology was not neutral; it obscured the fact that the use of chemical inputs to amend soils, control crop pests, remove species considered invasive or harmful, and reduce uncertainty by standardizing landscapes and environments, contributed greatly to the contamination of soils, groundwater, and waterways. If the term "insecticide" in agriculture dates from 1838, it was replaced in the middle of the twentieth century (1959 in France) by the evocative term "pesticide," which came from English. Though the term was readily adopted into French, it was curiously unsuitable, even if in hindsight it was curiously apt. In French, "pest" does not refer to "parasite" but rather to an epidemic disease; in French, "la peste" translates as "the plague." Subsequently, the term "phytosanitary" spread. Plant sanitation was a sufficiently neutral term: phytosanitary products encompassed insecticides as well as "herbicides" (around 1930) and "fungicides" (1912). Sanitation, as far as it was suggestive of scientific management, had little impact on the imagination; insect-, herb-, and fungus-killers, however, provoked other ideas.[94]

Between 1950 and 1985, global grain production increased 2.6 times. Though it was unevenly distributed, this explosion was only possible thanks to the massive use of chemical fertilizers—nitrogen, phosphoric acid, and potash (the famous NPK trio). Their application increased from 4 million to 17 million metric tons between 1900 and 1950, but then to 130 million metric tons by the end of the 1980s. Fertilizers were joined by a considerable increase in irrigation and the use of pesticides, the quantity of which was multiplied by twenty during the same period.[95] The rise of petrochemicals played a major role in this change because the use of hydrocarbons made it possible to market new synthetic substances. The use of feed additives for livestock, such as acrolein, also deserves mention. A very toxic substance produced from a petroleum derivative, acrolein was used to produce a protein component for livestock called methionine. It was used in feed and in anticoccidials whose development permitted more intensive forms of poultry farming.[96] Particularly in the United States and Great Britain, these innovations provided the context for a quantitative intensification of antibiotics that were used widely and on a massive scale in livestock farming.[97]

The use of chemical fertilizers and pesticides required that land be reorganized first. Old polycultures gave way to monocultures, and the number of farmers, who still represented half of the population in the United States in 1920, decreased drastically. Mechanization also played an important role; the spread of tractors—300,000 worldwide in 1920, predominantly in the United States—increased to 3 million by 1940, 6 million a decade later, and 16 million by 1970. Modernization had a snowball effect: since industrial monoculture was more susceptible to pests and soil nutrients declined more rapidly, the need for more pesticides and fertilizers established a vicious cycle to compensate for the problems of these new gray soils. The extent of agricultural chemical pollutions was derived, therefore, from these developments. Agricultural pollutions increased more quickly in the United States and Canada than in Europe, and faster in Europe than in the rest of the world.[98]

Prior to 1939, the United States had already experienced health scandals related to insecticides, especially from the use of arsenicals, but the Second World War ushered in a real turning point.[99] As early as 1949, California farmers spread parathion; in 1951 they sprayed demeton; malathion in 1953; and metacide in 1956. The danger of these organophosphorus products came from the action of an enzyme that disrupts the nervous system and causes heart problems and cancers.[100] Among the main pesticides used after 1945, the famous DDT became prominent due to leftover stocks from the Second World War.[101] Although the hazard of pesticides gradually emerged as a global political problem—and was addressed at the international level—much of the early information on this issue originally came from California in the 1950s–1960s, where it was first and foremost regarded as an occupational health concern. In 1949, in an incident near Marysville, twenty-five farm workers became ill after working in an orchard sprayed with parathion. The California State Department of Public Health then identified 300 cases of poisoning by agricultural chemicals. Geography and history combined to make California a major site for the production of fruits and vegetables; 50 percent of the country's oranges, and 100 percent of the olives, avocados, and apricots were produced there. The introduction of new species, accompanied by the introduction and proliferation of insects, pushed farmers toward chemical control of the environment. Twenty percent of all pesticides used in the country were applied in California. The workforce, often made up of immigrant laborers, struggled to be heard, even though Mexican workers in California's orange groves conducted a "death march" in the 1950s to denounce their working conditions.[102] Outside California, thousands of more or less serious chemical-related accidents also occurred between 1945 and 1970 among the African American workers of the cotton plantations of the southern States.[103]

In Europe, and especially in France after the Liberation, the entire agricultural industry was in dire straits. Shortages persisted; the French government was forced to reinstitute wartime bread rations in 1946. In comparison, North American agriculture became a model of efficiency praised by agronomists such as René Dumont in his *Leçons de l'agriculture américaine* (1949). "Routine agriculture," Dumont warned, "withdrawn into an autarchic and Malthusian position would ruin the entire country."[104] French agriculture changed rapidly: the country saw a reduction of one million agricultural workers between 1946 and 1954, its share falling from 36 percent to 27 percent of the active population. The First Plan of 1947, which relied on the American Marshall Plan funds, enabled the modernization of old practices and the widespread mechanization of agriculture. The number of tractors increased from 137,000 in 1950 to 558,000 in 1958 and 1 million around 1965, while the equipment for spraying the fields with pesticides spread. Initially, pesticides were applied through human and animal traction, but sprayers pulled by tractors followed. Within a few years, DDT and HCH (hexachlorocyclohexane) replaced the old arsenical products (arsenates of lime and alumina).[105] Agricultural modernization and intensive use of chemical substances were especially beneficial to the larger farmers, while smaller farmholders were swept away. In France, Charles de Gaulle's rise to the presidency in 1958 further accelerated these changes. The creation, in 1960, of land development and rural settlement corporations (SAFER), responsible for acquiring land, and the Common Agricultural Policy (CAP) of the European Economic Community (EEC), which entered into force in 1962, promoted the adoption of pesticides as a modernizing factor in production.[106] The state's technical services, including the National Institute for Agricultural Research (INRA, 1946), encouraged the use of these products. So did the largest farmers' organizations, such as the National Federation of Agricultural Holders' Unions (FNSEA, 1946) and the National Club of Young Farmers (CNJA, 1957). Jean Bustarret, a specialist in hybridization and improvement of plant species, an inspector and then director general of INRA from 1949 to 1972, was fascinated by the American example. He became an ardent promoter of pesticide products, launching first the National Federation for the Defense of Farming in 1956, and the Weed Control Committee, which included representatives of the industrial world—Pechiney, for example, which in 1953 developed a weed killer for flax.[107]

The situation outside the Western world was quite different, because chemical products remained expensive and required more concentrated operations on smaller farm plots. As a result, their use was more limited. Beginning in 1953 in the USSR, Khrushchev initiated an ambitious plan to increase Soviet agricultural productivity and fight against undernourishment. Khrushchev's plan involved cultivating the Kazakh steppes, as well as parts of Siberia to extend cultivated areas from 9.7 million to 28.7 million hectares (about 24 million to 71 million acres) within a decade. The initiative failed due to soil erosion and the rapid degradation of soil fertility because of overexploitation.[108] To compensate for this failure, the country began a rapid conversion to chemistry in the 1960s; the amount of chemical fertilizers produced tripled between 1965 and 1975, from 31 million to 90 million metric tons, placing the country at the head of world producers. Pesticides such as DDT were also used in increasing quantities on large collective farms. Dispersed by helicopters and planes, they contaminated the environment. Moreover, Soviet pesticides were less effective and more polluting than those used in the West.[109]

In the global South, economic means dictated that most small farms were excluded from chemical modernization. However, large cereal, sugar, and cotton fields in Latin America—as well as those devoted to export crops inherited from colonization in Asia—gradually adopted chemical treatments. In great secrecy, some very toxic pesticides were used on plantations. Nemagon, the commercial name for a powerful insecticide developed in the 1950s by Dow Chemical and Shell Oil, was liberally sprayed to destroy the microscopic nematode worms that attacked the roots of banana trees. Used in large quantities in banana plantations in Central America, the Caribbean, and the Philippines, Nemagon raised major health concerns after 1975 when it was banned in the United States. Nevertheless, tens of thousands of agricultural workers in Southern Hemisphere countries were victims of cancer, blindness, infertility, and other medical problems directly related to its use.[110] In the same spirit as in the West, and with strong encouragement from the central government—which garnered support from American foundations (the Ford Foundation, for example)—the Green Revolution and its tons of pesticides also affected India beginning in the 1960s. Pollutions were accompanied by considerable social changes, which upended the traditional peasantry and made them wholly subsidiary to agroindustrial powers. In China, resistance to this transformation was stronger, and the use of pesticides remained limited before 1976 and the death of Mao Zedong.[111]

The health and environmental problems linked to the intensive use of pesticides only became a concern for agricultural chemistry's proponents once insect resistance to chemicals emerged. Once the chemical industry developed new molecules, DDT's use declined. As in the Lyon region in 1966, organochlorines were chased out of the market in favor of increasingly complex organophosphorus compounds.[112] Denunciations of the new products, initially marginal, rose nevertheless. In 1952, Roger Heim, director of the Museum of Natural History, warned against "the large-scale use of certain chemical substances, insecticides, fungicides, rodenticides, herbicides, [which] may lead to very regrettable alterations of natural balances."[113] In 1962, the American biologist Rachel Carson published her famous *Silent Spring*, in which she condemned the ravages of pesticides on animals. It was also at this time that the first movements for modern organic farming appeared, mainly in Germany and England. Inspired by the agronomic principles theorized in the interwar period by the Austrian philosopher Rudolf Steiner, the German Ehrenfried Pfeiffer, the Englishman Albert Howard, and the Japanese Masanobu Fukuoka, advocated organic farming (*l'agriculture biologique* in French). The practice began to acquire momentum in the 1950s and 1960s as a reaction against the growing use of chemistry.[114] In France, the French Association of Organic Agriculture and the Nature and Progress Association were created in 1961 and 1964 respectively; in this context, in 1972 the French government chose to reform a 1943 law on the approval of toxic substances in agriculture to adapt it to new health and environmental risks caused by agricultural pollution. This reform, which was also inspired by the 1969 American decision to ban the import of French milk and cheese containing high levels of DDT, was part of a worldwide movement. In 1970, Sweden banned DDT, followed by the United States and France in 1972, and most Western countries thereafter. But DDT remains in daily use on other continents.[115]

* * * *

On the eve of the 1970s, the miracle and the hopes of the postwar period began to turn into a nightmare. Warnings against pesticide products multiplied in a manner similar to those that had accompanied widespread motorization and the rise of mass consumption. Governments were compelled to intervene, and companies were made to change their practices and markets. Nevertheless, the attraction of consumption limited the scope of these disputes. After a few years of economic crisis following the oil shocks, the image of a glorious and prosperous era emerged as the true expression of the times; it is no accident that the term "Glorious Thirty" was coined in 1979. However, an analysis of the setbacks to production and consumption forces a shift in gaze and change of focus in order to reconsider this period. The benefit of greater hindsight might justify rebranding the "Glorious Thirty" as the "Polluting Thirty" or the "Devastating Thirty."[116] Since the First World War, economic and scientific transformations have been considerable; the world of 1975 hardly resembled that of 1914 at all. While the pollutions that accompanied this evolution changed in nature, the same logics of expertise and oversight remained. At the same time, a change in regulatory and political resistance to environmental contaminations corresponded with the change of scale in the pollutions themselves.

10

The Politics of Pollution

Wars, energy systems, and consumer societies all reconfigured the polluted world and its political management. Although regulations were still rooted in nineteenth-century modes and practices—with some localized variations, a combination of judicial recourse and administrative prevention and controls spread across different levels of jurisdiction—methods of intervening to protect contaminated spaces adapted to industrial excess and new means of international mobility, all while struggling to keep up with new, toxic products and the unanticipated hazards they posed. If the rise of white lead embodied the manner in which authorities demonstrated their bias in favor of industrialization in the previous century, asbestos became—after 1914—one of the new flagship products that came to symbolize the fragility and inadequacies of existing regulations. Asbestos was initially regarded as a modern miracle; this fibrous mineral silicate was first considered a symbol of purity because it was the perfect fire retardant, when fire was one of the greatest scourges of large industrial societies. Asbestos made its way into workplaces, offices, schools, and houses. Estimates suggest that 174 million metric tons of asbestos were used in one way or another over the course of the twentieth century. It was responsible for extensive contamination and hundreds of thousands of deaths.

The early history of asbestos as a miracle product is a little oversimplified. At the beginning of the twentieth century, a work inspector named Denis Auribault reported on the dangerous nature of this substance in a study that warned against "the pernicious action of its dust," and the suspected deaths of at least fifty workers in a factory in the Condé-sur-Noireau commune in the Calvados department in France. The link between exposure to asbestos fibers and deaths in certain professions had already been established.[1] In the United Kingdom, the first cases of death caused by asbestos were documented in the 1920s. Numerous articles,

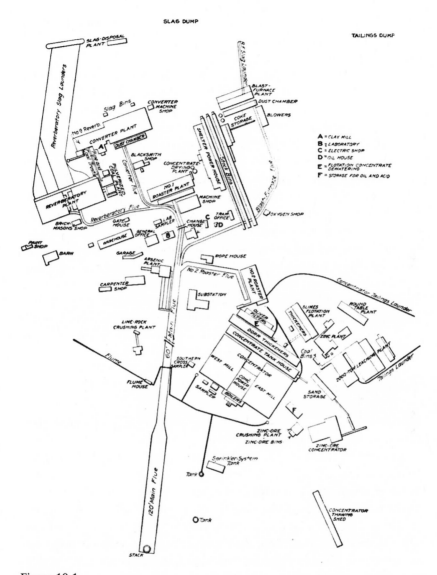

Figure 10.1

Anaconda Copper Factory at Washoe, 1916. Like many chemical industries, copper refining was extremely polluting, because of the acid and sulfurous gases that devastated the surrounding environment. The industrial scale of these refineries took on gargantuan proportions on the eve of the twentieth century. At the Anaconda Copper Company-operated Washoe Mine in Montana (United States), an expansive network of laboratories, workshops, and storage facilities was organized around the flue pipes (center), which converged on Anaconda Smelter Stack—which, at 178 meters, was then the tallest chimney in the world.

notably in German and English medical journals, dealt with the fiber's threat to the lungs. However, asbestos was not banned anywhere until the end of the twentieth century.

More than any other product, asbestos clearly demonstrates how a complex network of control and influence lay behind the management and regulation of polluting substances throughout the century. In continuing to apply nineteenth-century legal and administrative policies, industrialized states proved incapable of banning or controlling the spread of toxic products, as industrial lobbies gained influence and international trade intensified. As a rule, hazardous products were only eliminated when a more profitable substitute became available. Antipollution policies did exist, of course, as did various measures to protect the natural world, but their enforcement was sporadic and limited. However, by the 1970s, the new industrial and health risks of the "age of extremes" radically transformed the nature of pollutions. New hazards and better understanding of old hazards meant that concerns about contaminations were a more significant part of discourse for societies and states and forced legislators to revise regulatory systems. New environmental policies evolved, as disputes and ecological organizations grew in number. In this framework, the 1970s were a decisive turning point.

Weak Regulations

Prior to the 1960s, policies to limit and control pollutions were fragile. Even though international organizations such as the World Health Organization (WHO, 1945), the Food and Agriculture Organization (FAO, 1946), the International Labor Organization (ILO, 1919, reformed 1946), and the International Union for the Conservation of Nature (IUCN, 1948) were working to better understand the phenomenon, pollutions had for a long time remained a local or a state prerogative. The four principal regulation methods inherited from the nineteenth century were common law, administrative procedures, technological promise, and confidence in dilution. These endured with varying degrees of significance, depending on the time period and the country.

In the United States, both common law and local regulatory bodies became progressively less important, giving way to greater intervention from the federal government. In 1915, the Supreme Court declared that it was no longer unconstitutional for governing authorities to tackle industrial pollutions, a decision that set a precedent.[2] Such legal changes notwithstanding, the advent of the First World War weakened environmental initiatives and the conflict marked a pivotal turning point as the fight against smoke pollution lost momentum. All countries involved in the war witnessed a weakening of antipollution movements. In 1916, for example, the Supreme Court of the German Reich deemed that polluting the River Ruhr was a normal part of industrialization. The previous year, a plaintiff arguing over the destruction of his orchard had had his case dismissed from court; the judge ruled that it was irresponsible to endeavor to manage an orchard in an industrial city. Pollution had become an inevitable phenomenon, a natural part of industrialization.[3] The United States had seemed on the brink of resolving smoke pollution problems prior to 1914, but the funding to do so no longer existed during the Great Depression of the 1930s. In fact, any progress on the matter hardly went beyond the local setting. For example, in St. Louis between 1937 and 1950, the city's smoke commissioner, Raymond Tucker, succeeded in introducing a municipal ordinance that contributed to significantly reducing urban pollution by favoring the use of coke over its dirtier counterpart, coal. A similar improvement appeared to take place in Pittsburgh as a result of neighborhood protests that pushed local authorities to adopt measures to control industrial smoke.[4] The first piece of federal legislation regarding pollution addressed waterway pollution in the nineteenth century. It was not until 1955 that any federal legislation regarding air pollution was passed. The National Air Pollution Control Act, followed by the Clean Air Act in 1963, was particularly significant because it offered the first legal opportunity for the federal government to intervene in this domain.

In Europe, state intervention and public power varied according to social and political contexts. In France, municipal decrees on smoke fell into disuse during the Great War, but were reinstated thereafter to control pollution by the large suburban plants then in full development. The deadly fog in Meuse in December 1930 provoked numerous reactions in Belgium and abroad. Foreign reports of this event, however, relieved industrialists of all responsibility in favor of a more fatalistic vision. In 1933 this fatalism was further fostered by the Belgian prime minister, who wrote to the mining community that "even if it is important to prevent the atmosphere from being polluted by sulfur dioxide from factories, … we cannot advocate measures that would undermine the industry and take away their livelihoods."[5] Nevertheless, one consequence of the toxic fog was the Morizet law—named after André Morizet, a socialist senator and mayor of Boulogne-Billancourt—passed in 1932 to govern air pollution in France in response to increasing industrial smoke. Its effect, however, was limited; far from improving the situation, it instead hindered actions undertaken by municipalities by entrusting the prefects to manage the problem. Moreover, the law only applied to large industrial plants that were given a three-year period in which to conform to prefectural decrees.[6] The problem of smoke pollution was then overshadowed by both the tensions surrounding interwar politics and the great economic crisis of the 1930s. In France, as in Belgium and Germany, authorities hardly got involved, although certain large cities such as Lyon did take some initiative to protect the health of locals by negotiating limited compromises with industrialists. Globally, the political and social emergencies of the time, coupled with totalitarian dangers and mass unemployment, meant that concerns over pollutions were relegated to the margins of public debate.[7] The prefect in the Rhône department showed that this was indeed the case, in a report written in 1938 that noted "several factories ... have had to postpone the installation of smoke controlling devices, due to the economic slump and the precarious state of their treasuries."[8]

In France as in Great Britain, national laws governing air pollution were not introduced until the 1950s, in the wake of the deadly fogs in Donora, Pennsylvania in 1948 and London in 1952. The Great Smog of London accelerated the government's impetus to act. Hugh Beaver was chairman of the Committee on Air Pollution (1953–1954) that produced the Clean Air Act of 1956. Unlike previous laws, this Act contained a particular clause to include household fires, and introduced areas in some towns and cities where only smokeless fuel could be burned.[9] Yet another spectacular smog—this time in New York in 1953, which resulted in 200 deaths—brought about the Air Pollution Act in 1955 in the United States. These smog episodes had a residual effect around the world: in France, an interministerial committee was put in place and various events were organized to discuss the problem.[10] The Morizet Law was replaced by a law "on air pollution and smells" promulgated by the new Gaullist government in 1961. But this national legislation was largely inefficient, and Paris was really the only city concerned that cars were still not considered part of the problem.[11] This law was followed by a second, pertaining to water, its management, allocation, and preservation from pollution (1964).[12]

Generally, authoritarian regimes around the world accorded very little attention to pollution, and repressed complaints by residents. This was the case under Mussolini in the 1930s to protect the aluminum industry. Similarly, in Franco's post-civil war Spain, pollution remained a secondary concern. The country had long suffered political isolation and economic stagnation and it was not until economic growth began in the

1960s—accompanied by a hike in pollutions—that some laws were put in place by the central government to regulate air pollution (notably 1972, 1974, and 1975). Worldwide, environmental concerns usually came from opponents of a particular governing regime, and thus after Franco's death in 1975, environmental issues could finally be heard in the political arena at the national level as well as in autonomous regional administrations.[13] In Germany, before the Nazis' rise to power, the Weimar Republic was faced with the troubling fact that inflation and its negative social repercussions had led to a decline in the fight against smoke pollution, the result of a sort of official lethargy. The German bureaucracy was incapable of circumscribing pollution and most efforts were put toward improving furnace efficiency and installing purification techniques and filters to curtail toxic soot.[14] The status quo prevailed with Hitler's rise to power, even though some have thought that they could interpret Nazi policies as being ecologically significant, with the adoption of a law to protect animals as early as 1933, or the many speeches given on the protection of nature.[15] In fact, no great measures were taken to reduce pollution at all, with rearmament and preparations for war actually increasing pollutions at that time.[16]

In the East, the emerging Soviet Union and Cold War ideologies weighed heavily. After the 1917 revolution, and in spite of continuing internal troubles and the threat of civil war, Soviet authorities quickly focused on sanitation and public health issues. Health services were drastically overhauled, and beginning in the 1920s several decrees were made by the People's Commissariat for Labor to protect cities from industrial nuisances. For example, parcels of land were allocated for industry in order to situate it farther from populated areas. Moreover, Soviet scholars engaged in a series of measures and evaluations to calculate the rise in, and effects of, industrial pollutants. Various industrial health institutes were created, and numerous publications also discussed pollution as a concern. In 1935, the first Pan-Soviet conference dedicated to the fight against air pollution was held in Kharkiv, in northeastern Ukraine, followed by a second in 1938. There, representatives from the country's major health institutes and laboratories considered methods to determine air quality, and measures to be taken to protect the population. The backdrop to these explorations, of course, was the country's massive industrialization during which workers' rights were violated, so one might question the sincerity or centrality of such initiatives. Also around this time, specialized groups of health inspectors in numerous cities in the USSR were tasked with controlling air pollution.[17] The Second World War and the reconstruction period that followed it, under the rule of the Stalinist dictatorship, once again slowed down progress on pollution control initiatives. However, the 1940s saw Soviet authorities implement environmental legislation and standards for "maximum admissible concentrations" of toxic substances in air, soil, and water that were more stringent than those in the West. In 1949, for example, new factories could not be built, nor older factories reopened, unless health regulations were applied. A substantial amount of scientific work showed the harmful effects of pollutants on human health, and the Russian law of October 25, 1960, for the protection of nature included measures to combat pollution, comparable to those found in Western laws. In 1963, health monitoring became law, by federal order, in all republics of the USSR.[18]

During the Cold War period, in the authoritarian regimes of the East, like the GDR, pollutions and environmental problems were described and interpreted as the manifestation of the capitalist crisis and its obsession with profit. The technical progress associated with the socialization of the means of production must naturally solve the problem. In 1954, a law was passed in the GDR for the protection of nature, but it made no reference to pollution; pollution controls were adopted in the 1960s. At that point, scientific research and study on pollution commenced in earnest. This included an extensive inquiry into SO_2 emissions that was launched in 1965 and examined emission levels from 1,000 industrial plants in East Germany. In 1966, a decree from the Cabinet stipulated measures to be taken to limit air pollution and, at the end of the decade, other inquests attempted to evaluate industry's pollutant waste and its cost for municipalities, agriculture, and forestry. In the absence of pollution abatement measures, polluters were required to compensate victims: agricultural producers and urban dwellers. However, rather than penalizing polluting industries and requiring that they install preventive measures in their works, companies found they could anticipate and plan for damage payments and integrate compensation into their budgets and management. As a result, the installation of filters moved slowly; instead, local residents adapted to the presence of nuisances and to being compensated for them.[19]

This glut of legislation should not lead us into falsely thinking that environmental regulations became a priority. They remained on the back burner until the 1970s, and in the meantime tended to be very general, stopgap measures to address specific problems; in no way did they trouble themselves with hunting for the root of pollution problems or targeting dirty industrial sectors. In fact, the administrative framework

Table 10.1
European air pollution laws between 1956 and 1976

Year	Country	Title of law
1956	Great Britain	Clean Air Act
1960	USSR	Law for the protection of nature, including industrial pollution
1961	France	Law regarding the struggle against atmospheric pollutions and foul odors
1964	Federal Republic of Germany	Technical directives for air quality
1964	Belgium	Law for the struggle against air pollution
1965	Federal Republic of Germany	Clean Air Act
1966	Italy	Law against pollution
1967	Czechoslovakia	Clean Air Act
1968	Great Britain	Clean Air Act
1970	German Democratic Republic	Law for the protection of nature
1970	Ireland	Decree on the observation of air pollution
1970	Netherlands	Law for protection against atmospheric pollution
1971	Belgium	Decree for the purity of the air
1971	Italy	Decree for emissions thresholds
1973	Denmark	Environmental Protection Law
1974	Federal Republic of Germany	Federal law for protection against emissions and technical directives regarding the air
1975	Belgium	Royal decree for the struggle against air pollution for SO_2 and suspended solids
1976	France	Law regarding classified facilities on environmental matters
1976	Luxembourg	Law against atmospheric pollution
1976	Hungary	Law for the protection of the human environment

that regulated industrial establishments was still a distant legacy of the 1810 law. In France, legislation had been reformed with a law on December 19, 1917, which gave more power to third parties through a system of increased sanctions against polluting industries.[20] At the same time, however, the new law introduced the declarative regime, which allowed some industries to work outside the legislation's administrative scope. Technical improvement remained the focal point of French pollution measures. As a result, when European cooperation was enhanced, the fight against pollution was generally taken over by sectors responsible for promoting industrialization. The European Coal and Steel Community (ECSC, 1952) searched for ways to reduce industrial dust—but through technical means, since its overriding objective was to support the European coal and steel industries, to enable them to modernize, optimize their production, and reduce costs.[21] It was not until the 1970s that the European Economic Community began to deal with the environmental problems caused by the industry.[22]

As was the case with asbestos, industrial companies were infrequently subjected to binding regulations. Accusations of responsibility were difficult to prove amidst the vague and general notions on the protection of nature and air quality. This left public authorities essentially powerless to stop the spread of harmful substances; policies for the protection of public health failed on numerous occasions. In France, for example, the Eternit Company was founded in 1922, and from 1929 onward it provided 50 percent of asbestos cement (the principal use of the substance) consumed in the country, with production increasing tenfold between 1937 and 1974. Within forty years, the company was one of the world's leaders in the sector, becoming an enormous international aggregate of factories and mines in France (mainly Corsica), in its colonies such as Algeria and Senegal, and also in India, China, and Brazil. Over the span of fifty years, this "industry of death" contaminated environments, local populations, and workers at every step in the production process.[23] In 1962, the European Commission warned its member states of the dangers of asbestos, but no specific regulation to limit its use was introduced in France before 1975.

Constructing Uncertainty

The close relationship between states and increasingly powerful companies and their experts is a decisive factor in illuminating the continuing weakness of environmental regulations. Legislation and regulations are made in complex power relationships and are the product of discourse and acts, influences, and expertise that lead to decision-making. Within this rubric, the influence exerted by large multinational companies, whose power gradually exceeded that of the states, extended in tandem with their size. The rise of these large organizations with branches around the world—engaged in manufacturing in several countries—illustrates the quintessential challenges facing governments who would regulate against pollutions more stringently. In addition to their legislative influence through lobbying interests, these multinational companies are surprisingly nimble, and in the face of sterner pollution controls they relocate quickly and easily in search of raw materials and cheap labor. At the same time, these gargantuan corporate powers dispose of waste and pollutions from their production activities with impunity. Big American companies were especially responsible for initiating this internationalization; their practices contributed to the continuing exploitation of poorer countries and the constant displacement of pollutions, while trying to frame the debate and influence the way they were regarded. In the world of production-oriented and modernizing engineers, these corporations were subjected to constant challenges.[24] Faced with attacks and denunciations, companies invented the decontamination market and disseminated their technological solutions. In 1960, the Trade Union Chamber of the French Iron and Steel Industry created a Commission for the Fight Against Air Pollution within its Steel Technical Association. Some sectors, such as oil, established their own measures to counter criticism against their emissions and environmental misdemeanors. In 1949, for example, the French information department of the Esso company created the quarterly magazine *Pétrole Progrès*; equivalent journals appeared in the main industrialized countries. Richly illustrated, this periodical lauded the infinite potential of fossil fuels and celebrated the wealth of oil nations, all while carefully concealing the industry's ecological transgressions. When the topic of pollution arose more vigorously at the end of the 1960s, *Pétrole Progrès* focused on minimizing it by absolving the petroleum sector of responsibility and shifting the debate's attention toward the future of the "clean combustion engine."[25]

Industrialists' strategies to deny or obfuscate the reality of the pollutions they created were precociously and particularly well highlighted in the American case.[26] Indeed, many polluters acquired their techniques directly from the tobacco industry, which became a master at deflecting and misdirecting attacks against its products. The historian Robert Proctor has demonstrated the existence of a deliberate policy of generating scientific ignorance in popular, mainstream discourse about the effects of smoking. As scientific studies began to correlate lung cancer with smoking cigarettes as early as the 1930s in Germany, tobacco multinationals organized a defense based on uncertainty surrounding the medical results. In the early 1950s, when American scientists confirmed the link between smoking and cancer, representatives of four major corporations formed the Tobacco Industry Committee for Research. Their goal was to frame the tobacco industry in a more positive light and confront the negative backlash. The Committee for Research challenged the scientific evidence upon which critical studies were based; it intimidated researchers and policymakers; funded and produced "independent" research; lauded the industry's "clean" products; muddied the chain of causality; and infiltrated regulatory authorities. These and its other strategies were varied, energetic, and forceful.[27] The primary ambition was to challenge science as a final arbiter of truth and producer of formal proof. The tobacco industry's deliberate construction of ignorance—a practice since dubbed agnotology—bought it time by spreading doubt about the reality of the deadly effects of the products it marketed. Polluting industries followed the tobacco industry's lead, rejecting claims about the hazards posed by pollution or their products, and blocking any intervention that threatened their production model.[28]

The making or construction of uncertainty seems to apply to all of industry's harmful effects. When acid rain emerged as a problem in the 1960s, industry was accused of destroying American forests because of its introduction of smoke filtration systems in factories after the London and New York smogs in 1952 and 1953. Ironically, by filtering out the fuel particles, which had the power to dissolve the acidity by chemical combination, the sulfur dioxide suddenly increased then condensed in the atmosphere, falling to the ground in rainfall. Local pollution problems turned into regional problems. Whereas between 1969 and 1975, a great many articles addressed this problem, explicitly linking industrial pollution and forest destruction, industrialists followed the manual provided by the tobacco industry, claiming that the scientific knowledge was insufficient to provide certainty on the issue or justify regulatory measures. Similarly, the destruction of the ozone layer, caused by chlorofluorocarbon (CFC) pollution and identified in 1970, witnessed the same strategies deployed the following year. Manufacturers asserted that the hole found over Antarctica was due to natural variations. Almost verbatim, challenges to the abnormal occurrence of epidemic diseases around chemical plants were framed in the same way.[29]

The Manufacturing Chemists' Association (MCA) was frequently at the heart of networks that merged industrial expertise and lobbying. Founded in 1872 by sulfuric acid manufacturers, the MCA experienced considerable growth around 1900 after the American chemical industry took off: in 1902, DuPont moved toward chemical production, while Dow Chemical and Monsanto—founded in 1889 and 1902, respectively—and the multinationals Hooker Electrochemical and Union Carbide helped turn the United States into the world's leading producer of chemicals. After the First World War, as the chemical industry experienced a new boom, the MCA mobilized to counter growing suspicions that many of its products were toxic hazards.[30] In the 1920s, confronted with claims that tetraethyl lead, the basis of the new leaded gasoline, posed an especially toxic threat, the MCA deployed its network of expert researchers: Robert Kehoe, a university professor, and Anthony Lanza, who headed the Industrial Hygiene Division of the Public Health Service and was a consultant for the Metropolitan Life Insurance Company, dismissed concerns and vouched in turn for the product's safety; both also insisted that tetraethyl lead's economic benefits be taken into account—and, above all, that chemicals should be deemed harmless until unequivocal proof of their toxicity was established. In 1964, thanks to its effective lobbying strategy, the American industry produced 224,000 metric tons of leaded gasoline, three times more than in the 1940s.[31] Debates over the safety of CFCs, renamed "freons" by the industry, followed a similar trajectory. Thanks to the money from the MCA, Kehoe founded the Kettering Laboratory in 1930, which established itself as the most influential research institute on the toxicity of industrial chemicals. However, historical studies show that these influential experts and the groups that backed them were perfectly aware of their products' deleterious effects. Cover-ups and obfuscations masked the real hazards in order to save time and avoid regulations. When a serious industrial incident occurred in 1930–1931 (the Hawks Nest Tunnel disaster: as a result of digging a tunnel, nearly a thousand workers—mostly black—died of silicosis), industrialists created the Air Hygiene Foundation (AHF), which became a particularly effective lobbying body to deny the responsibility of industries in pollution and worker mortality. In the development of legislation, industries often ensured that their interests were preserved. In the United States, the 1947 Federal Insecticide Fungicide and Rodenticide Act, the 1948 Water Pollution Act, and the 1955 Air Pollution Act were each diluted—their bite removed—at the behest of industry. Because of the chemical industry's growing influence, and despite numerous reports that demonstrated the toxicity of products (notably from the oncologist Wilhelm Hueper), until the late 1960s manufacturers tended to enjoy legal success when their products were challenged. Thanks to its experts and growing lobbying clout, industry succeeded in blurring the meaning behind scientific results and imposing uncertainty about responsibilities, demanding additional time, and highlighting technical improvements as the inevitable solution. This was the paradigm—from the 1920s through the 1950s—for lead compounds in paint (considered by the Lead Industry Association, founded in 1928, as an acceptable risk to be weighed against their profits), and from the 1950s to the 1970s for plastics, particularly vinyl chloride (VCM) and polyvinyl chloride (PVC). For the latter, historians Gerald Markowitz and David Rosner uncovered how the chemical industry concealed from the general public what it knew about the hazardous nature of its products and their carcinogenicity right into the 1970s.[32]

The case of asbestos, however, was the most edifying. This magic fiber had been a primary health concern; warnings prompted numerous branches of environmental research. At the same time, however, its manufacturers adopted commercial strategies to disseminate their products while banishing or disregarding their health and environmental impacts. With the opening of many mines and factories, world production rose

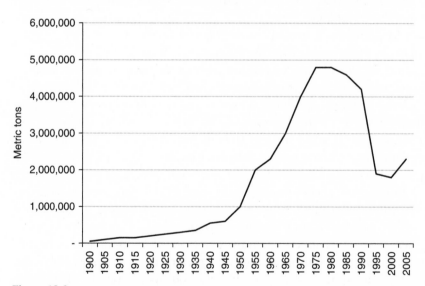

Figure 10.2
Global asbestos production, 1900–2005. *Source:* based on Jock McCulloch and Geoffrey Tweedale, *Defending the Indefensible: The Global Asbestos Industry and Its Fight for Survival* (Oxford: Oxford University Press, 2008), 14.

from 31,000 (1900) to 570,000 (1940) and 2.2 million metric tons (1960), reaching its peak in the 1970s: nearly 3.5 million metric tons were produced annually, mainly in Canada (1.5 million metric tons), the USSR (1 million metric tons) and South Africa.[33]

The growth of the industry ran parallel with a growing body of medical knowledge that warned about the product's harmfulness. As early as 1898, a British factory inspector identified workers' pathologies associated with asbestos production, followed by similar information in 1906 from the French labor inspector Auribault. In the 1920s, several medical descriptions pointed to asbestos as a pathogen and in 1931 the British government introduced dust control across the industry. In the 1930s, doctors in Germany and Japan reported that the fiber was a primary catalyst for lung cancer. During the war, Germany even recognized cancer of the pleura, a thin membrane around the lungs, as an occupational disease. In 1955, an epidemiological survey of South African mines confirmed the indices and extended the risks to surrounding populations: mesothelioma—or "asbestos cancer"—was highlighted in these studies. And yet, after 1945 production continued to grow; governments strongly encouraged asbestos use, which enjoyed growth in fields as diverse as construction, agriculture, and the electrical and automotive industries.[34] Some sites were particularly affected. The small town of Casale Monferrato in Italian Piedmont was one important example. A first asbestos plant was installed in 1906, after which it became the largest Eternit plant in the world. Asbestos was presented as a symbol of development and prosperity. Its dust was everywhere: in schoolyard slabs and corrugated roofs. Before it was closed and destroyed in 1986, the plant had long contaminated the environment and claimed thousands of victims.[35] The paradox of inaction can be explained by the "counter"-activism inspired by industrialists that took advantage of the trust society and government placed in industry to be good citizens and neighbors. To counter suspicions against asbestos, manufacturers activated their research networks by creating various institutes. The Air Hygiene Foundation, renamed the Industrial Hygiene Foundation in 1941, commissioned scientists like Lanza to produce counter-reports and promote the notion of an acceptable exposure threshold, which became an essential tool for governing toxic substances and resolving the disputes they raised. Canada was a lodestone for the asbestos example.[36] Although the hazard of asbestos was no longer in doubt after 1952–1953, industrialists ensured further increases in production by hiding and manipulating toxicity data. This obfuscation has since been established.[37] Even after the scandals and revelations of the 1970s, denials persisted, and it was not until the late 1970s that the major industrialized countries timidly began to regulate asbestos, via a supposed "controlled use."[38]

Inasmuch as it constituted a potent force, industry did not have a monopoly on toxic expertise. The deepening gaze on pollution was also shaped by the development of toxicology and ecological science. Historian Christopher Sellers has shown that post–Second World War scientific environmental thinking has its roots in the industrial toxicology of the corporate world.[39] Moreover, between the founding of the British Ecological Society in London in 1913 and the media recognition of this science in the 1970s, ecology changed from a simple branch of botany and zoology into a full discipline, complete with a coherent lexicon and accepted systems and praxis among a growing body of practitioners. Ecology designated both knowledge about the natural world and its functioning, but also a set of techniques and knowledge enabling a more rational and effective exploitation of the environment.[40] The study of natural imbalances and interactions between substances and living species was at the heart of this science. In 1935, Arthur Tansley defined ecosystems as "the basic units of nature on the face of the earth," elements that are constantly changing under the impact of human intervention.[41] There was tension, however, in the world of ecologists. Following the work of the American Eugene Odum and his famous treatise *Ecology*, published in 1963, a branch of militant ecology—typified by the bestsellers of biologists Rachel Carson and Barry Commoner—specialized in highlighting disturbances introduced by human activities into ecosystems.[42] For other ecologists, on the other hand, anxious to justify the scientific character of their discipline while resisting any militant (or political, anti-industrial) perspective, ecology was first and foremost a natural science, "neither the science of pollution nor the science of the environment, but that of the global march of the living world."[43]

In addition to the new ecological science, medicine remained a key source for toxic expertise. In industrialized countries—in Los Angeles, London, and Paris—many doctors began investigating the effects of air pollution, and helped to build a case against pollution by reinforcing its public visibility. However, the situation varied greatly between the United Kingdom—where regular measurements of air quality were required starting in the interwar period—and France, where this type of scientific measure was slow in coming. In the 1950s, French doctors learned about the work of their colleagues in the English-speaking world—such as Professor Paul Kotin of Los Angeles—who identified early on the potential for cancer risks from prolonged exposure to carcinogens at low doses.[44] From the 1950s onwards, scientific measures and assessments of pollution became more pronounced, as shown by the creation in 1958 of the Association for the Prevention of Air Pollution (APPA), which opened branches in the major French provincial cities—including Lyon and Marseille—but also at particularly polluted sites such as Lacq. However, scientific surveys spread slowly; their regulatory influence was also slow. Industry was more agile: in France, spokespersons for major polluters quickly integrated themselves into the bodies responsible for investigating pollution. Major energy companies, such as EDF and the Charbonnages de France, were represented on APPA's board of directors; industrialists began organizing their own association in 1960. The Comité d'Action Technique Contre la Pollution Atmospheric (CATPA) [Committee for Technical Action Against Atmospheric Pollution], whose first president, the engineer Louis Armand, was also the president of the highly polluting Houillères de Lorraine, was also joined by a more technical body, the Interprofessional Technical Center for Air Pollution Studies.[45] In fact, even if the medical professions were divided and not very homogeneous, the medical vision was often imbued with a technological approach that was based on chemistry and engineering sciences, relegating epidemiology to the background. That hierarchy hindered the accountability surrounding factories' pollution.[46]

Mobilizations

Ideas about nature—pristine or polluted—evolved a great deal during the twentieth century. Incongruous interpretations persisted, nevertheless. On the one hand, nature was viewed as an infinite, inexhaustible resource, whose very presence justified human exploitation and sacrifice on the altar of industry. On the other, nature was perceived as fragile and limited; pollution and overexploitation threatened its sustainability. Such incommensurable readings of nature had profound impacts on how pollutions were understood, especially as "pollution" entered the popular lexicon during the 1960s. Between the two world wars, pollution continued to be a sign of prosperity, full employment, and modernity, a continuation of ideas created in the great industrial basins of nineteenth-century Europe. Chimneys spewing black plumes were a common—and positive—feature of publicity that championed growth. That vision did not change until the late 1960s, when the factory began to be associated with aggression and harm, and blue skies gradually supplanted the images of darkened industrial clouds as a more desirable aesthetic.[47] Former senior officials, such as Philippe Saint-Marc, who were active in planning policies after 1945, turned to environmental issues. In 1972, Saint-Marc drew up a widely publicized Charter of Nature, which collected 300,000 signatures, before publishing in 1975 a book on pollution that received a great deal of attention.[48]

Even if many scientists were bound to the protection of big and polluting industries, many other scientists denounced polluting practices. Whistleblowers contributed significantly to popular media's access to questions of industrial risks and pollution throughout the twentieth century.[49] In the USSR, some scientists warned against the scale of environmental pollution and destruction produced by forced industrialization. In spite of the repression they endured, they formed the embryo of a community of naturalists who pushed for the preservation of nature.[50] In the United States, where scientists enjoyed greater social standing and the support of a powerful media regime, some of them worked extensively on the recognition of pollution and were vocal about changing regulations. For example, while the oncologist Hueper contributed to the banning of many chemicals whose role in the onset of cancer he denounced in the 1920s, Dr. Alice Hamilton, the first female professor at Harvard University, became a prominent public intellectual as her work on public health in the workplace galvanized important changes in labor laws. Hamilton was also a catalyst in launching professional epidemiology as a critical means of analyzing environmental pollutions. She denounced the use of white lead in paint and tetraethyl lead in gasoline by combining clinical examinations, toxicology, and epidemiological field investigations from 1910 onwards. She also showed that the mortality rate in most American industrial sectors was higher than their European counterparts.[51] At the same time, in Great Britain, the first Medical Inspector of Factories, Thomas Morrison Legge, popularized the notion of "industrial diseases," and contributed to a certain recognition of the relationship between industrial diseases and pollutions. The Taylorist organization of British industry ensured, however, that much of his legacy was forgotten after his death in 1932.[52]

During the Glorious Thirty, the industrial boom provoked resistance to and questioned the benefit of controlling pollutions. Henry Fairfield Osborn's pioneering work, *Our Plundered Planet*, offered some resistance and rationale. The Frenchman Roger Heim also weighed in with *Destruction and Protection of Nature*, a few years later, in 1952.[53] The work of these two renowned scientists testified to the important role that naturalists played in the fight against pollution. But the most decisive milestone was the work of biologist Rachel Carson. In the late 1950s, inspired by cancer research—including the work conducted by Hueper—Carson focused on the impact of pesticides, which led her to demonstrate unusual bird mortalities as a result of pesticides in the environment and the spread of these products in agriculture. In her resounding book, *Silent Spring* (1962), she accused the chemical industry of misinformation and explained how pesticides contributed to human cancers. The book quickly became a touchstone for the prohibition of toxic substances in the environment; in short order it was translated worldwide. Carson was strongly attacked by the chemical industry—especially by DuPont de Nemours, the main manufacturer of DDT—but she nevertheless managed to convince the public and policymakers. Political momentum against pollutions began to hold sway; Carson's work was instrumental in the 1972 ban of DDT in the United States, which was followed by concomitant bans in most Western countries.[54]

At the same time in the United States, probably the country most affected by industrial pollution between 1945 and 1970, other scientists introduced new charges against pollutions. In the late 1950s, the geochemist Clair Patterson began to study the concentration of lead on the earth's surface and showed that it had greatly increased since the dawn of the industrial age. Leaded gasoline, insecticides, food can solders, water service pipes, kitchenware glazes, and paints were especially responsible for its spread. In 1965, Patterson published an article in which he tried to warn the public about the hazards associated with lead additives and launch a public opinion campaign against their industrial use. Naturally, Patterson ran up against stern opposition from the lead industry, and he fell victim to their lobbying and smear tactics. In 1971, he was excluded from the National Research Council research group that investigated lead in air pollution. Nevertheless, Patterson's warnings prompted the Environmental Protection Agency (EPA) to announce in 1973 the reduction and elimination of all lead additives in consumer products.[55]

During the 1960s, pollution was an inextricable feature of technological discourse. Authors such as Ernst Schumacher and Ivan Illich developed a concerted case for appropriate, user-friendly, and clean technologies (as distinct from authoritarian and polluting technologies).[56] This debate enjoyed considerable intellectual fervor: innumerable publications and experiments fostered transformative practices and in the face of the flood of consumer goods, people turned to simpler, handmade tools. Many advocates contended that these "soft technologies" would break free of their "marginal ghetto" and conquer the world.[57] Large-scale systems required a kind of centralized authoritarianism in their manufacture and implementation. Their control demanded standardization and their widespread adoption and acceptance. In contrast, smaller, unstandardized technologies represented a kind of social chaos: a decentralization of control, power, and wealth accumulation. As a result, large-scale systems opposed democratic technologies: ones that were soft, appropriable, and produced little pollution. During this period, technological gigantism was widely denounced in numerous contexts; in airports, for example whose construction inspired more and more opposition and conflict, as their number increased after 1950. In addition to noise, airports' environmental impacts introduced new branches of expertise to assess the extent of their nuisance. Major airport construction projects in Roissy (near Paris) and London, as well as in Tokyo and Osaka, encountered considerable opposition. In 1975, an OECD report compared airport emissions to factories, creating pollution, nuisances, and environmental risks through their expansion.[58]

Some economists also denounced the fallacy of perpetual growth. Whereas economic theory inherited from the nineteenth century did not reckon with growth's deleterious effects on nature, new interpretations wrestled with the question of negative externalities, their costs, and their role in market disruptions. In the context of capitalism's crisis during the Great Depression of the 1930s, pollutions emerged not only as a cost, but a cost that distorted the realities of economic growth, multiplied the negative externalities, and favored the polluters to the detriment of others. For example, the British economist Arthur Cecil Pigou proposed introducing taxes on pollutions in order to restore market equilibrium and "price truth" by internalizing the external effects in the price of products.[59] Similarly, in 1950, the German-American economist Karl William Kapp showed how the activity of private enterprises induced social costs not taken into account by the prevailing economic theories and their price systems. Air and water pollutions, the decline in biodiversity, and the deterioration of working conditions and living environments never appeared on the balance sheet beside the production methods that birthed them. Instead, those costs were carried by society.[60] This debate gained traction at the end of the 1960s. In 1967, the English economist Ezra J. Mishan published *The Costs of Economic Growth*, an ambitious treatise that publicized Nicholas Georgescu-Roegen's and Herman Daly's heterodox theories of the bioeconomy. Pollutions were analyzed as a brake on further development, a threat to the economic system itself.

In a few years, the question received widespread analysis from economists; even critics had to contend with pollution as a burden to growth. Some economists, such as Jean-Philippe Barde, sought to construct an environmental economy that would take the form of an extension of the dominant neoclassical economy. On the contrary, others believed that the scale of pollution and the environmental issue made it necessary to deconstruct this analytical framework in order to build a new economic science. This was the point of view of René Passet, Georgescu-Roegen, and Bertrand de Jouvenel, who proposed replacing political economy with political ecology. Their work stressed the importance of analyzing limits to growth and adopting lessons from the life sciences to construct what some economists called a bioeconomy, what others called ecological economics.[61] Ignacy Sachs proposed building an environmental political economy that would make it possible to identify the winners and losers of the ecological game and the deterioration caused by the circulation of pollutions.[62] The publication in 1972 of the famous Club of Rome report contributed to the media and political visibility of these debates, which had previously been restricted to specialists.[63] *The Limits to Growth* and the entry of ecological economics into mainstream discourse forced many economists to turn to the environment and integrate pollution into their analysis models. In the United States, Wassily Leontief was one such example. Ambitious studies analyzed and modeled the economic costs, influence, and impact of pollutions. Jacques Theys, then a young sociologist, embarked on a study on the impact of automobile pollution. He was not alone in treading new paths—modeling air pollution in Lorraine, studying the hierarchies of costs associated with pollutions, and so forth, spurred many inquiries into pollutions and the economy during the 1970s, although curiosity and commitment to addressing these concerns ebbed in the following decade.[64]

Post–World War II globalization meant that this evolution in economic thought did not affect only Europe and the United States. As Japan rapidly converted to chemical agriculture—after its introduction by the American military—farmers readily adopted pesticides and herbicides as solutions to increasing yields and reducing intractable challenges to healthy crops. At the same time, Japanese doctors and consumers raised concerns about the chemicals' effects. Pursuing the first moves toward agriculture without chemical inputs since before the war, the Buddhist doctor Giryo Yanase unsuccessfully warned authorities about the dangers of the new chemicals in the 1950s. Similarly, Masanobu Fukuoka, a training microbiologist employed in the Plant Inspection Division of the Yokohama Customs Office, resigned from his post to protest against agricultural pollution. He subsequently turned to cultivating the land; in his 1975 book, *The Revolution of One Strand of Straw*, he invented "natural agriculture." The book's denunciation of scientific agriculture and its pollutions was based on a unified concept of man and nature nourished by Zen, Taoist, Shinto, and Buddhist traditions.[65] During the 1970s, movements for organic farming were emerging; the writer Sawako Ariyoshi supported them by publishing in the major daily *Asahi* a serial novel titled *Fukugō Osen* (Complex Pollution). It was also at this time that the system of teikei—or direct sales—was introduced by urban middle-class consumers to access healthy food products.[66]

In addition to scientists and consumers, certain professional societies or community groups who were victims of pollutions organized to defend themselves against environmental aggressions. In 1920, French beekeepers published the newspaper *L'Abeille de France* (*The French Bee*); after 1945 it appealed to the state against what it perceived as the new agricultural chemistry's transgressions. "On all sides," it proclaimed in 1946, "we hear the cries of alarm that have been rightly raised by beekeepers against the increasing use of agricultural insecticides."[67] French beekeepers demanded standards, labeling, and education to teach better spraying methods. But beekeeping—considered an archaic, anachronistic practice, an unprofitable activity mainly practiced on a small scale by amateurs and, as such, devalued by the weight of large agricultural producers—was voluntarily sacrificed in favor of increasing rapeseed yields. In the same way, fishermen's complaints were constantly resisted and marginalized after 1945 in the name of industrial production. As with beekeeping, industrial prosperity was a question of scale: big industries enjoyed big concessions from regulators. Sport and recreational fishing lacked the same kind of political clout.[68] However, in keeping with nineteenth-century historical traditions, fishermen's associations often played a pioneering role, learned to assess links between pollutant discharges and fish mortality, and were empowered to manage the damages and interest paid by river polluters. Meanwhile, the number of fishermen rose to nearly three million in France in the 1960s.[69]

From Inequalities to Environmental Injustices

At the intersection of regulatory inertia and polluting industrial activities, the question of environmental inequalities emerged. Urban dynamics, shaped by the automobile and large development projects, disrupted ecosystems and led to new spatial and social reconfigurations. The impacts of pollutions were played out on various scales, which increasingly sacrificed vulnerable classes and regions.[70]

Although it is difficult to draw up a complete overview of industrial infrastructures and how their relocations related to pollutions—since there are other factors, such as the cost of labor and changes in transport infrastructure—there is irrefutable evidence to suggest that some areas were sacrificed by hosting concentrations of the most polluting industries. This was particularly the case in the United States, where segregation is especially strong; Chicago and Pennsylvania's industrial cities serve as poignant examples. In an attempt to understand the social dynamics of environmental inequalities, Andrew Hurley traced the manner in which industrial pollutions exacerbated inequalities in Gary, Indiana. After the establishment of US Steel in 1906, Gary began to grow rapidly. At the beginning of the twentieth century, Gary was a small town on the shores of Lake Michigan, a few miles from Chicago. US Steel built the largest steel plant in the world; Gary became one of the most polluted cities in the United States. In 1939, an industrial front of more than 10 kilometers was built along the lake. The heavy pollution that came from its works unevenly affected the population. It mainly affected the workers—almost all of African American origin—but pollutions and their harm trickled out from there: to the working-class neighborhoods around the factory, composed of immigrants, then to communities populated by white workers, and finally to middle-class, white neighborhoods. Gradually, from 1945 to 1980, avoidance strategies were implemented by the richest sectors of the population, much to the detriment of the poorest who remained close to the sources of pollution. In Hurley's reading, the rise of environmental consciousness was a class-specific process.[71] If such research empirically demonstrates environmental injustice as an historical phenomenon, it is not easy to document it precisely. There are many examples, such as the "Love Canal" case revealed in 1976: after the closure of the Hooker Chemical plant in 1953, working-class subdivisions and a school were built on the site of a former toxic landfill.[72] As the chemicals leached into peoples' homes, contaminations became commonplace. Air quality measurements in the Louisiana chemistry corridor (Cancer Alley), between Baton Rouge and New Orleans, also indicate the incidences of higher levels of pollutions that align tidily with class geographies: the poorest populations live in the most polluted places.

Many works confirm that populations who benefit the most from industrial production suffer the least from its pollutions. In the same

Table 10.2
Rate of pollution scaled geographically and stressing the exorbitant exposure
burden suffered by the African American population in Convent, Louisiana

Location	African-Americans as percentage of population	Discharges of pollutants (kg per person)
United States	12.1	3.2
Louisiana	30.8	9.5
"Cancer Alley" Parishes, Louisiana	36.8	12.2
St. James Parish, Louisiana	49.6	163.3
Convent, Louisiana	83.7	1032.8

Data drawn from the 1995 Toxic Release Inventory. *Source:* based on Gerald Markowitz
and David Rosner, *Deceit and Denial: The Deadly Politics of Industrial Pollution*
(Berkeley: University of California Press, 2002), 265.

vein, those most exposed to environmental hazards tend not to benefit
from its riches. There is a long-standing relationship between inequality and pollutions. The management of nuisances follows a differentiated topography that subjects the most vulnerable social groups to the
most deleterious environments; wealthy classes, who enjoy greater social
and geographical mobility manage to segregate themselves from harm
through the construction of income-based enclaves in safer fringes of the
city, industrial reorganizations, and access to open spaces.[73] This phenomenon can be analyzed at different levels because municipal, regional,
and national variations offer different perspectives, which have become
all the more pertinent since the post-1945 globalization of the economy.
Indeed, environmental inequality trajectories transcend even national
boundaries: in colonial and postcolonial relationships it became possible to transfer the most unhealthy and dangerous industries to countries
which do not set such stringent environmental standards of production
as the Western (or Northern) countries now do.

Developing countries frequently surrender their subsoil to the control
of large Western groups, whose mining operations—under looser controls than they might experience at home—have become critical sources
of highly polluting activities and bereft lands. For example, the Belgian
Congo—which became Zaire after independence (and now the Democratic Republic of the Congo)—is the archetype of a colonized country
pillaged for its mineral wealth and the continuity of that exploitation

after decolonization. In this model, unsafe and unsavory extraction practices sacrifice workers' health and the health of neighboring populations. From the beginning of the twentieth century, Belgian, British, and American companies set out to conquer the many geological riches of its best-endowed region, Katanga, where tons of copper, cobalt, gold, diamonds, manganese, uranium, tungsten, tin, and tantalum were extracted. The cruelest form of extraction capitalism was deployed in Katanga's mines: semi-forced labor in appalling conditions. The mines' African workforce grew from 8,000 miners in 1914 to 42,000 in 1921, while the other regions of the Congo employed close to 100,000 miners under conditions that approximated slavery. Exploitation intensified during the Second World War. The number of laborers rose from 500,000 to nearly 1 million between 1939 and 1945; after the war, Congo became the world's third largest producer of copper, contributing to the devastation of the environment.[74] The conditions in the Congo were hardly an exception; similar conditions existed in most of the raw material extraction areas of the developing world. The uranium mines of the French African empire (Niger, Madagascar, Gabon), for example, offered comparable conditions.[75] In Vietnam, specific rights were accorded to extractive international corporations that permitted the exploitation of resources in complete disregard of workforce health and the pollution caused by the extraction and processing of ores. From 1889 to 1930, mining law (derived from the Mining Act of 1810) was amended to ensure maximal resource exploitation despite significant environmental damage, including to the water system. In addition to inert waste, the devastation of mining regions came from heavy metal dusts, considerable volumes of cyanide, hydrochloric acid, and quicklime discharges. Mortality rates linked to these pollutions were considerable; independence did little to reduce the high death rates in the mines.[76] Mining typically occurred out of sight of larger populations, but it was also not limited to colonial or postcolonial spaces; removed from centers of power and attention, mining was also prominent in developed countries. In the United States, mines caused irremediable pollution that affected the health of local populations and irreversible damage to ecosystems.[77]

In South Africa, the apartheid regime after 1948 took race, class, and environmental exploitation to its logical extremes, which translated into stark environmental inequalities. Pollutions were concentrated where vulnerable populations lived. In spite of environmental laws that followed in the technical spirit of those passed in Europe or the United States—in 1956 (water law) and 1965 (law on air)—the South African variants established differentiations on their stringency and enforcement that followed spatial lines. Public health and ecosystem protection were prioritized in some areas and rejected in poorer, black areas. As a result, in spite of numerous regulations, pollutions persisted because they were knowingly permitted in areas of social and political vulnerability. Apartheid ensured that these inequalities, and the severity of the pollutions, were kept more or less invisible, at least until the beginning of the 1970s, when the scale of pollutions and their deleterious effects affected the country as a whole.[78]

The geographic dispersion—of pollutions and class—reinforced social inequalities and their dynamics. Environmental well-being was long considered a postmaterialist issue, a luxury afforded only to more affluent sectors of society. The historical record suggests otherwise. Pollutions also appeared in the preoccupations of the workers and the working class, in both urban and rural settings. This pioneering field of environmental history research draws attention to the existence of many historical social struggles by subordinated groups keen to protect the environments in which they lived.[79] It is not always easy to reconstruct popular representations of pollution. Voices of the grassroots instigators were frequently drowned out by the deluge of propaganda (advertising, pro-chemical demonstrations, and the like) that denied or dismissed their suffering, while the pollutions were hailed as symbols of progress. However, several life stories and autobiographies illustrate the manner in which the poor and disenfranchised were not at all indifferent to the environmental nuisances that surrounded them. In France, the study of workers' writings published during the interwar period shows a strong anxiety about the contaminated air, and the desire to regain contact with an unsoiled nature. For example, Maurice Lime recounted the story of the life of a child in Metz during the Great War. He was the son of a railway worker who became a miner: "From here to the Moselle [River]," Lime wrote, "the mines first—then the steelworks—sully the Orne [River] and kill its fish, while the emissions from the blast furnaces kill the vines and make the forests languish." In *Faubourg*, published in 1931, Lucien Bourgeois presented the trajectory of a young working-class couple who settle in the suburbs, where they "breathed the pure air with joy, because the air from the town from which they came had been foul." In the story of his youth in Lyon in the 1920s, the libertarian writer and worker Georges Navel also described the "ugliness of working-class neighborhoods," with "their high chimneys and line of leprous facades, … shutters, shop fronts: blackened by smoke." The proletarian and socialist writer Albert Soulillou recalled the landscapes of the industrial suburbs of Paris, home of the "ugliest plain in the world on the side of La Courneuve and Stains." Soulillou described the riverways sacrificed to industry: a "disgusting brook which killed its beautiful grasses" and which, once arrived at Saint-Denis, "came out black, thick, bloody, ashamed, so ashamed that it was no longer moving."[80]

Even if trade unions and their struggles played a decisive role in improving the working conditions inside French factories—as recognized by the establishment in France of the Committees of Hygiene, Safety, and Working Conditions (CHSCT, 1947)—they seemed, on the other hand, to have been less interested in outside pollutions or the more global ecological stakes associated with industrial production. This observation can, however, be put into perspective as the concerns of the trade union organizations varied according to the context, the economic situation, the extent of the social conflict, or even the political opportunities that were presented. In many cases, union struggles cut across environmental struggles and created a real "labor environmentalism."[81] During the 1960s and 1970s, the United States experienced many conflicts led by oil, chemical, steel, and agricultural labor unions against the threats that waste and pollutions posed. In 1973, a big strike against Shell was supported by conservation organizations. Workers' health and the fight against pollutions made common cause in the face of a large company deemed to be predatory—both against the environment and against its labor force.[82] Indeed, American labor and environmental movements forged an early alliance; that bond only started to fray in the 1980s at the height of Ronald Reagan's presidency.[83] Alliances between environmental and worker organizations were also established in the fight against pesticides.[84] Anthony Mazzocchi, a prominent labor leader, embodied this rapprochement between environmentalism and trade unionism in the United States. Starting in 1953, his continuous and determined action to combine occupational health and environmental health within the chemical industries helped to force the subject of workers' and environmental health onto the government's agenda. In no small measure did his efforts (and those who worked with him) spearhead the move toward the monumental Occupational Safety and Health Act of 1970.[85] Similar forms of labor activism that mobilized against pollutions occurred in Italy, France, and Canada, where they also won some hard-earned results.

A brief case study of vinyl chloride illustrates the alliance that was built between occupational health and pollution. Internationally, the International Federation of Chemical Workers' Unions (and in particular its secretary Charles Levinson) had been raising concerns about PVC since the early 1970s, while in France doctors had expressed worry (for example, Groupe information santé), followed by the chemistry federations of the CGT and the CFDT. The protection of health in the workplace and the denunciation of pollutions were also raised by other trade unions, who increasingly challenged industry's tactics of monetizing risk. Occupational health was regarded as an employer's medicine. Instead, trade unionists wanted to invent another expertise, one that worked more closely with and directly for laborers. Such tensions came to a boil in a number of social and environmental struggles, such as those of Rhône-Poulenc in Vitry (1970) and Ugine-Kuhlmann in Pierre-Bénite near Lyon (1971), as well as that of migrant workers in the Penarroya lead factory in Lyon (1971–1972), and another that began in the asbestos company Amisol in Clermont-Ferrand in 1974. In all cases, these struggles led to demands for the invention of another expertise, in partnership with the workers.[86] In the early 1970s, after decades of fruitless dialogue, Japanese trade unions also succeeded in teaming up with neighboring communities to confront the chemical pollution that had caused Minamata disease. Rejecting industrial corporatism, unions ended up denouncing the chemical industry's misdeeds; they supported public complaints against pollutions, and pushed employers to recognize and pledge assistance for occupational diseases and to diminish industry's environmental transgressions. Indeed, Japanese society was more than ready to demonstrate greater regulatory awareness and responsibility for industrial pollutions. By the 1970s, a number of cases had become mainstream issues and causes of considerable concern: mercury poisoning at Minamata (1973), the cadmium river in the Toyama prefecture (1970), and the asthma "epidemic" suffered by the population in Yokkaichi (1972).[87]

1968: The Turning Point

The expansion of trade union mobilizations against pollutions was indicative of a groundswell that gained widespread support around 1968 and the years that followed. A plethora of disasters captured media and public attentions: the oil spill caused by the sinking of the Torrey Canyon off the coast of Cornwall in 1967; the massive insecticide spill in the Rhine in 1969, which spanned more than 600 kilometers and killed about 20 million fish, prompting outrage; and others. Both in the Soviet world and in the West, mobilization spread as vast campaigns against urban fumes, pesticides, and discharges of hydrocarbons, lead, or asbestos all emerged.

The first major mobilization of public opinion in the USSR—spurred by scientists and writers—occurred in the early 1960s against the risk of pollution in Lake Baikal.[88] In France, radio journalist Jean Carlier galvanized opinion to prevent a large tourist project in the Vanoise Park that threatened to defile the majestic alpine sites. Backed by the brand new French Federation of Conservation Societies (FFSPN), this green campaign was a success. Publications criticizing environmental threats attracted more and more attention; a veritable militant ecological press emerged in the major industrialized countries around 1970.[89] In the United States, the Earth Day celebrations systematically referred to air pollution; conferences on the subject attracted people from all over the country. Opinion polls showed that the proportion of those who considered air pollution a "very serious" problem grew from 28 percent in 1965 to 55 percent in 1968 in the United States.[90]

More than an awareness or awakening of ecological consciousness—which heralded pollutions' mainstream arrival in public discourse—the 1960s corresponded to a more explicit shift in the approaches toward environmentalism. Where earlier conservation sentiments tended to be framed in an elitist tradition and worked for local regulations, the new environmentalism was more militant and rejected the siting of polluting industries anywhere. More militant approaches were evident in new publications such as the French journal *Survivre et Vivre* (Survive and Live), founded by the mathematician Alexandre Grothendieck. Its contents mixed radical discourse and satirical denunciation of modern societies and their pollutions.[91] Beginning in 1970, the series of children's *Barbapapa* books, for its part, proffered a very ecological vision of society, and mounted an implicit (and often explicit) challenge to technical productivism; polluting industry was frequently portrayed as a mortal predator (see figure 10.3).

As new environmental issues transformed the political landscape, the movement's vocabulary underwent considerable transformation. The word "environment" itself changed meaning and forced its way into ever more discourse, especially among North American planners and forecasters confronted with the proliferation of waste and pollution.[92] The English verb "to recycle," coined in 1926, spread after 1945. Its French counterparts, "recycle" and "recycling" (*recycler* and *recyclage*) appeared in 1959 and 1960. Their use extended in both languages during the 1970s, precisely at a time when the amount of waste on a global scale was growing exponentially.[93] This change in language was also accompanied by a broader restructuring of the intellectual and political understanding of

Figure 10.3
Satirical denunciation of air pollution. *Source: Survivre et vivre*, no. 10 (October–
December 1971).

the environment and an institutional reorganization that initiated something of an "ecological revolution" in many countries.[94] The increasing
concern over pollutions and their effects led to the admittedly slow, punctuated, difficult, and marginal emergence of intellectual currents that can
collectively be understood as the incipient stages of a new and united
notion of political ecology. This was by no means a homogeneous and
unified theoretical corpus; rather it constituted a complex nebula ranging

from liberal and centrist positions to anarcho-libertarian approaches that assumed and sometimes combined—even within the same author's work—strands of nostalgic romanticism and anticapitalism.[95] In the intellectual and political ferment that accompanied the 1968 revolts, political divisions became strained; pollutions inspired new debate over the boundaries between nature and culture.

A new militant world emerged around pollution issues; many new organizations operated on a scale that reached well beyond local concerns. Greenpeace, a nongovernmental organization founded in 1971 in Vancouver (Canada), made important connections between its two complementary bases: the fight against pollutions and the protection of the environment on the one hand, and antinuclear militancy on the other. An increasing number of environmental groups were responding to the growing audience for their message. In 1971, Friends of the Earth was formed; a section of this international network quickly established in France on the initiative of the journalist Alain Hervé, who was also founder of the magazine *Le Sauvage* (1973–1980). Pollution's entry into mainstream politics also took place through more traditional electoral and institutional dynamics. In December 1971, the People's Movement for the Environment—the first explicitly environmentalist party—was created in Neuchâtel, Switzerland; it won eight seats in the 1972 communal elections. The Australian green party, United Tasmania Group, was also formed in March 1972. In the following years, green parties sprang up in most industrialized countries; their presence transformed the political landscape, bringing the issue of pollution squarely into the public arena.[96] In France, the "Ecology and Survival" movement was launched in 1973 by Solange Fernex, a pacifist and feminist. Its candidate in the 1973 legislative elections in Mulhouse was the first French green to stand for a national election (he won 2.7 percent of the vote). The following year, René Dumont, who was Honorary President of Friends of the Earth, became a green candidate for the presidential election. Dumont, who began as an agronomist but turned to Third Worldism and then to environmental issues in the 1960s, published *Utopia or Death* in 1973. It sold 150,000 copies. Dumont laid out his radical criticism of wasteful society, castigating the "selfish rich countries," the automobile as "symbol of our aberrations," the "mechanized industries" that ruin local crafts and disrupt ecosystems; pollutions occupied a central place in his work, which championed a radical and emancipatory conception of politics.[97] Dumont obtained only 1.32 percent of the votes cast, but he came to embody pollution's new visibility in media and in political affairs.

This politicization of pollution issues around 1970 led to a major reorganization of administrative and regulatory structures. Willy Brandt, Mayor of Berlin and Social Democrat candidate (SPD) in the German Federal Elections of 1961, insisted to members of his party in a convention speech that "the sky over the Ruhr region must be blue again!" Brandt's call, which initially aimed to extend the party's electoral influence beyond its usual circles of the working class, took important root. In 1965, a new air pollution act—albeit endowed with a weak means of enforcement—initiated a national framework for environmental institutionalization.[98] By the beginning of the 1970s, states were creating new administrations that were supposed to provide answers to now widely accepted environmental problems. A flurry of environmental legislation came out of the United States, under the presidency of Richard Nixon. On December 2, 1970, in the wake of the adoption of the 1969 National Environmental Protection Act followed by the organization of the first Earth Day, Congress established the Environmental Protection Agency (EPA), whose purpose was to coordinate government action on environmental protection, set up and manage environmental standards, support research, and inform the public about risks related to pollution. The creation of a motor vehicle certification system that imposed pollution emission thresholds and the 1972 prohibition of the use of DDT (except for products intended for export) constituted a first wave of the EPA's agenda. The agency also restructured the management of watercourse pollution in order to centralize actions of public authorities to enable riverine ecological restoration.[99] This type of public organization subsequently spread throughout the English-speaking world, where environmental protection agencies became commonplace.

New ministries of the environment were created in countries around the world after 1971; they became new tools of intervention at the heart of the state apparatus. They ensured that there existed a dedicated home and central authority within states for environmental issues; the environment ceased to be torn between several competing branches of government. From this institutional base came the slow emergence of a new specific legal domain. In France, for example, the regulation of establishments requiring environmental oversight had been previously carried out by the labor inspectorate. Such responsibilities were entrusted to the Mines Department at the end of the 1960s before being transferred to the new Ministry of the Environment, under Robert Poujade. But many soon realized that these changes were more talk than action, and that far from breaking with previous legislation, the reforms became part of their continuity. The French law of July 19, 1976, which reformed the regime of classified industrial installations, appeared to backtrack on labor's hard-won gains during the previous decades; the new law maintained preference for technical solutions and determined that industrial improvements adopt the best available technologies at an economically acceptable cost. Rather than punishing polluters, inspectors' powers were limited to encouraging them to modernize their production facilities—often while drawing on large state subsidies. In this respect, the new reforms marked a certain regression in French environmental legislation.[100]

Before the 1960s, Japan was a polluter's paradise, but the environmental crescendo of the late 1960s brought significant changes. In 1964, the city of Yokohama, near Tokyo, formalized emission standards with local industries, while involving the population in the development of regulation. In 1967, the governor of Tokyo set up a Pollution Research Bureau and, in 1969, promulgated an ordinance on pollution, with even stricter standards. Pressured by social and environmental movements, the Ministry of Health recognized the industrial diseases of Minamata in 1968, which paved the way for a process of recognition of victims' status. In an exceptional session of Parliament in 1970—called the Pollution Diet—fourteen environmental laws were approved, putting Japan at the global vanguard of regulations against pollution. Completing the system, an Environmental Agency was created in 1971; two years later, a law on compensation for pollution-related damage, backed by significant means. In less than ten years, Japanese legislation against pollution had radically changed perspective.[101] Outside the industrialized world, some institutional changes occurred in countries under Western influence. In 1976, the government of Félix Houphouët-Boigny in the Ivory Coast, a former French colony that became independent in 1960, created a Ministry for the Protection of Nature and the Environment; its existence was, however, very precarious and short-lived.[102] In 1974, China also attempted to address its pollution problem; a body of the Council of State set targets for reducing pollution, but its success was negligible.[103]

Inasmuch as national legislation seemed to dominate environmental policy at the end of the 1960s and into the early 1970s, the fight against pollution was prominent on the agenda of international institutions at the beginning of the 1970s. Certainly, the Cold War had already inspired some actions concerning environmental expertise and standards in numerous international organizations. The WHO, for example, had been dealing in particular with health problems caused by air pollution at least since 1963.[104] But one of the defining moments in the history of international environmental engagement came with the 1972 United Nations Conference on the Human Environment, known as the United Nations "Stockholm Conference," or "Earth Summit," which resulted in the establishment of the United Nations Environment Program (UNEP). The United States was pivotal in coordinating (and funding) this international reconfiguration of environmental action; it insisted on making the environment a thematic, central issue in North Atlantic Treaty Organization (NATO) discussions. Research for the famous Club of Rome report was based at the Massachusetts Institute of Technology (MIT), and its American authors were leaders of that collaborative team. The United States also initiated the Global Environmental Monitoring System (GEMS) in the mid-1970s.[105] In Europe, alongside the first dust programs of the ECSC (1958), the Council of Europe launched several initiatives such as the Strasbourg Conference on Air Pollution, which convened in 1964, and the adoption in 1968 of the Water Charter and the first international convention limiting the use of certain toxic detergents in product maintenance. For its part, the FAO organized the December 1970 Technical Conference on Marine Pollution and its Effects on Biological Resources and Fisheries in Rome. For the first time, an international organization managed to bring together scientists, engineers, lawyers, and administrators under the same roof to discuss the problem of marine pollution. For the biologist Jean-Marie Pérès, then director of the Endoume Marine Station in Marseille and a specialist in marine pollution, this initiative showed the considerable progress made in a very short time frame, compared to the time "when we considered that all that was dumped in the sea disappeared forever in its immensity, without harming anyone."[106] The International Maritime Organization also played an important role in the adoption of the first major conventions, including the Convention for the Prevention of Marine Pollution by Dumping of Wastes and Other Materials from Ships (1972), and the Convention for the Protection of the Mediterranean Sea against pollution, adopted in Barcelona in 1976.[107]

Although twentieth-century regulations largely followed in the spirit of those of the previous century, the new magnitude of pollutions and their increasing reach resulted in regulatory and legislative innovations to reduce nuisances. Especially after 1968, political ecology, social struggles, and the beginnings of a voluntarist international cooperation seemed to be changing the scope of environmental concerns and implementing major policies. The more systematic reprocessing of industrial wastewater made it possible to improve the ecological situation in certain rivers (such as the Rhine) and in more affluent countries; and urban pollution from sulfur dioxide undeniably decreased in cities. Yet these ecological gains were limited, tenuous, and temporary.[108] They were also unevenly distributed; reductions in pollutions frequently corresponded to a displacement more than real elimination. For the affluent nations, anyway, the spread of globalization became a valuable vehicle for relocating unwanted contaminants far and away from informed gazes. Britain's success in reducing sulfur pollution in the 1980s was due to the fact that Conservative Prime Minister Margaret Thatcher's neoliberal policies—which promoted free trade and import dependency—elected to abandon the coal industry as a means of weakening trade unions. During the 1970s, pollution was undeniably at the center of many debates. But, far from disappearing little by little thanks to wise regulations and technical progress, as many authorities claimed, pollutions were simply displaced, or reappeared in other forms at the end of the twentieth century.

Epilogue

Charging Headlong into the Abyss

Since the 1970s, pollution has become a fundamental category of analysis and action in contemporary societies. As political ecology has emerged, environmental contaminations have never been as numerous as they are today, leading to major public health problems. In a world now populated by more than seven billion people (compared to just four billion in the mid-1970s)—who continue to consume more and more—environments and bodies are saturated with toxic substances. Nothing seems to be able to slow down this process. Today, the word "pollution" is omnipresent and the semantics around the word continue to expand to keep pace with new forms of pollution: it can be "light," "visual," "noise," "electromagnetic," even "mental" or "genetic." Electronic mail servers are full of spam that clogs and contaminates the internet. In French, the word spam translates as "pourriels" (France) or "polluriel" (Quebec), and "pollupostage" has been created to describe the act of spamming. This metamorphosis of language reflects the globalization of the stakes in play; pollutions have emerged as one of the great obsessions of our time, a near constant and unavoidable topic in the media. Countless scientific publications now warn against the collapse of biodiversity, health risks, and global environmental imbalances.[1] Therefore, how can it be explained that, despite pollution's full-fledged integration into the mainstream political agenda, pollutions are increasing and diversifying and the modes of public intervention have changed so little? Faced with the many communities of experts and scientists—doctors, chemists, epidemiologists, sociologists, and geographers—who explore all the implications of the problem, what can the historian contribute?

Because our historical exploration stops in the years 1970–1975, a pivotal moment in the evolution of world economic structures and the conceptualization of the environmental sciences, this epilogue will attempt to highlight the principal elements of contemporary changes.

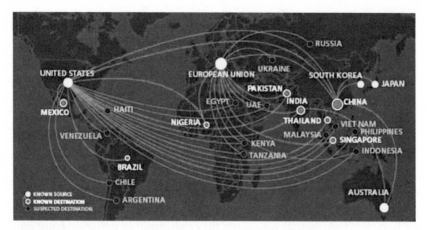

Figure 11.1
The global exchange of electronic waste, 2012. *Source:* based on Karin Lundgren, *The Global Impact of E-Waste: Addressing the Challenge* (Geneva: Organisation internationale du travail, 2012), 15. The spread of computing objects and networks is typically lauded as the triumphant condition of a dematerialized—or postmaterial—world that generates less pollution. However, this fallacy rests on the necessary disappearance of the increase in toxic waste that these tools create and reflects the inequality of global ecological exchanges. The waste trade is just one of the many examples that show that the burden of pollution is mainly borne by the poorest people in the poorest countries on the planet.

The global context has changed considerably: exponential population growth, the rise of neoliberalism in most of the world's economies, the spread of globalization, and the unprecedented importance of finance; and new information and communication technologies have induced structural changes in the industrial world, its geography, and hierarchies among sectors. At the same time, global environmental threats, such as climate warming or mass extinction of species, are changing the way we think about pollution.

This epilogue, which examines the present, is less intended to provide an exhaustive picture of pollution over the last forty years—one book would not suffice, and is it even possible in so few pages?—rather, it means to reflect, with regard to history, on recurrences and inflections, volumes of contamination, and the emergence of new polluting cycles.[2] It is clear that despite countless warnings from the 1970s, little has changed. Interest lobbies and the "merchants of doubt" have steadily grown stronger and more organized, creating a sclerosis of uncertainty, slowing down any effective measure. The requalification of the categories used to analyze and engage with pollutions is a first step to take into account, as well as the frenetic pace at which technology has developed over the last few decades to create new—and, as ever, unforeseen and unanticipated—risks and contaminations. But the preeminence of the old forms of pollutions, which prolong nineteenth-century trajectories of industrial capitalism, must also be underlined because they perpetually reiterate themselves in new places. Inasmuch as new hazards attract attention, polluting sediments cause more damage through accumulation than through substitution. In addition to the industrial and commercial lobbies, the organization of awareness and knowledge, which have never been so pointed and specialized, contributes to the confusion and reinforces this conundrum. For all its good intentions, increased specialization has meant losing sight of the forest for the trees. Could there be signs of hope for a world that would put the eradication of pollution front and center?

Requalifications

Many new concepts were forged, which characterized the new period of pollution that began in the 1970s. Both French and American sociologists Alain Touraine and Daniel Bell theorized about the dawning of postindustrial societies; others preferred the idea of postmodernity.[3] In either case, both invoked a paradigm shift that would see the subordination of material elements to immaterial elements (knowledge and information), and the organization of the economic system through complex networks rather than pyramidal hierarchies. In Western societies subject to waves of deindustrialization and driven by the ever-increasing weight of financial services and capital gains, the concept remains highly debated. At the current juncture, the American futurist Jeremy Rifkin defends the thesis of a third industrial revolution, based on the scientific and technological potentialities of our world (for example, so-called "intelligent networks") and their supposedly positive impacts on the environment.[4] The proponents of this third industrial revolution, and all those who extol digital capitalism, start from an apparently just observation: the laws of energy govern economic activity. But the current crisis marks the loss of impetus for past energy trajectories. Fossil energy and the "rare earths" that drove industrial civilization's economic success are depleted. The entropic debt from these extractions—resulting from past economic activity—is accumulating much faster than the biosphere is able to absorb it. But these approaches remain locked in a somewhat simplistic vision that technology can once again release civilization from its material shackles; they testify to a refusal to see the world as a holistic unit, or to confront the heart of the productive system. Because, at the global level, it is industry's ever-growing power and importance, and its effects that must be considered.

Another perspective held by the German sociologist Ulrich Beck in the mid-1980s was that of the "risk society." According to Beck, where old pollutions were perceptible to the senses—but marginalized by the urgency to manage and overcome resource scarcity—new pollutions were now largely hidden from perception. While past environmental risks could have appeared as the result of a lack of technology and hygiene, now their causes were blamed squarely on "industrial *over*production." Risks became "a *wholesale product* of industrialization, and [were] systematically intensified as [modernization became] global."[5] From Beck's risk society, we would have entered into an era of "reflexive modernity," increasingly aware of the threats provoked by human action. In this reassuring reading of our time, the last decades would have seen an unprecedented awareness of the dangers of pollution and an increase in actions to confront and reduce them.[6] But this analysis fails due to its historical myopia. The industrial and technoscientific changes of the past two centuries have not been unconscious; warnings and debates have continued to accompany them. Our contemporaries favored industrialization and power despite their inherent risks and threats, and at the cost of environmental pollution.[7] History shows how much environmental reflexivity was already present in previous centuries, while recent periods largely belie the hope that a new technological democracy might be found to solve its deleterious effects.

To reconcile industrial growth and pollution control, the concept of "sustainable development" has been coined. The term appears to have been used officially for the first time in the 1987 United Nations report drafted by the Norwegian Gro Harlem Brundtland, but the concept had already been debated for several years in environmental circles. It appears as early as 1980 in a joint report by UNEP and the World Wildlife Fund (WWF).[8] "Sustainable development" must respond to "the needs of the present, without compromising the ability of future generations to meet their own needs." It offers a wide vision in which everyone can recognize themselves, but it is also a kind of "paralyzing" oxymoron.[9] A propaganda tool for companies engaged in greenwashing, sustainable development is also regularly invoked by policymakers whose rhetoric lauds the virtues of compromise while cloaking their decisions in actions that are hardly ecological. As a result, the concept is often heavily criticized.[10]

Where "sustainable development" represents the model of good governance intent on reconciling ecology and capitalism, it in fact resembles more and more an empty promise used to maintain the status quo.[11]

It is in this context that the most important contemporary requalification of pollutions emerges around climate issues and debates. Indeed, the study and profiling of pollutions is now evolving in a world where global warming has emerged as the ecological emergency of primary significance. The greenhouse effect provoked by atmospheric gases—already brought to light by the French physicist Joseph Fourier in 1824 and which Arrhenius had hypothesized was exacerbated by carbon dioxide (1896)—began to mobilize the community of scholars in the postwar period. Consensus around the possibility that the large-scale combustion of fossil fuels could dramatically disrupt the planet's climate gradually emerged. Therefore, this harmless gas—even beneficial for the growth of plants—entered into the spectrum of polluting substances.[12] In 1965, a report submitted to the American President Lyndon Johnson on the risks associated with increased pollution pointed to the responsibility of human activities for CO_2 emissions.[13] Ten years later, one of the report's authors, the American geochemist Wallace Broecker, announced in the journal *Science* the imminent eruption of global warming. His article predicted that the symbolic CO_2 threshold of 400 ppm would be reached in 2010. At the time, global CO_2 levels were at 330 ppm. Broecker was not far off; 400 ppm was reached in 2013, an increase as important as that calculated for the period 1750–1975. Broecker also believed that "the present ... trend will, within a decade or so, give way to a pronounced warming induced by carbon dioxide."[14] The expression "global warming" was subsequently challenged; for some it was too blunt in blaming human actions for climatic harm. For a few years, the National Aeronautics and Space Administration (NASA) preferred "inadvertent climate modification," which allowed the phenomenon to be euphemized.[15] In 1979, UNEP and the World Meteorological Organization (WMO) considered the situation to be of sufficient concern to convene the first World Climate Conference in Geneva. What followed was the launch of a global research program involving the largest international scientific bodies.

Experts began to demonstrate the correlation between rising temperatures and CO_2 levels and other gases such as methane (CH_4). On June 23, 1988, as the United States was overwhelmed by a scorching early summer, James Hansen, director of a NASA lab, testified in front of a US Senate committee, and declared that global warming was not caused by the natural variability of climate but by human activities. He added that the trend was expected to increase, and that he was "99%" sure of his claims. The climate issue was born. In the same year, at the G7 Summit in Toronto, the concept of sustainable development was adopted and the Intergovernmental Panel on Climate Change (IPCC) was created. Consequently, many diplomatic negotiations were established to understand and stop the phenomenon: the United Nations Framework Convention on Climate Change (UNFCCC) was adopted in 1992 at the Earth Summit in Rio de Janeiro. That framework culminated in 1997 with the drafting of the Kyoto Protocol, which was supposed to reduce emissions of anthropogenic greenhouse gases to 1990 levels. Now meeting under the Conference of the Parties (COP), these major international meetings have consistently failed to install restrictive devices. If the 2015 Paris Conference (or COP 21) realized agreement and common cause amongst nations around a minimum agreement, none managed to significantly slow the increase in the concentration of greenhouse gases entering the atmosphere.[16]

The debates focused first on the reality of the phenomenon, then on the question of whether or not greenhouse gases should be considered as pollutants. The controversy was particularly acute at the beginning of the twenty-first century, pitting climate scientists against climate skeptics, environmental groups against oil lobbies. Many studies show how climate skepticism was bought and paid for, subsidized by the petroleum and coal industries, or closely linked to conservative think tanks. Some of its authors were former confederates of the tobacco and chemistry industries, commissioned to minimize the dangers of their products. As professional doubt merchants, they artificially maintained public ignorance in order to prevent any measure of reduction in greenhouse gas emissions. Obfuscating over whether or not carbon dioxide is a polluting substance, they have gained traction in grinding climate action to a relative halt.[17] In the fight against "real pollution," and because CO_2 is vital for plants, the Belgian chemist István Markó stated that "CO_2 is neither a poison nor a pollutant! It has never been and never will be."[18] Conversely, climate specialists and environmental movements claim it is the major source of pollution of the industrial age. The WHO, for its part, endorses the notion of "'climate pollutant" for CO_2 insofar as, beyond a certain threshold, it contaminates the environment and modifies the chemical composition of the atmosphere, resulting in health effects and ecological imbalances.[19]

As the case of greenhouse gases shows, the boundaries that define what are considered pollutions are—more than in the past—porous. For example, chemical contaminations, with poorly understood cumulative effects owing to the longevity of their molecular makeup, are on the rise: not only are yesterday's persistent organic pollutants still widely present in the environment (such as DDT or lindane, another organochlorine-based insecticide), but hundreds of thousands of new ones continue to be marketed. But how to evaluate the extent of pollutions, or evaluate changes when scales of contamination and contextual situations are so diverse? And how to evaluate effectively when every indicator hides as much as it reveals? Contrary to previous centuries mentioned in this book, fumes and blackened waters, perceptible by the senses, no longer constitute the most dangerous pollutions. Instead, the innumerable complex molecules whose minute doses and uncertain effects are more difficult to spot pose the more significant threat.[20] According to the toxicologist André Cicolella, the proliferation of synthetic products, developed to meet uses as diverse as cleaning and cosmetics, produces an "invisible scandal of chronic diseases."[21]

Endocrine disruptors serve as a useful illustration of this process of the contemporary requalification of pollutions just as they highlight certain continuities in the apprehension of toxic substances. In her 1996 survey, *Our Stolen Future*, the American zoologist, epidemiologist and environmental activist Theo Colborn, theorized and popularized the notion of endocrine disruptors to refer to any molecule or compound chemical agent that upset the hormonal system, resulting in abnormalities. Through her work, Colborn was especially concerned about such impositions on the reproductive system. *Our Stolen Future* asks: "What if pollution made people less intelligent?" Colborn suggested that the effect of chemicals on hormones could lead to concentration difficulties for children and even forms of aggression and violence.[22] This type of warning—initially rejected as part of a catastrophist imaginary or a "sociobiological drift" and condemned as a form of "demagogic dramatization" likely to drown the real issues—nevertheless managed to play an important role in the redefinition of chemical pollutants at the end of the twentieth century.[23] "Poison" is a little too black-and-white as a descriptive term; these new hazards are substances that can disrupt the body in subtle, discreet ways. The presence of endocrine disruptors in products as abundant as pesticides, plastics, and certain car emissions made them a public health problem, especially since they made clear that dose or exposure level was no longer the central metric for ascertaining harm; a very small dose could have a major effect. Epidemiologists and public health workers also needed to determine the moment of exposure (especially in utero) and the prospect of multiple exposures and their famous "cocktail effect."

In the very wide range of these new compounds, bisphenol A, synthe-sized at the end of the nineteenth century but mainly used by the plastics industry after the 1960s, offers an exemplary case. Now mass-produced—6 million metric tons a year—it is used in the manufacture of plastic and is found in many products of everyday life, from sunglasses to CDs to baby bottles. At the beginning of the twenty-first century, doubts began to cir-culate about exposure to low-dose bisphenol A. More in-depth studies and research led to Canada banning it in baby bottles in 2008, followed by a European directive in 2011. After bisphenol A, phthalates became the new focus of attention; more than 3 million metric tons are produced each year around the world. These synthetic molecules are present in plastics, cosmetics, paints, and clothes, among others; some of them are highly toxic to human reproduction.[24] More and more surveys confirm how invisible chemical contaminations at low doses are dangerous; after the reproductive system, they could affect the brain and damage cognitive and cerebral faculties, especially in children.[25]

Dioxin contamination also reveals new ways of understanding pol-lutions. Although already present during the Vietnam War, the issue of dioxins made a sensational debut in the public sphere with the Seveso disaster in 1976, when a toxic cloud escaped from a reactor of the Icmesa chemical plant, which manufactured trichlorophenol. The cloud spread across the Lombard plain in Italy. Locals experienced immediate health problems, including skin disorders such as chloracne. As well, 20,000 hectares (about 49,000 acres) of soils were contaminated, and 80,000 poisoned cattle had to be slaughtered.[26] Dioxins are "persistent organic pollutants" that remain in the environment for many years. Considered as carcinogenic substances by the WHO, the dioxin family actually com-prises several hundred different molecules, about thirty of which are particularly toxic. They are manifested during thermal processes, which is why they are regularly found in higher concentrations around incin-erators. Apart from accidental causes, dioxins infiltrate the food system, accumulating especially in animal fats. Since the Seveso accident, many cases of dioxin pollution have hit the headlines: the case of the dioxin drums at the Ciba chemical plant near Basel, Switzerland (1982–1985), cow's milk containing dioxins around an incinerator near Albertville, France (2001), contaminated greases delivered to feed producers in Ger-many (2011), and so on. Although dioxins are produced locally, they are spread all over the planet and are particularly concentrated in PCBs, whose pollutants are also recurrent due to the large quantity of stocks that require disposal.[27]

Invisible Molecules, Biotechnologies, and Digital Worlds

The dispersion of invisible contaminants into the environment did not only come from innumerable chemical molecules, but also from new genetic, nano-, and digital technologies. Often presented as solutions to the pollution of the old industrial world, the new technological regimes that emerged at the end of the twentieth century in reality introduced a new series of polluting cycles. Thus, from the 1980s, the first genetically modified organisms (GMOs)—that is to say, whose genetic heritage is modified by human intervention via transgenesis or insertion into the genome of one or several new genes—were developed by industry and seed companies, including Monsanto, in the name of the fight against world hunger and agricultural pollution. Since certain GMOs were designed to include pesticide substances, their adoption was supposed to lead to a more tempered and "rational" use of insecticides and herbicides.[28] After the first transgenic plant was obtained experimentally in 1983, an herbicide-tolerant plant appeared in 1987. In 1990 the first transgenic plants were marketed.[29] At the end of the 1990s, nearly 40 million hectares (about 99 million acres) of transgenic plants were cultivated worldwide. Transgenic-based agriculture expansion accelerated despite regulations and numerous controversies. By 2006, 102 million hectares (about 252 million acres) were cultivated, with nearly 200 million hectares (about 494 million acres) ten years later, especially in the United States, Argentina, Brazil, India, and China where GM fields have reshaped agricultural models. Because of cross-pollination from genetically modified plants that transmit their new genes to neighboring non-transgenic plants belonging to the same species, it is possible to speak of "genetic" pollution. In Canada, for example, there is almost no natural, uncontaminated rapeseed in cultivation. Touted as rational, technical solutions to fight against the use of chemicals, genetically modified plants ended up producing quite the opposite effect. Because many GMOs are specially designed to resist herbicides—which makes the spraying of fields with chemicals less problematic—more, rather than fewer, chemical applications took place.[30] New generations of GMOs produced drought-resistant plants, or trees with a modified cellulose/lignin ratio to meet the needs of biorefineries and the paper industry. GMOs ushered in a definitive specter of boundless proliferation, threatening to add to the old, "gray pollution (industrial pollution) an 'augmented' green pollution (genetic pollution)," leading to more controversy and resistance around these new products.[31]

With nanotechnology (a term coined by the Japanese physicist Norio Tanigushi in 1974), which produces elements less than one millionth of a millimeter—the order of magnitude of the distance between two atoms, and therefore smaller than cells—it is their size that becomes a source of pollution. At this infinitesimal level, biological barriers are easily crossed, a property currently exploited by the cosmetics industry. Nanotechnology emerged in the late 1980s, and followed a development parallel to that of GMOs, both chronologically and epistemologically. Obtained through molecular manipulation, these nanomaterials posed fundamental ethical questions. Many nanomaterials were recognized as toxic, inducing oxidative stress, cytokine inflammations, and cell necrosis. In addition, they could be absorbed by mitochondria and the cell nucleus; as a result, they could cause DNA mutations. In addition, by incorporating these minute metal particles into products without any hope of recovering or recycling them, there is no way to know where they will end up. For example, nanosilver contained in clothing for its antibacterial properties was leached into the environment when laundered, without any means of recovery. Worse still, titanium nanoparticles are ingested, since they are a food additive (E171) that whitens yogurt and adds shine to confectionery. Since 2007, a journal dedicated to these new risks, *Nanotoxicology*, has appeared regularly.[32]

Beyond these controversial pollutions, the incessant production and renewal of high-tech objects has increased pollutant waste and new, more or less hidden nuisances. These new digital tools, now ubiquitous, promised to open an era of less polluting practices by promoting the dematerialization and pooling of certain services, such as carpooling, but they have actually caused many side effects whose study has only just begun. Electromagnetic pollution (electro-smog) is recognized by many experts; electromagnetic fields produced by mobile phones and relay antennas are classified as possible carcinogens for humans by the International Agency for Research on Cancer (IARC). In May 2015, 190 scientists from 38 countries called on the UN and the WHO to take measures to limit exposure, confirming the reality of the phenomenon. Public authorities, however, felt reassured that monitoring emission norms and exposure thresholds would suffice to continue expanding the new digital world's reach and access. No prospect of restrictions was ever seriously envisaged.[33]

Computers were virtually nonexistent in 1970 but have now invaded the daily lives of the majority of the world's population. It would be a mistake to consider this an immaterial phenomenon because the computer is first and foremost made of silicon, plastics, and metals. Life-cycle specialists of material goods have been trying to assess how much energy and resources are needed to produce electronic readers, smartphones, and computers—which in a few short years have become mass consumer goods—and how much toxic residues they produce.[34] The number of mobile phones (first created in the United States in 1983) increased from 300,000 in 1985 to more than 200 million in 2006. Smartphones, virtually nonexistent at that time, are growing at an even more dramatic pace: 1.5 billion units were sold worldwide in 2016. To produce these quantities, the electrical and electronics sector consumes 12 percent of global gold demand, 30 percent of silver, 30 percent of copper, and up to 80 percent of rare metals such as ruthenium or indium, whose extraction and refining cause significant environmental damage. Some sectors, such as coltan (tantalum ore), feed wars in Africa, while "rare earths"—a group of metals with electromagnetic properties, such as scandium and yttrium—require removing vast quantities of the earth's crust for their extraction.[35] China currently dominates global production, resulting in the local release of many toxic elements, such as heavy metals or sulfuric acid, as well as radioactive elements. In mining sites, cancer incidence can reach 70 percent, all while local agricultural yields drop. For example, in Baotou, the rare earth capital, factories process 70 percent of the world's production and dump thousands of liters of toxic waste into a lake that today extends over 10 square kilometers.[36]

Another facet of the digital world's pollutions that should be taken into account is the industry's energy consumption, which is already reaching the levels of air transportation. While the equipment's energy efficiency is progressing, like nineteenth-century furnaces the rebound effect offsets any positive results: the exponential growth of data production and the proliferation of screens and general digital usage are increasing the demand for energy and the consumption of materials.[37] Moreover, data centers consume massive quantities of energy to process queries and run data storage servers, which need to be constantly cooled, which also consumes energy. In France, data centers consume 10 percent of the country's electric energy. Though individual usage may seem trivial, a simple internet request on Google equals the consumption of a 12-watt light bulb for 2 hours; sending a 1-megabyte attachment to ten correspondents requires the energy needed to move a car 500 meters. In an hour, around the world, 10 billion emails are sent, which corresponds to 50 gigawatt hours, or 4,000 Paris-New York round-trips by plane.[38]

Waste from cutting edge technology has also become a significant source of ecological pollution and exchange which is unequally distributed around the world. Debris from the conquest of space also deserves mention, albeit (unfortunately) much too briefly. Since the launch of the first satellite, Sputnik 1, by the USSR in 1957, the number of objects sent into space has skyrocketed, especially since the 1980s with the rise of telecommunications. They are either left to drift in orbit around the earth—unwanted satellites traveling at nearly 30,000 kilometers per hour—or they eventually fall back to earth. According to NASA, in 2013 these debris consisted of 29,000 objects of more than ten centimeters, 670,000 of more than one centimeter, and 170 million objects that were more than one millimeter.[39] At the Seventh European Space Debris Conference held in Darmstadt, Germany from April 18 to 21, 2017, scientists once again expressed concern about the significance of this debris, which poses safety problems because of collision risks.[40]

High-tech waste is mainly the result of daily consumption. Due to the constant renewal of equipment and its increasing obsolescence (750 million mobile phones are discarded each year), the amount of electronic waste has been growing at a steady pace of between 3 percent and 5 percent per year since the 1990s. This waste contains compounds that are extremely toxic to human health and the environment. Mercury, lead, cadmium, and cyanide, for example, are all present and the workers responsible for their disposal or recycling are particularly exposed. For example, it is estimated that the 315 million computers that became obsolete between 1997 and 2004 contained more than 500 million kilograms of lead. The recycling of this waste is the subject of international trade, described by some as "toxic colonialism."[41] Waste, which comes from electronic devices that are mainly used by consumers in rich industrialized countries, is sent to Asia, Africa, and Latin America for recycling or storage in landfills that cause significant pollution. In 2007, nearly 80 percent of US computer waste was exported to Asia. While the Basel Convention—signed in 1989, and ratified by nearly 190 countries (but not the United States)—officially bans this traffic, millions of metric tons are transported each year to open dump sites, such as those in the suburbs of Accra, the capital of Ghana, as recently filmed by German documentarian Cosima Dannoritzer in *The E-Waste Tragedy* (2014), the tagline for which asserted that "the illegal recycling of electronics is a toxic business on a global scale." China and India are currently importing most of the world's electronic waste. The Guiyu dumpsite, opened in the late 1970s, is now the largest in the world; in 2005, it employed

100,000 people, but that number has undoubtedly grown since then.[42] Many environmental organizations denounce these pollutions and the social, health, and environmental ravages they cause. These waste sites, though, serve as symbols of a globalization of pollution and hardening of environmental inequalities in recent decades.

Globalization and Unequal Ecological Exchange

Unequal exchange and expanding transfer distances have continued to be a part of polluting dynamics. With globalization, the process has become even more entrenched and compounded; as a result of globalization's new mobilities and their capacity to radically alter the scale of these transfers, it is now possible to speak of an unequal ecological exchange on a global scale. In the cities of the global North, it is undeniable that the atmospheric concentration of certain well-known molecules—such as sulfur dioxide, carbon monoxide or lead oxides—has decreased. Such successes have reduced the volume of acid rain. On the other hand, emissions of nitrogen oxides and sulfur dioxide have increased significantly elsewhere; in China they are nearly ten times higher than in developed countries. And while fine particles have increased everywhere, especially because of automobile traffic, this increase is mainly affecting the countries in the global South.

Historical analysis shows that the new pollutions of the so-called postindustrial world constitute additions (rather than substitutions for older contaminants) to the planet's pollution burden. The notion that new, postindustrial technologies and their pollutions work as improved substitutions comes from a form of myopia induced by globalization that unevenly redistributes risks. Globally, lifestyles the world over demand ever more resources, in spite of the considerable disparities between countries and social groups. Emerging economies—and their hunger for the resources and affluence enjoyed by the globe's fortunate "haves"— demonstrate this uncontrollable snowball effect on the world-system's horizon.[43] Each year, the symbolic date of exceeding the threshold for the consumption of resources that the planet can renew comes a little earlier: according to the current calculation method, the Earth's biocapacity was exceeded in 1976 on November 19, it came on October 25 in 1986, October 4 in 1996, August 24 in 2006, and August 8 in 2016. If all humans lived as the inhabitants of Australia, it would take 5 and a half Earths to provide for their needs, 3 Earths if they lived as in France, 2 Earths if they lived as in China, and 0.7 Earths if they lived as in India.[44]

In cities in the North that are heavy consumers of goods and services, infrastructures are sufficiently developed to evacuate waste and provide a degree of urban hygiene, which leaves the impression of a world without pollution, even though consumption is based on a productive chain (including mines, transport, and industry) that causes many contaminations. At the other end of the chain, there are still shantytowns, where about one-third of the world's urban population lives—far more than in the 1970s—and where countless modern proletarians live in and with wildly insalubrious conditions. The lack of sanitary infrastructure, the pervasiveness of toxic industries, and uncontrolled car traffic defy imagination throughout Central Asia and sub-Saharan Africa. At the beginning of the twenty-first century in a mega-city of 10 million inhabitants like Kinshasa, there is no mains drainage; in Manila, only 10 percent of dwellings are connected to it. Mike Davis observes that most of the big cities of the Third World have "a Dantesque district of slums shrouded in pollution and located next to pipelines, chemical plants, and refineries: Mexico's Iztapalapa, São Paulo's Cubatão, Rio's Belford Roxo, Jakarta's Cibubur, Tunis's southern fringe, southwestern Alexandria, and so on."[45] The "discharge syndrome," which characterized many nineteenth-century industrial cities in Europe, existed on a considerable scale in the global South's slums in the 1980s. In a Bangkok shantytown, chemical plant explosions in 1989 and 1991 caused the deaths of hundreds of inhabitants and extensive pollution. The city of Cubatão, near São Paulo, constructed on ancient banana plantations, reached 100,000 people in the early 1980s and was home to huge industrial complexes that produced nearly 40 percent of the country's steel and fertilizers. It became a "valley of death": the infant mortality rate was ten times higher than the rest of the state. Soot and suspended particles were so prevalent that birds and insects simply disappeared and vegetation died inexorably. Recently, however, the situation has begun to improve.[46]

For much of the current pollution regime, China—the new workshop of the world—provides a terrifying illustration, demonstrating that the old pollutions inherited from past practices have not disappeared. In just a few years, the country has become the most polluted in the world. In 1979, China launched the "socialist market economy," and opened up to international trade. It engaged in rapid industrialization, particularly in labor-intensive industries such as textiles and low-quality manufactured goods. Its growth rate was close to 10 percent and the Special Economic Zones (SEZs) offered multinational companies tax incentives and low-cost docile labor that instigated a massive relocation of production to these territories. The social and ecological cost was significant.[47] China accounts for nearly 40 percent of global coal consumption per annum, which supplies 68 percent of its energy needs and 80 percent of its electricity. In 2008, the country became the world's largest emitter of greenhouse gases, responsible for nearly 23 percent of total emissions. In industrial areas, coal pollution became so bad that it caused serious health problems, rendered agricultural land sterile, and forced the authorities to displace the population. In thirty years, China's industrialization and agricultural modernization has contaminated 20 percent of its arable land—or 20 million hectares (about 49 million acres)—and 40 percent of its rivers. Some 300 million rural people consume water with heavy metal levels dangerous for health. Air pollution is just as dramatic; acid rain is increasing, and incidences of lung disease are off the charts, a situation that could kill 1.6 million people every year in the country. The media regularly reports on the magnitude of the smog and the distress of the population in the face of what is regarded as an uncontrollable phenomenon.[48] Even the Chinese Ministry of the Environment admits the extent of the problem. In 2013, after the Ministry ignored journalist Deng Fei's map of the main polluted sites in China, it published a list of contaminated villages where the proportion of people who were sick (especially with cancer) was particularly alarming. There were nearly 450 such "cancer villages," all located in industrial areas.[49]

Beyond China, the whole of Asia is the victim of massive pollutions caused by continuous industrial expansion. Since the 1990s, global coal consumption has steadily increased, particularly because of the construction of electric power plants. By 2013, the Asia-Pacific region was billowing more than two-thirds of the 36 billion metric tons of carbon dispersed into the atmosphere by the burning of coal and hydrocarbons (a doubling since 1975).[50] The so-called Asian brown cloud—a huge sheet of soot particles and polluting gases produced by industry, road traffic, and also agricultural practices such as slash-and-burn and waste incineration—has been a prominent feature of satellite photographs for twenty years. It is arguably the largest pollution in the world, extending from Pakistan to India and China, disrupting the local climate and threatening already fragile populations.[51] Water pollution is another major problem in the region: large dams (such as the Three Gorges on the Yangtze in China), overconsumption, and massive agricultural and industrial waste emissions result in river salinization and pollution.[52]

Since 2007, the American NGO Pure Earth has published a list of the most polluted sites in the world. Sub-Saharan Africa and Southeast Asia consistently rank first as the most contaminated regions.[53] The relocation of wastes and pollutions to Southern countries goes hand in hand with the uneven distribution of risks. One such example was a large pesticide factory built in 1978 in Bhopal, the capital of the Indian state of Madhya Pradesh with a population of 300,000 and located 600 kilometers south of New Delhi. In 1984, following a series of technical failures, a toxic cloud escaped and spread over an area of 25 square kilometers, causing thousands of deaths and hundreds of thousands of illnesses.[54] More recently, fires in textile factories in Bangladesh and Pakistan have resulted in hundreds of casualties, demonstrating the inadequate security conditions in these countries. In China, there was a 98 percent increase in industrial accidents between 2010 and 2013, especially in mining.[55]

In order to fully grasp the magnitude of regional differences in terms of pollution and risk, it is necessary to distinguish between places of consumption and places of production, especially since a large part of the toxic production in the South exists to meet the demand from rich countries. In other words, it is necessary to refine the analysis according to social classes and their consumption patterns: thus shifting the blame for the causes of pollution solely from emission sources to include the consumerist practices that inspire them. Consumption is the real source of industrial pollution. As journalist Hervé Kempf notes, it is the rich who destroy the planet and externalize the damage. If China has become the world's leading polluter, it is because it has transformed itself into a huge workshop serving the consumers of the North; its toxic productions are the result of meeting the needs of North American and European consumers more than from satiating its own domestic market.[56]

The outsourcing of pollution is particularly apparent in the treatment of waste. Even if it was practiced before, the 1970s saw the export of hazardous waste become a truly international activity: Mexico began to accommodate American waste on a massive scale, Southeast Asian countries agreed to stockpile waste from Japan, and Morocco and some other African countries received waste from Europe and the United States. By the end of the 1980s, the international trade in toxic waste amounted to millions of metric tons per year.[57] Some cases illustrate this phenomenon, and some of them are difficult to believe. In 1987, for example, the media reported on a barge loaded with residues from a toxic incinerator in the United States that was sailing aimlessly around the Atlantic for lack of a country ready to receive it. Such incidents accompanied by pressure from various protest movements spurred the first protocols and conventions aimed at regulating this toxic and illegal trade. Yet, as with computer and electronic equipment residues, exports from the North to the South continue to grow at a steady pace, even though most of the recipient countries do not meet the sanitary standards that are in place in the countries of origin. Thus, a blatant, worldwide environmental injustice, which is even explicitly criminal in the case of some banned poisons in developed countries, is being promoted.[58] This was the case, for example, in Côte d'Ivoire in 2006 during the *Probo Koala* disaster. The *Probo Koala* was an oil tanker belonging to Trafigura, which transported 581 metric tons of highly toxic hydrocarbon wastes—a mixture of petroleum, hydrogen sulfide, phenols, caustic soda, and organic sulfur compounds—to the African coast. The cocktail of contaminants produced deadly gases; tens of thousands of people were poisoned.[59] More generally, the toxic exchange is more mundane. Diesel fuels sold on the African continent are much more toxic than those in Europe, even though it is Western petrochemical companies that organize this trade.[60]

Thus, most developed countries continue to favor—as in the past but on a whole new scale—the outsourcing of pollutions, via the relocation of the most polluting factories or the export of dangerous substances, without ever calling into question their potential impact elsewhere. Out of sight: out of mind. The transport revolution made possible by gigantic container ships enables this type of global traffic, while itself becoming a source of pollution through its diesel consumption or the abandonment of ships in Southern countries.[61] Whether it is the highly polluting activity of tanning or heavy industries such as metallurgy, polluting factories have left Europe for countries with less stringent environmental, health, and labor regulations.[62] The whole world is now organized to manage this differentiation of risk and the ecological inequality of global production. The problem of environmental injustice has been renewed on a global scale.[63]

Recrudescence of an Energy-Consuming Industrial World

The new types of pollution and the new logic behind where to locate polluting industries did not replace the old dynamics; they added to them. Even today, it must be remembered that most of the electricity produced in the world still comes from fossil fuels: 40 percent of electricity is still produced from coal and about 75 percent of coal is used for electricity.[64] Coal remains a major fuel, especially when the price of oil starts to climb; for its part, nuclear energy was called into question after the Fukushima accident in 2011. Coal remains the basis for industrial development in

China and India. The extraordinary growth of India's coal exploitation has left ecologically ravaged enclaves in its wake: destroyed forests, ground subsidence, air and water pollution. In the Meghalaya Basin, the multitude of artisanal mines (dubbed "rat holes") are worked by migrant workers from Bangladesh and Nepal. Here too, the land is devastated by this exploitation.[65]

Despite the development of renewable energies—regardless of their permanent growth—neither coal nor oil production has declined. Indeed, the contrary is true. Rather, since the 1990s, the exploitation of oil sands and shale gas deposits have become prominent features of the global energy regime. With the depletion of oil reserves, the exploitation of these unconventional hydrocarbons has become financially attractive, even under conditions previously considered difficult. Oil sands in Canada, Venezuela, and Madagascar are driving a relative boom of activity. Exploitation of the northern Alberta (Canada) oil sands began in 1967, but production accelerated markedly after 2002. The oil (or tar) sands are composed of a mixture of sand, water, clay, and crude bitumen, whose extraction causes much greater environmental damage than conventional oil. To extract the pure material, it is necessary to heat the bitumen with solvents. The surrounding region is literally devastated, and doctors have noted the appearance of rare cancers.[66] Shale gas, which is a natural gas contained in clay layers rich in organic matter, can only be exploited by hydraulic fraction that requires the injection of water at high pressure and various chemical additives, including lubricants, biocides, and detergents. Its large-scale operation began in the 2000s, especially in the United States. In addition to groundwater contamination, it is also an industry that emits high levels of greenhouse gases while adding to seismic risks. The American landscape is now dotted with areas polluted by this exploitation, especially in the pioneering states of North Dakota and Montana.[67]

This extension of the hydrocarbon range does not detract from the more traditional oil pollution. The years 1975–2000 were marked by major oil spills, which led to increased regulations on supertankers and the containment of their cargoes. Double hulls became mandatory. Accidents were often highly publicized: the *Torrey Canyon* in 1967, the *Amoco Cadiz* off the coast of Brittany in 1978 (223,000 metric tons), and the *Exxon Valdez* off Alaska in 1989 (37,000 metric tons of oil spilled). Many other, less publicized accidents occurred off the African and Latin American coasts with even greater spills. For example, in Trinidad and Tobago the *Atlantic Empress* spilled 280,000 metric tons in 1979; the *ABT Summer* in Angola lost 260,000 metric tons in 1991. There is a long list of such accidents, which should not overshadow the chronic, much larger volume, estimated by the WWF at 4.5 million metric tons per year.[68] In addition, leaks or accidents on platforms or oil fields are widespread and frequently underreported. The 2010 explosion of the Deepwater Horizon oil platform, operated by British Petroleum in the Gulf of Mexico, caused the worst oil spill in history: 835,000 metric tons of oil escaped into the waters, and contaminated more than 2,000 kilometers of coastline.[69] Less publicized—because it remained largely invisible—a huge methane leak from the Southern California Gas Company near Los Angeles at the end of 2015 also reflected the fragility of energy infrastructure. Similarly, some landscapes were affected by chronic leaks and releases that destroyed whole regions, such as the Niger Delta in Nigeria.[70] War-related oil spills are also common, particularly those affecting the Near and Middle East regions since the 1980s: the Iran-Iraq War from 1980 to 1988, the first Gulf War in 1990–1991, the invasion of Iraq by US-led troops in 2003, and ongoing conflicts in Syria and the Middle East. For example, in 1991, when it withdrew from Kuwait, the Iraqi army sabotaged a large part of the country's oil wells; more than a million metric tons of oil spilled into the sea, precipitating immense contamination of coastal areas and significant health problems for the populations.[71]

The oil boom is still current, and it feeds the ever-growing petrochemical industry, a part of which is plastics. If plastic consumption has shown signs of stagnating in Europe, it has exploded around the rest of the world. Less than 50 million metric tons of plastics were manufactured each year in the late 1970s; forty years later, that number has swelled to more than 300 million metric tons. Millions of metric tons end up in the oceans. This undecomposed waste constitutes the origin of the vast "plastic continent"—which obviously is no continent at all—discovered in the 1990s, and dubbed the Great Pacific Garbage Patch.[72] When he made his documentary about the Pacific Ocean's Midway Islands (*Midway Journey*, 2011), one of the most isolated archipelagos in the world, photographer Chris Jordan was struck by the amount of plastic waste that was there. It was responsible for high incidences of albatross mortality; the birds ingested the plastics. He returned with photographs that caused stupefaction: birds, whose entire innards were filled with plastic detritus.[73] More worryingly, the ocean is now shrouded—up to 10,000 meters deep—with persistent organic compounds (such as PCBs) and plastic microparticles that are undetectable to the naked eye. The latter come from the residues of laundering synthetic-fiber clothing, and brake wear and car tires.[74]

The contemporary extraction of metalliferous resources follows the same trajectory as hydrocarbons, as it continues to grow in tonnage: however, mining excavation requires digging ever deeper for fewer rich mineral deposits. For example, for copper, the average metal content in the ore dropped to 0.5 percent at the beginning of the twenty-first century, and this downward trend is consistent with all metals, resulting in more aggressive extraction processes employing stronger chemicals, and greater energy expenditure.[75] Mining is completing the economic colonization of the world. Inside the Arctic circle, the Soviet exploitation of nickel and apatite (a type of phosphate) has resulted in extreme levels of pollution that continue to increase with the Russian productivist frenzy. Mercury and methylmercury are released with the tailings, and at levels that endanger the health of Inuit populations and threaten several marine species. Even some of the most remote places on Earth are now contaminated, although they have never contributed to this pollution.[76]

Since it is not possible to review all the metals, a survey of aluminum might suffice as an illustrative example. While this sector was sometimes presented as clean after 1980 because of the adoption of new pollution controls, the ecological ravages have continued—and have spread internationally. Releases of red mud from alumina production regularly trigger large-scale disasters, including at old sites such as Gardanne, which is still contaminating the Calanques National Park.[77] This is also the case in Hungary, where as a result of the collapse of one of the basins, the Ajka aluminum plant released a million cubic meters of red mud in 2010. This plant was established in 1942 by the Nazis, privatized in 1993, and employed 1,100 people at the time of the accident.[78] Dust is another polluting, long-term byproduct of aluminum's production. As is often the case, it is an industry that is becoming more international: aluminum operations are beginning to be outsourced to poor countries, where the risks are less visible. Aluminum production has grown considerably in Africa (Cameroon, Ghana, Guinea), South America (Brazil, Jamaica, Surinam), India, and China, which, along with Russia, has become a world-leading producer.[79] In 2015, the Rio Doce disaster in Brazil, one of the worst in the history of the country, created a toxic mudslide that spanned 500 kilometers. Tens of thousands of retention ponds where extractive waste was stored threaten to contaminate the environment and destroy the lives of local residents.[80]

In the agricultural sector, the cycle of chemical pollutions also continues to spread, leading to accelerated soil degradation and declining yields. The quantity of chemical fertilizers used in the world rose from 30 million metric tons in 1960 to 150 million metric tons in 1990, and 175 million metric tons by 2011—an increase that mainly occurred in Asia (+450 percent between 1970 and 2000), South America (+200 percent) and Africa (+100 percent).[81] Similarly, in countries that were already committed to chemical agriculture, the quantities of pesticides consumed did not decrease between the 1980s and 2010, but rather tended to stagnate or even increased slightly, despite considerable regional fluctuations. In 2007, 2.5 million metric tons of pesticides were consumed worldwide, even though the efficacy of the products used had increased significantly (by a factor of 10 compared with 1945), meaning that in fact ten times less was needed to produce the same effect, but this was clearly not the case in practice. Europe consumes 3.9 kilograms per hectare (about 3.5 pounds per acre), compared with an average of 1.5 kilograms worldwide. At the beginning of the twenty-first century—and mainly on the strength of its vineyards, which were a major consumer of herbicides, insecticides, and especially fungicides—France remained the world's third largest pesticide user with about 110,000 metric tons per year, behind only the United States and Japan.[82]

Committed to the "Green Revolution," many countries in Asia, Africa, and Latin America modernized their agriculture by adopting high-yielding plants that required more fertilizer and pesticides. Between 1972 and 1985, pesticide imports increased by 261 percent in Asia, 95 percent in Africa, and 48 percent in Latin America. Asia was still a small consumer before 1970, but it became an important outlet for large international groups that set up factories with the support of authorities who promised loose regulatory oversight. While some agricultural chemicals were progressively banned, many other substances were introduced during the 1970s and 1980s: pyrethroids, then atrazine (finally banned in France in 2002), and neonicotinoids—which affect the central nervous system of insects (Gaucho, produced by Bayer, for example, which was partially banned in France in 2009)—as well as systemic (or "total") herbicides whose active substance is glyphosate. The best-selling glyphosate in the world is Monsanto's Roundup, marketed since 1975, although relicensing of its use is currently under discussion in the European Union. At the same time, new techniques, such as seed coating, allow permitted substances to become increasingly invasive.[83] The situation seems more ambiguous than ever. Pesticide use in France has become much more stringent: aerial spraying is strictly controlled, and on January 1, 2017, local authorities banned the use of chemical sprays in public spaces. After 2019, pesticide products will no longer be sold to individuals. However, in spite of the progress of organic farming, agriculture at large remains dependent on pesticides whose consumption continues to grow each year—by nearly 6 percent between 2011 and 2013 and 9 percent in 2014, according to the official statistics of the Ministry of Agriculture.[84]

Demand for fossil fuel energy for transportation has not decreased either. Since the 1970s, modes of transport have steadily multiplied and the global ecological footprint of the automobile continues to grow.[85] In this sector, the question of pollution is still mainly thought of in technical terms (references to "clean cars," for example), despite the strong criticisms castigating automobile civilization formulated since the 1970s. In France, the number of cars grew from 24 million in 1985 to 38 million in 2015. Around the world, nearly 100 million vehicles are produced a year, mainly in China, the United States, Japan, and Germany. One billion cars were in circulation in 2010. According to a French free newspaper—one mainly subsidized by advertisers—current trends indicate that this number could swell to 1.7 billion by 2035; the article concluded that the car must remain "an irreplaceable instrument of freedom of movement and that—made cleaner, smarter, more efficient, and possible to share—it will find new life."[86]

Catalytic converters have ensured a reduction in polluting emissions such as carbon monoxide and nitrogen dioxide. Filters on diesel engines have reduced pollution by fine particles. But all these technical improvements are mitigated—counterbalanced—by constant increases in the size of vehicles, their power, the myriad added onboard electronic devices such as GPS and air conditioning, as well as by the greater distances traveled.[87] It is also too often forgotten that the technologies expected to reduce pollution—such as the catalytic converter—rely on the use of precious metals (including palladium and platinum), and that these devices displace pollution more than they remove it.[88] Efforts to develop less polluting engines and fuels make it difficult to seriously rethink whether all of these objects are truly necessary or if we could be using them all in a more frugal manner.[89] Many official reports attempt to assess the health and financial costs of air pollution generated by transportation; they would be $900 billion for the OECD area alone.[90] Since the 1970s, motorization has also been extended to all the middle class populations of the world, to the great benefit of state affluence and manufacturers, who continue to see the automobile as an indicator of economic prosperity.[91] In India, cars are now one of the main causes of urban pollution in the country. At the end of the 1990s, the transportation sector already accounted for more than 40 percent of the petroleum products consumed in the country; pollutions from nitrogen oxide and airborne particles were derived mainly from modes of transport. It was only in 1990 that the first emission standards were adopted in the country.[92] In the last twenty years, India has also become a major car manufacturer, and the Tata company is striving to reinvent Ford's promise of cars for all, in Southern countries. Almost nonexistent in the 1970s, the Chinese car fleet reached 40 million vehicles in circulation at the beginning of the twenty-first century. Massive pollution of Chinese metropolitan areas has gone hand in hand with this spectacular increase. Their peaks in pollution are regularly reported by the media.[93]

Alongside land vehicles, air transport must also be considered. It was still in its embryonic stages and reserved for a small elite in the 1960s (110 million passengers), but the number of passengers then doubled every 10–15 years to 1 billion in 1987, 1.5 billion in 2003, and up to 3.7 billion people in 2016. It is expected to rise to 7 billion by 2030, which will have significant ecological impacts, particularly in terms of CO_2 emissions.[94] As with cars, in the face of such growth, authorities and manufacturers are relying on the "technological leaps" that are supposed to inaugurate "sustainable" aviation. Thanks to better traffic management, the development of more efficient aircraft and engines, and the use of agrofuels, manufacturers are expecting reductions in pollutions from aeronautical travel and transportation.[95]

The explosion of greenhouse gas emissions due to widespread mobility and motorization, along with the growing focus on the planet's climate issues, have given new vigor to the nuclear trajectory. Initially the construction of the first nuclear power plants in the late 1960s met with significant apprehension with regard to radiation. These fears diminished however, while the number of reactors in the world increased rapidly. After the oil crisis, the Messmer plan in France—named after the Prime Minister and adopted in March 1974—prompted a vast, unprecedented, and unparalleled construction program.[96] In the early 1990s, there were about 400 reactors in the world, but then construction stagnated.

In 2012, the nuclear sector represented 12.9 percent of the electricity produced worldwide and 4.4 percent of the energy consumed. The United States had the largest number of reactors in operation. France ranked first in nuclear power's share of the nation's energy mix: 37 percent of its energy consumption and 78 percent of its electricity. Japan has also become a major nuclear country—the third largest in the world—with 54 reactors in operation that produced 29 percent of its electricity before 2011. The late 1970s, however, marked the beginning of major questions surrounding nuclear safety, after several major accidents—at Three Mile Island in the United States in 1979, and especially at Chernobyl in the USSR (Ukraine) in 1986 and Fukushima in 2011 in Japan. As early as 1957, the USSR had experienced a nuclear accident in Kyshtym (in the Urals), at the Mayak nuclear complex. An explosion was the source of a large radioactive contamination.[97] The Chernobyl disaster, however, was the critical disaster at a crucial time that likely quelled nuclear power's potential growth. It called into question the cost–benefit merits that nuclear purported to have in its favor. From the meltdown, a radioactive cloud passed over Europe; the effects were felt several thousand kilometers away. Determining a precise balance sheet of the disaster is impossible to evaluate and subject to controversy, but between 25,000 and 125,000 "liquidators," required on the site to plug the breach, would have died rather quickly, more than 200,000 would have suffered disability, and between 14,000 and 985,000 people could have died from radioactive contamination. A few years after the fact, the UN and WHO estimated that 10 million people could potentially suffer the consequences of these radioactive pollutions, many of them through blood or thyroid cancers. The area around the plant will remain contaminated for hundreds of years; Pripyat, which at that time had a population of 50,000, has become a ghost town.[98] The Fukushima disaster reactivated the same fears and questions, not least because Japan's safety protocols and measures were amongst the most stringent in the world. Here again, reports court controversy and many studies are still in progress, but the radioactive pollution is proven, and it continues to spread (especially in the sea) through the plant's cooling water systems, which remain unconfined. Radioactivity levels are higher than initially suspected and continue to raise alarm bells.[99]

Faced with these disasters, it is impossible to extricate the issues surrounding nuclear risk from the pollution debate. Many regard nuclear power as a source of "green" or "clean" energy, because it emits fewer toxic emissions for flora and fauna, the environment, and those living near power stations. In France, this position is embodied, for example, by the engineer Jean-Marc Jancovici.[100] However, nuclear energy does emit CO_2 throughout its cycle, during the extraction of uranium and its transformation into fuel and during the construction and dismantling of reactors. There are also numerous insidious radioactive contaminations from mining waste, authorized discharges, or chronic pollution related to waste management and reactor dismantling, which are generally disguised by the establishment of "a dose management system."[101]

Governments must defend their positions on nuclear power in the face of antinuclear criticism in the age of global warming.[102]

The issue of radioactive waste remains more relevant than ever. Until 1982, although it was a controversial practice even within the engineering community, 100,000 metric tons of radioactive waste were dumped at sea in concrete containers; 75 percent of it came from the United Kingdom. The practice was subsequently prohibited by the London International Convention (1993), but discharges into the sea by pipeline still happen, such as at the nuclear fuel reprocessing facility at La Hague, at the end of the Cotentin peninsula in France. First rendered operational in 1966—and initially for military use—and attached to a waste storage facility, the plant converted to civil engineering in the 1970s and treats used fuel from Germany, the Netherlands, Switzerland, Belgium, and Japan, to separate out the radioactive elements. The recovered plutonium is then transported to the Marcoule plant to produce MOX, a nuclear fuel from waste reprocessing. In 2014, France had 1,460,000 cubic meters of radioactive waste. Burial projects under thick clay layers at great depth—as on the site of Bure in France—are subject to intense scrutiny and controversy. Justified as the only viable solution for some, it is widely considered a huge threat and a short-sighted strategy that simply delays or postpones the prospect of risk from hazardous contaminations.[103]

The Construction of Impotence

The regulation of pollutions continued to follow the responses, methods, and practices that typified the regimes established in the nineteenth century. Expertise remained the primary criterion for legitimizing political decisions and new regulations—an inchoate bundle of commands and constraints that did little to diminish contaminations, even though marked progress in measuring instruments and scientific knowledge should have yielded better results. Toxicology, a discipline that explores the risks associated with the exposure of living organisms to foreign substances—particularly at the forefront of alerting people to the dangers of pollution—was often diverted in favor of a "public health from on high" that tended to dismiss the severity of risks, such as the responsibility for pollution in the explosion of cancer. In addition, manufacturers continued to insist that the importance of industrial secrecy absolved them from having to publicize toxicity studies.[104] As for epidemiology, it was marginalized because it struggled to establish concrete evidence, especially with the immediacy that many toxic hazards required; also, to produce proven results in epidemiology implied accepting the principle of large-scale human experimentation and the inherent dangers that it entailed.[105] As in the past, the credibility gap between scientists and the public challenged scientific expertise and authority; more and more secular groups challenged the monopoly of the scholarly and administrative expertise that followed in the footsteps of such nineteenth- and twentieth-century trusts.[106]

Yesterday as today, inaction is obviously never absolute. Some toxic products have been banned, either totally or partially, while their uses have been subjected to more and more controls. This is the case with asbestos, banned in 2005 throughout the European Union (in 1993 in Italy, 1994 in Germany, 1997 in France and Belgium, and 1999 in the United Kingdom), even if its legacy of health effects persists and will continue to be felt in the coming decades: in France alone, between 61,000 and 118,000 deaths were attributable to exposure to asbestos between 1955 and 2009, and 75,000 additional victims are expected by 2050.[107] Asbestos use continues to be allowed outside Europe but it is mainly produced in Russia, China, Kazakhstan, and Brazil, with a volume of over 2.5 million metric tons a year; this figure is (still) steadily rising. Canada, an historic producer and exporter, only slowed production after 2011 and would not cease production until 2018.[108] Another emblematic product is chlorofluorocarbons (CFCs), used in refrigeration and air-conditioning equipment, for which the Montreal Protocol on Substances that Deplete the Ozone Layer was reached in 1987, and which all countries signed in 2009. Since global production of CFCs was halted in 1994, a drastic planetary reduction in emissions constitutes one of the great success stories in the struggle against global contaminants. A UNEP report estimates that two million skin cancers are prevented each year by this ban, and the ozone layer stopped receding in 2016. As ever, however, the task is not yet complete; among the gases introduced to substitute for CFCs in manufacturing, hydrochlorofluorocarbons (HCFCs) will not be completely banned for decades to come, and hydrofluorocarbons (HFCs)—even if they do not contravene the Montreal Protocol by destroying the ozone layer—are potent greenhouse gases. The Kigali Amendment of 2016 was signed in Rwanda as an addition to the Montreal Protocol. Its focus took aim at outlawing harmful CFC substitutes, but without really knowing what to replace them with, even as the use of air conditioning continues to spread.[109] In general, when a toxic product is banned, it is because the market can provide a ready substitute. This was the case with DDT, banned in the agricultural practices of several countries since the 1970s, a ban that the Stockholm Convention (2001–2004) sought to generalize definitively, but which India ignored.[110] This was also the case for leaded gasoline, which was banned in countries from the 1970s to the present (although some countries still allow it), or the vinyl chloride contained in aerosol sprays, banned in France since 1976.[111]

But apart from these few emblematic cases, the continuing proliferation of toxic products and the inadequacy of institutional responses dominate narratives of pollutions. The social and political construction of impotence in the face of pollutions is a major phenomenon of contemporary public policies. It can be explained by the continuity of the historical dynamics described throughout this book, such as industrial lobbying, pressures on the labor force, and a series of structural biases that ensure that any regulatory process does not hinder the bottom line: a devoted and dedicated commitment to economic growth. Most of the data used as a basis for the regulation of toxic substances and radiation are produced by industry itself. In addition, many of the regulators' experts have unclear relationships—open conflicts of interest but also more subtle connivances—with the industry they are supposed to regulate. Powerlessness can also be due to the ever more intimate relationship between industry and research. Some mercenary scientists artificially maintain a level of doubt, while private or public-private funding makes some knowledge difficult to produce. "Merchants of doubt" and interest lobbies remain active, generating uncertainty, which prevents any effective or ambitious measure.[112] The gap between alarmist speeches and actions favoring "business as usual" is now emerging as a further obstacle and source of regulatory impotence. The influence enjoyed by the big polluting industries has probably never been so important; scandals reveal the close links between industry and the scientific community. One example among many: in 2017, Michel Aubier, a renowned French pulmonologist and author of numerous reassuring declarations on the dangers of air pollution from diesel, received a six-month suspended sentence for lying about his ties with the oil giant Total since the 1990s.[113]

Opposition to environmental regulations is not just a matter of interest and lobbying in its classical form, but also a process of building ignorance that is developing in an increasingly subtle and surreptitious way. Establishing uncertainty and ignorance in the stakes of these disputes is of central importance to the "merchants of doubt" and their industrial employers, as shown for example in the case of glycol ethers; their invisibilization is eminently political.[114] All the same, the way in which public health problems are framed around pesticides remains rooted within the same confrontation and paradigm of uncertainty that pits a government committed to data and instruments against skeptics determined to undermine what the data actually say.[115] Within this macabre—and all too common—refrain, science's capacity to serve as an asset in ascertaining truth and salubrity is fraught with difficulties. Instead, science has become an instrument of indecision, inaction, and even ignorance. Agnotology— the active construction of ignorance—was already present in the nineteenth century, but its centrality in toxic debates accelerated after 1945; it remains more decisive than ever in our thinking about contemporary pollutions.[116] Similar findings can be made on the issue of workers' health related to emerging chemicals; professional risk management systems are designed by and for experts, which makes it very difficult for employee representatives and citizens to take ownership of them. If such controls become, for all intents and purposes, invisible through a lack of knowledge or access to knowledge, the status quo prevails.[117] At the same time, authorities preserve an aura of ignorance surrounding past pollutions by reconverting old industrial sites and giving new meaning to the scars they leave on the landscape. Industrial heritage policies, initiated during the 1970s in Europe and the United States, were designed—basically—to remove traces of old polluted areas. Indeed, the industrial patrimony that transforms remaining sites into glorious monuments comes from a strategy of concealing degraded spaces sacrificed to industry. Following in the hallowed footsteps of the buildings of the rural proto-industry of the early modern period and the metallic structures of the nineteenth century, it is now possible to witness incipient efforts in the patrimonialization of the contemporary chemical industry.[118]

The construction of impotence takes many forms, which explains how contaminations continue to grow in quantity and severity, despite the many devices and laws that have been introduced to attempt to stem the phenomenon. Namely, regulatory devices are legion, and they have evolved away from protecting the polluters the way they did in the nineteenth century. Environmental law—and pollution law, more specifically—has emerged as a new branch of law that has increasingly come to include the criminal sphere; notions of environmental crime and even ecocide have become common.[119] Moreover, the recent period has been marked by the burgeoning of environmental administrations and the strengthening of international cooperation, leading to some successes and many disappointments.[120] But a golden rule persists: the law remains ineffective without the political will of the state. As early as the 1980s, reports and surveys highlighted the great inefficiency of environmental law, a "baroque edifice" marked by uncertainty and contradictions.[121] In the case of French establishments classified as industrial, the number of inspectors has remained structurally insufficient, and their numbers grow much less quickly than the factories that require oversight. In 1978, France had 273 inspectors, 515 in 1988, 581 in 1996, before their numbers grew rapidly—to more than 1,000—at the beginning of the twenty-first century. But even this expansion failed to keep pace with industrial growth; the inspectors were responsible for more than 500,000 classified establishments.[122] In Great Britain, customary law and its pragmatism remain at the heart of regulations and, as in France, inspectors favor negotiations; criminal sanctions remain rare.[123]

The increasing complexity of regulations at all levels—local, national, and global—and further institutional entanglements also contribute to making well-intentioned actions ineffective by rendering decisions and initiatives useless.[124] International conventions on the environment, concluded mainly under the auspices of the United Nations or its agencies, wait years before being ratified—if they are at all.[125] Moreover, most conventions are heavily adapted to comply with the interests of, or the pressures exerted by, the major industrial stakeholders. For example, the climate conventions and the fight against greenhouse gases are based on the carbon market—an emissions trading system that the industry has managed to impose—rather than on taxation. As a result, the extensive environmental commodification introduced in the nineteenth century is allowed to continue, with polluters paying compensation while being allowed to operate their factories.[126]

REACH (Registration, Evaluation and Authorization and Restriction of Chemicals), a recently formed agency for European chemical oversight, constitutes another attempt at regulation at the supranational level that encountered difficulties in its application, not least because of the numerous exemptions and derogations it introduced. Entered into force in 2007 after lengthy negotiations, its objective is to record, evaluate, and better control chemicals manufactured, imported, and placed on the European market. But the central paradox of this regulation came from the fact that new formulations arrived on the market far more quickly than the new policing system could assess their toxicity. Rather than catching up with the delay in evaluating new chemicals, the situation became all the more chaotic and out of control. As a means of managing its sprawling obligations, the regulations were applied only to new or imported substances whose volume exceeded 1 metric ton per year per company. In other words, many molecules escaped attention, especially nanomolecules, those that were very light, or those for which the duration of exposure poses a much more significant threat than the dosage.[127]

The proliferation of environmental law and regulations to circumscribe pollution, therefore, changed scale, but its evolution has never been a wholly linear and continuous process. Further, as in the past, the most polluting industries do not cease to create new methods to resist, obfuscate, or soften environmental standards or reduce opposition. While many international and supranational regulatory efforts express a degree of earnest goodwill, many countries have refused to adhere to pollution control policies when they threaten to hinder their economic growth, even when a consensus is slowly being reached that no such economy–environment dichotomy exists. In China, for example, the role of the State Environmental Protection Administration (SEPA) is expanding, with the number of employees doubling between 1985 and 1995 to reach 88,000 civil servants. However, the results are mixed: at the end of the 1990s, only twelve pollutants had been subjected to controls, and only the industrial sector expressed any concern over nitrates and pesticides as major sources of contamination.[128] The authoritarian organization of the Chinese regime and the priority accorded to the rise in production constantly marginalize environmental initiatives. In 1998, a Legal Assistance Center for Victims of Pollution was launched, on the initiative of academics and environmental activists. The following year, the Communist Party imposed a temporary moratorium on industries in Beijing to limit pollution on the occasion of the celebration of fifty years of the People's Republic.[129] But regular warnings of poor air quality and smogs of unprecedented size and density have lead to a veritable "airpocalypse." China's continuing dependence on coal is the main cause of this heavy chronic pollution. If Chinese authorities regularly promise—in vain for the moment—ambitious plans of action to convert to a transformative "green" economy, it is legitimate to doubt the success of these half-measures that never counter the drive for production.[130] In 2014, according to the OECD, the economic costs of the health effects of Chinese air pollution would have reached $1,400 billion (and $500 billion for India).[131]

The example of China shows that, without really breaking with the past, environmental law and regulations depend very much on the political voluntarism of decision makers who are caught up in a mesh of factors combining representations, cultures, political power relations, and social equilibriums. For example, when the Mexico City Council decided in 2014 to remove polluting cars from the city to improve the air quality that impaired the health of its millions of inhabitants, it provoked a strong outcry that stalled any action.[132] In the absence of real political will, such achievements are a far cry from stated objectives. In 1975, the French planned to treat and eliminate 80 percent of effluents before they returned to the environment by 1990, but they never came close to reaching this target. A report published in 1988 noted that 95 percent of water pollution offenses remained unreported or unaddressed.[133] More recently, the EcoPhyto plan, adopted in France in 2008 after the Grenelle Environmental Round Table—a multijurisdictional dialogue on national environmental challenges—targeted a 50 percent reduction in the use of pesticides over the next ten years. That goal was a complete failure, since consumption (as previously noted) continued to grow.[134] The fight against pollution remains subordinate to economic and financial interests and priorities, struggles for influence, and the interests of those in power. The "polluting industrial" complex is increasingly colonizing states and governments to shape regulations to their advantage—when it is unable to organize its efforts to cleverly cheat and circumvent existing standards, as the 2016–2017 Dieselgate case, involving many European car manufacturers, suggested.[135] To make matters worse, many political leaders refuse to treat pollution as a serious problem, and remain ignorant of the issues. During George W. Bush's presidency (2001–2009), the American administration actively set back the national fight against pollution: downsizing the EPA, freezing industrial waste management programs, dismantling air and water quality laws inherited from the 1970s, and expressly denying climate change and the science behind it.[136] Even though President Barack Obama's administration undeniably paid greater attention to environmental issues, the election of Donald Trump as US President in 2016 reactivated a policy of denial that risks calling into question a large number of international agreements and research and action programs.[137]

Powers and Asymmetries

In the fight against pollution, history shows in the end how sociopolitical power relations are fundamental to understanding the motives behind the struggle. Around the 1970s, environmental movements managed to establish a series of environmental concerns on the political agenda. But aside from this brief environmental moment, the historical narrative has witnessed the recurrence of asymmetries, between a dominant doxa favorable to industrialism—whether it purports to be a new, "green" industrialism or remains orthodox in its polluting practices—and the environmental groups that alert the public and denounce contaminations while being constantly marginalized. The contemporary era appears to have fallen in lockstep with historical continuity, even if local contexts provide some nuance, and despite a new global environmental awareness that seems to be emerging little by little.

Environmental failures and the glacial pace of regulation are due to the dominant economic imagination embodied by experts and decision-makers anxious to boost growth by rejecting anything that could hinder it, because of the additional costs that this would impose on companies that must be competitive in the globalized economic system. Multinational corporations, already powerful after the war, have seen their influence increase during the last decades, because they benefit from the acceleration of mobility and the circulation of capital. These conglomerates, with subsidiaries around the world and factories in several countries, are fluid entities, relocating freely to find raw materials and cheap labor. That mobility plays a decisive role in the fate of environmental regulations. While global enterprises were denounced in the 1970s for their predatory actions and their pollutions—intellectuals such as Schumacher and Illich strongly attacked these organizations, whose practices they considered counterproductive—in the 1980s an intellectual offensive renewed and rehabilitated their public policy standing. The "new spirit of capitalism," carried by an aggressive managerial discourse and indoctrinated executives, extolled the figure of the entrepreneur as forward-thinking pioneer, while Marxism was reduced to crisis, its place in social criticism in tatters. In a period that demanded optimism and was typified by a permanent celebration of financial success and consumption, enterprise became an unassailable totem.[138] Numbering 7,000 in the early 1980s, there are now roughly 80,000 multinationals, employing 75 million people and accounting for nearly 70 percent of global trade flows. While companies in some Southern countries are becoming more international, most multinationals are from the North and contribute significantly to the uneven growth of pollution. They are widely denounced for their role in the world's social and ecological crisis. The American linguist Noam Chomsky sees in it "private tyrannies" and "totalitarian institutions" that exercise illegitimate power outside of any democratic control. Nevertheless, their growing power and influence has been unyielding. Multinationals embody a new type of empire based on hierarchical social relations, the domination of nature, and legal impunity.[139] The business world boosts itself by stifling its opposition, especially investigative journalism. Thus, to counter the revelations of whistleblowers about their secretive and shameful practices that dodge taxation or the law, the European Parliament approved a "directive on business secrecy" in April 2016 for the benefit of these large companies. The underground influence of these lobbies helped to block any reform project and any attempt to impose limits on polluting emissions.[140]

In this framework, which prioritized a business-as-usual mentality, policies with an environmental objective also had to try hard to preserve competitiveness and the pursuit of profit, thus leading to the tragic legacy of "sustainable development" rhetoric. The so-called Porter hypothesis—coined in 1991 by a Harvard professor of management—explains that environmental regulations can both reduce pollution and not threaten corporate profits. According to this thesis, good and flexible standards push corporations to innovate to reduce nuisances, which favors productivity gains and in the long run the expected benefits would be greater than the costs of depollution.[141] In keeping with the long-term dynamics explored in this work, this thesis ultimately continues to protect polluters by creating an ever more favorable framework for the deployment of their activities. While the Porter hypothesis claims to propose means of controlling industry, in the case of global warming the economists who dominate the disciplinary field continue to invoke perfect markets, the rationality of agents, an infinitely generous Nature, as well as profuse rhetoric on the legitimization of a carbon market and industrial investment in polluting sectors, offset by other front-line actions.[142]

In recent years, "ecological modernization" as a new paradigm has also been emerging. The term refers to new approaches to the relationship between the economy and the environment, based on managerial, rationalist, and technical conceptions. It makes the protection of the environment a condition for future growth and a source of economic opportunity.[143] According to its proponents, the scale of pollution and global environmental imbalances have become such that only complete control of the planet's physical processes would be able to counterbalance their effects. From here, legions of responsible, informed, pragmatic, and optimistic people draw the future's paths and propose their solutions, which are mainly technical. Published in the United States in 2015, the "Ecomodernist" Manifesto, signed by 18 scientific leaders, is a striking example of this strategy. For these "ecopragmatists," as they choose to call themselves, pollutions are essentially a problem of innovation and the need for better technical management around the world. Global climate change, the depletion of the stratospheric ozone layer, ocean acidification, air and water pollutions: all could be solved by more innovations to "decouple human well-being from environmental destruction" while regreening the earth and retaining the high-tech comfort of the wealthiest. Coal and oil are praised for their ability to preserve forests; pesticides and tractors are celebrated because they increase yields and avoid further agricultural expansion; nuclear energy is presented as the best solution available in the face of energy shortages.[144]

Political elites, industrialists, and scientists are betting on technological breakthroughs and innovation as a solution to the threats posed by pollutions.[145] In the final agreement from the Climate Conference held in Paris in December 2015, the words "pollution," "energy," "oil," "coal," and "ecology" are absent; on the other hand, mention of "techniques" and "technologies" is omnipresent. The consensus on climate risks does not call into question the materialism of lifestyles but promotes new technologies and vast technoscientific projects in climate engineering.[146] In this framework, geoengineering constitutes a fountain of youth for technological fervor. It is promoted by multinationals and a few major public institutions as offering technoscientific solutions by imagining, for example, systems that harness pollutions. Geoengineering advocates artificializing everything to control everything, to the point of injecting massive amounts of sulfur into the atmosphere to compensate for the effect of greenhouse gases, as proposed by Paul Crutzen, recipient of the Nobel Prize in Chemistry.[147] Following nineteenth-century scientific rationale, pollutions conceptually remain a uniquely scientific problem. Many reports intended to guide political choices promise the advent of "e-agriculture," which is cleaner and more efficient, thanks to big data and robotics.[148] Oxymorons proliferate: the "green economy" is rich in such promises of a better, greener, richer future. Ecodesign must produce materials designed from the outset to avoid pollutions. The future imaginary will introduce "microspheres capable of cleaning the blood of victims of chemical, radioactive, or biological pollutants," or miraculous coatings capable of absorbing dangerous substances, like a "cleansing tar that ingests nitrogen molecules" or "pollution-eating microbes" capable of cleaning the soil and water.[149] For their part, ecologists and biologists are studying "ecological solutions to diffuse pollutions." While full restoration of an ecosystem is impossible, there are more and more natural strategies for decontaminating polluted landscapes. "Soil phytoremediation" will allow the use of plants and microorganisms to "extract, contain, deactivate, and degrade soil contaminants." Many call for the development of "ecotechnologies" and a "restorative science" to reactivate the technicist paradigm once in existence. The specialists recognize, however, that all these depollution techniques are insufficient because there are limits and thresholds beyond which any repair of the environment becomes illusory.[150]

Other prophets announce the advent of new, "inexhaustible and clean" energy trajectories. Jeremy Rifkin has heralded the advent of a "hydrogen economy" that would break with the old nuisances of the industrial age.[151] For the past ten years, there have been countless publications, projects, and investments to promote the development of the hydrogen regime, presented as the "only way" forward in "the fight against air pollution," according to Pierre-Étienne Franc, Vice President of the advanced Business and Technologies (aB&T) unit with the Air Liquide group, and an influential lobbyist for the sector in European institutions.[152] Geologists, for their part, champion the infinite potential of "natural hydrogen," which would not require any fossil energy in its production, and would thus offer a "nonpolluting" energy, even if the authors recognize that such a promise is still far from proven.[153] The fight against pollution has become an omnipresent marketing and advertising argument for selling new products, obtaining financing, and attracting the interest of public authorities and citizens; the craze for renewable energies (solar, wind, and so on) is sometimes part of this logic.

Faced with this steamroller that combines "economic realism," optimism vis-à-vis the virtues of the market as well as human ingenuity, and liberal and entrepreneurial ideas, movements that center their approach around the environment and pollution and insist that economic decisions should be constrained by the planet's ecological rhythms are invariably marginalized and ignored. This marginality does not mean that their warnings go unheard, but rather that their influence on regulations is very limited and always on the fringes of decision-making. Indeed, apocalyptic fiction films and documentaries, journalistic investigations, and committed essays rehash the problems with the status quo, condemning in each country the responsibility of industrialists and their connivance with policymakers. In France, journalists such as Hervé Kempf, Fabrice Nicolino, and Marie-Monique Robin—like Jean-Pierre Rogel in Quebec—are amongst the most vocal critics of the enduring status quo.[154] It has become fashionable to consider ecology as a new, widespread doxa of the future and that resistance to polluting practices could become a central tenet of contemporary life. It is true that antipollution militancy has acquired a form of legitimacy, inaugurating "hybrid societies"—dark green or light green depending on the extent of the movement in each country. At the same time, the institutionalization of organizations and ecological militancy continues along different paths.[155] Everywhere in the world, environmental organizations are emerging, green parties are growing on the political landscape, and international NGOs are extending their fields of action to fight against pollutions and alert public opinion to their dangers.[156]

For all that, however, pollution and contamination have not declined over the past forty years, with some local and product-specific exceptions. Multiple standardization strategies are at work to disqualify political ecology and make pollutions tolerable if they can be controlled. The idealists with a vision of infinite growth and happy globalization—blind to the ecological consequences of their policies—are constantly striving to absorb criticism and succeeding in transforming the radical and apocalyptic discourse into resigned acceptance.[157] More broadly, the question of pollutions is, however, euphemistic in its media framing: the solution entrepreneurs insist on small individual gestures and good practices without ever calling into question the global organization of the world and its productive and consumerist model.[158]

While pollutions accentuate inequalities and global injustices, their regulation requires a radical reshaping of power and expertise. The insistence on innovation alone in order to reduce the materialism of the economy and invent "clean" and inexhaustible energies never permits reimagining sociopolitical organizations and lifestyles, which would be necessary to live in a world without pollution.[159] In the face of modernizing promises and multiple strategies of depoliticization and dissimulation, other avenues involve considering trajectories of less materialism or energy sobriety as the only methods capable of reducing the ecological footprint of contemporary societies and the quantity of polluting substances released. Recent scenarios demonstrate there are models that inspire hope. They involve rethinking production and consumption structures, like the path proposed by the négaWatt association, which is paving the way for 100 percent renewable energy by 2050, provided it is given the freedom to develop.[160] Similarly, Jason Corburn's study of "street science" shows how Brooklyn's community movements for asthma, air pollution, and lead contamination are promoting the recognition of environmental inequalities and galvanizing a democratization of the construction of the problem of pollutions.[161] Rather than accentuate the technoscientific artificialization of the world in order to try (at best) to control pollutions that are considered inevitable, new projects involve turning away from the promise of infinite growth in favor of more sober and equitable ways of life.

*   *   *   *

The deleterious effects of human activities on the environment turned from local nuisances into global pollutions. The historical exploration of this mutation has shed light on its sources and implications, giving meaning to a phenomenon that now looks like a runaway race into the abyss. Regarded from the point of view of the social sciences, this diagnosis must be shared with the other sciences: physical, biochemical, ecotoxicological, medical, legal, and economic. The planet is growing warmer, seas are becoming more acidic, species are disappearing, bodies are being altered: looking at all of this from an historical point of view makes it possible not to sink into a state of stupefaction, or to only envision

Figure 11.2
Project: "Пляшем" (Dancing); author: Pasha Cas; curator: Rash X; photo: Olya Koto; location: Temirtau, Kazakhstan, 2016. Displayed on the side of an apartment building in the industrial city of Temirtau in Kazakhstan, this work by street artist Pasha Cas evokes Henri Matisse's famous painting *Danse*. The image warns of a world that idolizes wealth derived from oil, around which people turn with an illusion of being sated. Meanwhile, in the background, lead pollution from the city's metallurgical plants is five times higher than authorized standards.

an inevitable process on which we could not act effectively. Instead, we can understand that pollution is above all a social and political fact that relies on the ideas of progress that must be discussed at the time of mass contaminations. It is true that enduring pollutions, their persistence in synthetic or nuclear molecules, their geographic dispersal to the most isolated areas, from the Arctic to space, and the unprecedented fragmentation of residues pose a particularly difficult problem and make any resolution seem obscure and daunting. It is possible—indeed, important—to hope, however, that a long-term reflection that points out the inflections, the pivotal moments, and the inertia, as well as the means of action and the balance of power, might offer a key to understanding and opening a door to the emergence of new social and political configurations, able to face contemporary environmental challenges by ushering in a more just and less toxic world.

Notes

Foreword

1. Caroline Shields, ed., *Impressionism in the Age of Industry* (New York: Del-Monico Books, 2019).

2. In this fascination, the Impressionists were not alone. The British landscape painter Joseph Mallord William Turner's 1844 painting of the Great Western Railway, *Rain, Steam, and Speed* evokes similar themes. In the same general period, in 1876, the American poet Walt Whitman would herald the steam locomotive as "Type of the modern—emblem of motion and power—pulse of the continent." Turner, *Rain, Steam, and Speed—The Great Western Railway*, 1844. Oil on canvas, 91 x 121.8 cm. National Gallery, London, Turner Bequest, 1856. Whitman, "To a Locomotive in Winter," *Leaves of Grass* (New York: Bantam Classics, 1983), 375.

3. I have been unable to find a reliable citation for this popular quotation. The oft-quoted line in Spanish read: "El diccionario se basa en la hipótesis, obviamente no probada, de que los idiomas están formados por sinónimos equivalentes." My instinct is that the quotation may be something Borges said rather than wrote. On Borges and translation, see Ana Gargatagli and Juan Gabriel López Guix, "Ficciones y Teorías en la Traduccíon: Jorge Luis Borges," *Livius* 1 (1992), 57–67. I am grateful to József Szabó and James McDonald for their efforts in trying to track down the original "Borgesism."

4. James Woods, "Movable Types: How *War and Peace* Works," *New Yorker* (19 November 2007). Woods's discussion was inspired by a new, quite literal translation of Leo Tolstoy's *War and Peace*, by Richard Pevear and Larissa Volokhonsky. Drawing on Pevear's introduction, Woods stresses that translation's role is not simply a matter of transferring "meaning from one language to another," but more "a dialogue between two languages."

5. Jorge Luis Borges, "The Homeric Versions," in *Selected Non-Fictions*, edited by Eliot Weinberger (New York: Penguin Books, 1999), 69–74. Quotation is from page 69.

6. Damian Carrington, "Air Pollution Deaths are Double Previous Estimates, Finds Research," *Guardian* (12 March 2019).

Introduction

1. Daron Acemoglu, Philippe Aghion, Leonardo Bursztyn, and David Hemous, "The Environment and Directed Technical Change," *American Economic Review*, vol. 102, no. 1 (2012): 131–166.

2. John Copeland Nagle, "The Idea of Pollution," *UC Davis Law Review*, vol. 43, no. 1 (2009): 1–78.

3. Mary Douglas, *De la souillure: Essai sur les notions de pollution et de tabou* (Paris: Maspero, 1971 [1966]).

4. Patrick Fournier, "De la souillure à la pollution: Un essai d'interprétation des origines de l'idée de pollution," in *Le Démon moderne: La pollution dans les societes urbaines et industrielles d'Europe*, edited by Christoph Bernhardt and Geneviève Massard-Guilbaud (Clermont-Ferrand, France: Presses universitaires Blaise-Pascal, 2002), 33–56.

5. Adam W. Rome, "Coming to Terms with Pollution: The Language of Environmental Reform (1865–1915)," *Environmental History*, vol. 1, no. 3 (1996): 6–28.

6. Edward Frankland, "Les eaux de Londres," *Revue des cours scientifiques de la France et de l'étranger*, no. 3 (19 December 1868): 34–40.

7. For example, in France in two articles in *Bulletin de la Société libre d'émulation du commerce et de l'industrie de la Seine-Inférieure, année 1873* (Rouen: Henry Boissel, 1874), 70, 72, and 197.

8. The Lancet, *Pollution de la Tamise par la vidange* (Paris: J. Cusset, 1883); André-Justin Martin, *Congrès internationald'hygiène et de démographie de 1889: Rapports sur la protection des cours d'eau et des nappes souterraines contre la pollution des résidus industriels* (1889).

9. Peter Thorsheim, *Inventing Pollution: Coal, Smoke, and Culture in Britain since 1800* (Athens: Ohio University Press, 2006).

10. Christoph Bernhardt and Geneviève Massard-Guilbaud, "Écrire l'histoire des pollutions," in *Le Démon moderne*, edited by C. Bernhardt and G. Massard-Guilbaud (Clermont-Ferrand, France: Presses universitaires Blaise-Pascal, 2002), 9–30.

11. Guy Debord, *A Sick Planet,* translated by Donald Nicholson-Smith (London: Seagull Books, 2008), 77.

12. Stéphane Foucart, *La Fabrique du mensonge: Comment les industriels manipulent la science et nous mettent en danger* (Paris: Denoël, 2013); Naomi Oreskes and Erik M. Conway, *Les Marchands de doute: Ou comment une poignée de scientifiques ont masqué la vérité sur des enjeux de société tels que le tabagisme et le réchauffement climatique* (Paris: Le Pommier, 2012 [2010]).

13. Attempts at synthesis do exist. See, for example, Adam Markham, *A Brief History of Pollution* (London, Earthscan, 1994).

14. Grégory Quenet, *Qu'est-ce que l'histoire environnementale?* (Seyssel, France: Champ Vallon, 2014).

15. John R. McNeill, *Something New under the Sun: An Environmental History of the Twentieth Century* (New York: Norton, 2000) which was not translated into French until 2010: *Du nouveau sous le soleil: Une histoire de l'environnementmondial au xxe siècle* (Seyssel, France: Champ Vallon, 2010); Joachim Radkau, *Nature and Power: A Global History of the Environment* (Cambridge: Cambridge University Press, 2008 [2000]); Stephen Mosley, *The Environment in World History* (London: Routledge, 2010).

16. Geneviève Massard-Guilbaud, *Histoire de la pollution industrielle (France, 1789–1914)* (Paris: EHESS, 2010), 325.

17. Christophe Bonneuil and Jean-Baptiste Fressoz, *L'Événement anthropocène: La Terre, l'histoire et nous* (Paris: Le Seuil, 2013); Jason Moore, *Capitalism in the Web of Life: Ecology and the Accumulation of Capital* (London: Verso, 2015).

18. Alf Hornborg, *Global Ecology and Unequal Exchange: Fetishism in a Zero-Sum World* (London: Routledge, 2011), 48.

19. Stefan Giljum and Roldan Muradian, "Physical Trade Flows of Pollution-Intensive Products: Historical Trends in Europe and the World," in *Rethinking Environmental History: World-System History and Global Environmental Change*, edited by Alf Hornborg, John R. McNeill, and Joan Martinez-Alier (Lanham, MD: AltaMira Press, 2007), 307325.

Part I

1. Fernand Braudel, *Civilisation matérielle, économie et capitalisme (xve-xviiie siècle)*, 3 vols. (Paris: Armand Colin, 1979); Immanuel Wallerstein, *The Modern World-System*, vol. 2, *Mercantilism and the Consolidation of the European World-Economy (1600–1750)* (Cambridge: Cambridge University Press, 1980).

2. Serge Gruzinski, *Les Quatre Parties du monde: Histoire d'une mondialisation* (Paris: La Martinière, 2004); Gruzinski, *L'Aigle et le dragon: Démesure européenne et mondialisation au xvie siècle* (Paris: Fayard, 2012); Sanjay Subrahmanyam, "Connected Histories: Notes towards a Reconfiguration of Early Modern Eurasia," *Modern Asian Studies*, vol. 31, no. 3 (1997): 735–762.

3. Christopher A. Bayly, *La Naissance du monde moderne* (Paris: Éd. de l'Atelier/Éd. Ouvrières, 2007 [2004]), chap. 1; Kenneth Pomeranz, *Une grande divergence: La Chine, l'Europe et la construction de l'économie mondiale* (Paris: Albin Michel, 2010 [2000]).

4. John F. Richards, *The Unending Frontier: Environmental History in the Early Modern Centuries* (Berkeley: University of California Press, 2003); Carolyn Merchant, *Death of Nature: Women, Ecology, and the Scientific Revolution* (San Francisco: Harper & Row, 1980).

5. Robert B. Marks, "The Modern World since 1500," in *A Companion to Global Environmental History*, edited by John R. McNeill and Erin Stewart Mauldin (Oxford: Wiley-Blackwell, 2012), 57–78.

6. Jan De Vries, "The Industrial Revolution and the Industrious Revolution," *Journal of Economic History*, vol. 54, no. 2 (1994): 249–270.

7. Jan Luiten Van Zanden, ed., *The Long Road to the Industrial Revolution: The European Economy in a Global Perspective (1000–1800)* (Leiden: Brill, 2009); Patrick Verley, "La révolution industrielle. Histoire d'un problème," in *La Révolution industrielle*, edited by P. Verley (Paris: Gallimard, 1997), 13–120; Verley, *L'Échelle du monde: Essai sur l'industrialisation de l'Occident* (Paris: Gallimard, 1997).

Chapter 1

1. Pierre Quef, *Histoire de la tannerie* (Paris/Namur: Wesmael-Charlier, 1958); Peter C. Welsh, "A Craft That Resisted Change: American Tanning Practices to 1850," *Technology and Culture*, vol. 4, no. 3 (1963): 299–317.

2. Jean-Pierre Leguay, *La Pollution au Moyen Âge* (Paris: Éd. Jean-Paul Gisserot, 1999).

3. Mary Douglas, *De la souillure: Essai sur les notions de pollution et de tabou* (Paris: Maspero, 1971 [1966]); Michel Serres, *Le Malpropre: Polluer pour s'approprier?* (Paris: Le Pommier, 2008).

4. Madhav Gadgil and Ramachandra Guha, *This Fissured Land: An Ecological History of India* (Berkeley: University of California Press, 1992).

5. Gerald Groemer, "The Creation of the Edo Outcaste Order," *Journal of Japanese Studies*, vol. 27, no. 2 (2001): 263–293; Rama Sharma, *Bhangi, Scavenger in Indian Society: Marginality, Identity and Politicization of the Community* (New Delhi: M.D. Publications, 1995); Vijay Prashad, *Untouchable Freedom: A Social History of a Dalit Community* (New Delhi: Oxford University Press, 2000).

6. Patrick Fournier, "De la souillure à la pollution: Un essai d'interprétation des origines de l'idée de pollution," in *Le Démon moderne: La pollution dans les societes urbaines et industrielles d'Europe*, edited by Christoph Bernhardt and Geneviève Massard-Guilbaud (Clermont-Ferrand, France: Presses universitaires Blaise-Pascal, 2002), 33–56.

7. Bertrand Gille, ed., *Histoire des techniques: Technique et civilisations, technique et sciences* (Paris: Gallimard, 1978).

8. Gérard Gayot, *Les Draps de Sedan (1646–1870)* (Paris: EHESS/Terres ardennaises, 1998); Jean-Michel Minovez, *La Puissance du Midi: Drapiers et draperies, de Colbert à la Révolution* (Rennes, France: Presses universitaires de Rennes, 2012).

9. Paul Delsalle, *La France industrielle aux xvie–xviie–xviiie siècles* (Paris/Gap: Ophrys, 1993), 247–257.

10. Jean-Pierre Darcet, *Description d'une magnanerie salubre* (Paris: Huzard, 1838); David Jenkins, ed., *The Cambridge History of Western Textiles* (Cambridge: Cambridge University Press, 2003), see chap. 12.

11. François Jarrige, "Quand les eaux de rouissage débordaient dans la cité. Essai sur le mode d'existence d'une nuisance (France, xviiie–xixe siècle)," in *Débordements industriels: Environnement, territoire et conflit (xviiie–xxie siècle),*

edited by Thomas Le Roux and Michel Letté (Rennes, France: Presses universitaires de Rennes, 2013), 137–154.

12. Brenda Collins and Philip Ollerenshaw, *The European Linen Industry in Historical Perspective* (Oxford: Oxford University Press, 2003).

13. Sylvain Olivier, "Rouissage et pollution des cours d'eau en Languedoc méditerranéen au xviiie siècle" in *Pollutions industrielles et espaces méditerranéens (xviiie–xxie siècle)*, edited by Laura Centemeri and Xavier Daumalin (Aix-en-Provence, France: Karthala, 2015), 29–44.

14. André Guillerme, *Les Temps de l'eau. La cité, l'eau et les techniques (nord de la France, fin iiie-début xixe siècle)* (Seyssel, France: Champ Vallon, 1983), 168–169; Leonard Rosenband, *La Fabrication du papier dans la France des Lumières: Les Montgolfier et leurs ouvriers (1761–1805)* (Rennes, France: Presses universitaires de Rennes, 2005 [2000]).

15. Brenda J. Buchanan, ed., *Gunpowder, Explosives and the State: A Technological History* (Aldershot, UK: Ashgate, 2006); Jan Lucassen, "Working at the Ichapur Gunpowder Factory in the 1790s," *Indian Historical Review*, vol. 39, no. 1–2 (2012): 19–56 and 251–271.

16. Wayne D. Cocroft, *Dangerous Energy: The Archaeology of Gunpowder and Military Explosives Manufacture* (Swindon, UK: English Heritage, 2000).

17. Denis Woronoff, *Histoire de l'industrie en France (du xvie siècle à nos jours)* (Paris: Le Seuil, 1996), 105–143; Robert Delort and François Walter, *Histoire de l'environnement européen* (Paris: Presses universitaires de France, 2001); Fernand Braudel, *Civilisation matérielle*, vol. 3, *Le Temps du monde*.

18. Braudel, *Civilisation matérielle*, vol. 3.

19. Dean T. Ferguson, "Nightsoil and the 'Great Divergence': Human Waste, the Urban Economy and Economic Productivity (1500–1900)," *Journal of Global History*, vol. 9, no. 3 (2014): 379–402.

20. Susan B. Hanley, "Urban Sanitation in Preindustrial Japan," *Journal of Interdisciplinary History*, vol. 18, no. 1 (1987): 1–26.

21. Vijay Prashad, "The Technology of Sanitation in Colonial Delhi," *Modern Asian Studies*, vol. 35, no. 1 (2001): 113–155.

22. Guy Dejongh, "New Estimates of Land Productivity in Belgium, 1750–1850," *Agricultural History Review*, vol. 47, no. 1 (1999): 7–28; Isabelle Parmentier, "Résidus de consommation, tri sélectif et recyclage à Nivelles au xviiie siècle," *Bijdragen tot de geschiedenis*, vol. 84, no. 4 (2001): 399–417.

23. André Guillerme, *Les Temps de l'eau*.

24. Chloé Deligne, *Bruxelles et sa rivière. Genèse d'un territoire urbain (xiie-xviiie siècle)* (Turnhout, Belgium: Brepols, 2003), 107–108; Emily Cockayne, *Hubbub: Filth, Noise and Stench in England (1600–1770)* (New Haven, CT: Yale University Press, 2007); Thomas Le Roux, "Une rivière industrielle avant l'industrialisation: la Bièvre et le fardeau de la prédestination (1670–1830)," *Géocarrefour*, vol. 85, no. 4 (2010): 193–207.

25. Pierre-Denis Boudriot, "Essai sur l'ordure en milieu urbain à l'époque préindustrielle: Boues, immondices et gadoue à Paris au xviiie siècle," *Histoire, économie et société*, vol. 5, no. 4 (1986): 515–528; Boudriot, "Essai sur l'ordure en milieu urbain à l'époque préindustrielle: De quelques réalités écologiques à Paris aux xviie et xviiie siècles. Les déchets d'origine artisanale," *Histoire, économie et société*, vol. 7, no. 2 (1988): 261–281.

26. Reynald Abad, "Les tueries à Paris sous l'Ancien Régime ou pour-quoi la capitale n'a pas été dotée d'abattoirs aux xviie et xviiie siècles," *Histoire, économie et société*, vol. 17, no. 4 (1998): 649–676; Margaret Dorey, "Controlling Corruption: Regulating Meat Consumption as a Preventative to Plague in Seventeenth-Century London," *Urban History*, vol. 36, no. 1 (2009): 24–41.

27. Brian W. Peckham, "Technological Change in the British and French Starch Industries (1750–1850)," *Technology and Culture*, vol. 27, no. 1 (1986): 18–39.

28. Line Teisseyre-Sallmann, "Urbanisme et société: l'exemple de Nîmes aux xviie et xviiie siècles," *Annales ESC*, vol. 35, no. 5 (1980): 965–986.

29. Eva Halasz-Csiba, "Le tan et le temps: Changements techniques et dimension historique du tannage en France (xive-xviiie siècle)," *Techniques et culture*, vol. 38, no. 1 (2001): 147–174; Jean-Claude Dupont and Jacques Mathieu, *Les Métiers du cuir* (Québec: Presses de l'université de Laval, 1981), 130.

30. Jocelyne Perrier, "Les techniques et le commerce de la tannerie à Montréal au xviiie siècle," *Scientia Canadensis, Canadian Journal of the History of Science, Technology and Medicine*, vol. 24, no. 52 (2000): 51–72; H. Depors, *Recherches sur l'état de l'industrie du cuir en France, pendant le xviiie siècle et le début du xixe siècle* (Paris: Imprimerie nationale, 1932), 85; Daniel Heimmermann, *Work, Regulation and Identity in Provincial France: The Bordeaux Leather Trades (1740–1815)* (New York: Palgrave Macmillan, 2014); Jean-Pierre Henri Azéma, *Moulins du cuir et de la peau: Moulins à tan et à chamoiser en France (xiie–xxe siècle)* (Nonette, France: Éd. Créer, 2004).

31. Robert Fox and Agustí Nieto-Galan, eds., *Natural Dyestuffs and Industrial Culture in Europe (1750–1800)* (Canton, MA: Science History Publication, 1999); Christine Lehman, "L'art de la teinture à l'Académie royale des sciences au xviiie siècle," *Methodos*, no. 12 (2012). Online: https://journals.openedition.org/methodos/2874#text (last accessed 7 November 2018).

32. Daniel Faget, "Une cité sous les cendres: Marseille et les pollutions savonnières (1750–1850)," in *Débordements industriels: Environnement, territoire et conflit (xviiie–xxie siècle)*, edited by Thomas Le Roux and Michel Letté (Rennes, France: Presses universitaires de Rennes, 2013), 301–315.

33. Fernand Braudel, *Civilisation matérielle*, vol. 1, *Les Structures du quotidien*, 421–435 and 607–615.

34. Georges Vigarello, *Le Propre et le sale: L'hygiène du corps depuis le Moyen Age* (Paris: Le Seuil, 1985); Vigarello, *Le Sain et le malsain. Santé et mieux-être depuis le Moyen Âge* (Paris: Le Seuil, 1993); Vigarello, "Le sain et le malsain," special issue of *Dix-huitième siècle*, 1977; Fournier, "De la souillure à la pollution."

35. Keith Thomas, *Dans le jardin de la nature: La mutation des sensibilités en Angleterre à l'époque moderne (1500–1800)* (Paris: Gallimard, 1985).

36. This premise follows the famous thesis proposed by the medievalist Lynn White in *Medieval Technology and Social Change* (Oxford: Oxford University Press, 1962). See also White, "The Historical Roots of Our Ecological Crisis," *Science*, vol. 155, no. 3767 (1967): 1203–1207.

37. Ferhat Taylan, *Mésopolitique: Connaître, théoriser, gouverner les milieux de vie (1750–1900)* (Paris: Presses de la Sorbonne, 2018).

38. Mark Jenner, "Environment, Health and Population," in *The Healing Arts: Health, Disease and Society in Europe (1500–1800)*, edited by Peter Elmer (Manchester, UK: Manchester University Press, 2004), 284–314; Gilles Denis, "Dégâts sur les plantes, des météores aux manufactures: De la rosée de miel (Stanhuf, 1578) aux gaz vénéneux (Candolle, 1832)," in *Ordre et désordre du monde: Enquête sur les météores, de la Renaissance à l'âge moderne*, edited by Thierry Belleguic and Anouchka Vasak (Paris: Hermann, 2013), 389–422.

39. Jo Wheeler, "Stench in Sixteenth-Century Venice," in *The City and the Senses: Urban Culture since 1500*, edited by Alexander Cowan and Jill Steward (Aldershot, UK: Ashgate, 2007), 25–38; Carlo Cipolla, *Contre un ennemi invisible: Épidémies et structures sanitaires en Italie de la Renaissance au xviie siècle* (Paris: Balland, 1992 [1978]).

40. Denis, "Dégâts sur les plantes."

41. Julien Vincent, "Ramazzini n'est pas le précurseur de la médecine du travail: Médecine, travail et politique avant l'hygiénisme," *Genèses: Sciences sociales et histoire*, vol. 89, no. 4 (2012): 88–111.

42. Mark Jenner, "The Politics of London Air: John Evelyn's *Fumifugium* and the Restoration," *The Historical Journal*, vol. 38, no. 3 (1995): 535–551.

43. Harold Cook, "Policing the Health of London: The College of Physicians and the Early Stuart Monarchy," *Social History of Medicine*, vol. 2, no. 1 (1989): 1–33.

44. Roselyne Rey, "Anamorphoses d'Hippocrate au xviiie siècle," in *Maladie et maladies. Histoire et conceptualisation. Mélanges en l'honneur de Mirko Grmek*, edited by Danielle Gourevitch (Geneva: Droz, 1992), 257–276; Charles-Louis Montesquieu, *L'Esprit des lois*, 1748.

45. Pierre Van Musschenbroek, *Essai de physique*, 2 vols. (Leiden: Samuel Luchtmans, 1739, 1751 [1726]); quotation is from vol. 2, 615–617.

46. Pierre Darmon, *L'Homme et les microbes (xviie–xxe siècle)* (Paris: Fayard 1999); Jean-Pierre Peter, ed., *Médecins, climats et épidémies à la fin du xviiie siècle* (Paris/La Haye: Mouton, 1972).

47. James C. Riley, *The Eighteenth-Century Campaign to Avoid Disease* (London: Macmillan, 1987); Vladimir Janković, *Confronting the Climate: British Airs and the Making of Environmental Medicine* (New York: Palgrave Macmillan, 2010).

48. Alain Corbin, *Le Miasme et la jonquille: L'odorat et l'imaginaire social (xviie–xixe siècle)* (Paris: Aubier, 1982); Jean-Pierre Goubert, "Le phénomène

épidémique en Bretagne à la fin du xviiie siècle (1770–1787)," *Annales ESC*, vol. 24, no. 6 (1969): 1562–1588; Sabine Barles, *La Ville délétère: Médecins et ingénieurs dans l'espace urbain (xviiie–xixe siècle)* (Seyssel, France: Champ Vallon, 1999); Arlette Farge, "Signe de vie, risque de mort: Essai sur le sang et la ville au xviiie siècle," *Urbi*, no. 2 (1979): 15–22.

49. Mark Jenner, "Follow Your Nose? Smell, Smelling and Their Histories," *American Historical Review*, vol. 116, no. 2 (2011): 335–351; Cockayne, *Hubbub*; Elizabeth Foyster, "Sensory Experiences: Smells, Sounds and Touch in Early Modern Scotland," in *A History of Everyday Life in Scotland (1600–1800)*, edited by Elizabeth Foyster and Christopher Whatley (Edinburgh: Edinburgh University Press, 2010), 217–233; Douglas Biow, *The Culture of Cleanliness in Renaissance Italy* (Ithaca, NY: Cornell University Press, 2006).

50. Quoted in Wheeler, "Stench in Sixteenth-Century Venice."

51. Gilles Denis, "Normandie, 1768–1771: Une controverse sur la soude," in *La Terre outrage: Les experts sont formels!*, edited by Jacques Theys and Bernard Kalaora (Paris: Autrement, 1992), 149–157.

52. Thomas Le Roux, *Le Laboratoire des pollutions industrielles (Paris, 1770–1830)* (Paris: Albin Michel, 2011), 47.

53. Jean Nicolas, *La Rébellion française: Mouvements populaires et conscience sociale (1661–1789)* (Paris: Le Seuil, 2002), 155.

54. Edward P. Thompson, *Les Usages de la coutume: Traditions et résistances populaires en Angleterre (xviie–xixe siècle)* (Paris: EHESS/Gallimard/Le Seuil, 2015 [1991]).

55. Michael Stolberg, *Ein Recht auf saubere Luft? Umweltkonflikte am Beginn des Industriezeitalters* (Erlangen, Germany: Harald Fischer Verlag, 1994), 18–23.

56. Laurette Michaux, *Tanneurs et travail du cuir en Moselle du Moyen Âge au xxe siècle* (Metz, France: Archives départementales de la Moselle, 1989), 94; Arlette Brosselin et al., "Les doléances contre l'industrie," in *Forges et forêts. Recherches sur la consommation proto-industrielle du bois*, edited by Denis Woronoff (Paris: EHESS, 1990), 23; Kieko Matteson, *Forests in Revolutionary France: Conservation, Community and Conflict (1669–1848)* (Cambridge: Cambridge University Press, 2015), 83–86.

57. *The Decision of the Court of Session from Its Institution to the Present Time*, vol. 16 (Edinburgh: Bell & Bradfute, 1804), 13191–13193.

58. Dorothy Porter, *Health, Civilization and the State: A History of Public Health from Ancient to Modern Times* (London: Routledge, 1999).

59. Vincent Milliot, *Un policier des Lumières* (Seyssel, France: Champ Vallon, 2011); Milliot, ed., *Les Mémoires policiers (1750–1850): Écritures et pratiques policières du siècle des Lumières au Second Empire* (Rennes, France: Presses universitaires de Rennes, 2006); Joanna Innes, "Managing the Metropolis: London's Social Problems and Their Control"; Neal Garnham, "Police and Public Order in Eighteenth-Century Dublin," in *Two Capitals: London and Dublin (1500–1840)*, edited by Peter Clarke and Raymond Gillespie (Oxford: Oxford

University Press, 2001), 53–82 and 83–92; Jérôme Fromageau, "La Police de la pollution à Paris de 1666 à 1789" (PhD diss., Université de Paris 2, 1989).

60. Paul Slack, "Responses to Plague in Early Modern Europe: The Implications of Public Health," *Social Research*, vol. 55, no. 3 (1988): 433–453; Daniel Panzac, *Quarantaines et lazarets: L'Europe et la peste d'Orient (xviie–xxe siècle)* (Aix-en-Provence, France: Édisud, 1986).

61. Cockayne, *Hubbub*, 19; James Oldham, *The Mansfield Manuscripts and the Growth of English Law in the Eighteenth Century*, vol. 2 (Chapel Hill: University of North Carolina Press, 1992), 882–925.

62. Nicolas Des Essarts, *Tableau de la police de la ville de Londres* (Paris: 1801), 17.

63. Jean-Baptiste Robinet, *Dictionnaire universel des sciences morale, économique, politique et diplomatique*, vol. 28 (London, 1783), 186 ("Servitude").

64. Nicolas Delamare, *Traité de la police*, 4 vols. (Paris: 1713–1738).

65. Joseph-Nicolas Guyot, *Répertoire universel et raisonné de jurisprudence, 17 vols. (Paris: Visse, 1784–1785)*, 17:626; *Encyclopédie méthodique par ordre des matières*, vol. 8, *Jurisprudence* (Paris: Panckoucke, 1789), 278.

66. William Hawkins, *A Treatise of the Pleas of the Crown* (London: 1762), 1:199; William Blackstone, *Commentaries on the Laws of England* (Oxford: 1770), 3:217.

67. Joshua Getzler, *A History of Water Rights at Common Law* (Oxford: Oxford University Press, 2004).

68. Ernest Wickersheiner, "Fumées industrielles et établissements insalubres à Rouen en 1510," *Annales d'hygiène publique industrielle et sociale*, vol. 5 (1927): 567–575; William H. Te Brake, "Air Pollution and Fuel Crises in Preindustrial London (1250–1650)," *Technology and Culture*, vol. 16, no. 3 (1975): 337–359.

69. Pierre-Claude Reynard, "Public Order and Privilege: Eighteenth-Century French Roots of Environmental Regulation," *Technology and Culture*, vol. 43, no. 1 (2002): 1–28.

70. Le Roux, *Le Laboratoire*, chap. 1. Quotations are from page 50.

71. Antoine-François Prost de Royer, *Dictionnaire de jurisprudence et des arrêts*, vol. 3 (Lyon, France: 1781–1788), 744.

72. Jean Georgelin, *Venise au siècle des Lumières* (Paris: EHESS/Mouton, 1978), 33–36; Wheeler, "Stench in Sixteenth-Century Venice"; Braudel, *Civilisation matérielle*, 1:569.

73. Peter Brimblecombe, *The Big Smoke: A History of Air Pollution in London since Medieval Times* (London: Methuen, 1987).

74. Leona Jayne Skelton, "Environmental Regulation in Edinburgh and York (c. 1560–c. 1700): With Reference to Several Smaller Scottish Burghs and Northern English Towns" (PhD diss., Durham University, 2012); Cockayne, *Hubbub*, 206–214; Brimblecombe, *The Big Smoke*.

75. Jan De Vries and Ad Van der Woude, *The First Modern Economy: Success, Failure, and Perseverance of the Dutch Economy (1500–1815)* (Cambridge: Cambridge University Press, 1997), 274; Peter Poulussen, "Ennuis de voisinage et pollution de l'environnement," in *La Ville en Flandre. Culture et société (1477–1787)*, edited by Jan Van der Stock (Brussels: Crédit communal, 1991), 72–76.

76. Victor S. Clark, *History of Manufactures in the United States (1607–1860)* (Washington, DC: Carnegie Institution of Washington, 1916), 68–70.

77. Braudel, *Civilisation matérielle*, 1:579.

78. Imai Noriko, "Copper in Edo-Period Japan," in *Copper in the Early Modern Sino-Japanese Trade*, edited by Keiko Nagase-Reimer (Leiden: Brill, 2015), 10–31.

79. David Arnold, *Toxic Histories: Poison and Pollution in Modern India* (New York: Cambridge University Press, 2016), 177 and 189.

80. Le Roux, *Le Laboratoire*. Quotation is from page 11.

81. William Cavert, *The Smoke of London: Energy and Environment in the Early Modern City* (Cambridge: Cambridge University Press, 2016), chap. 5.

82. Le Roux, *Le Laboratoire*, chap. 3.

83. Oldham, *The Mansfield Manuscripts*.

84. Clark, *History of Manufactures*, 63–64.

Chapter 2

1. David Blackbourn, *The Conquest of Nature: Water, Landscape and the Making of Modern Germany* (New York: Norton, 2006); Fredrik Albritton Jonsson, *Enlightenment's Frontier: The Scottish Highlands and the Origins of Environmentalism* (New Haven, CT: Yale University Press, 2013).

2. Richard L. Garner, "Long-Term Silver Mining Trends in Spanish America: A Comparative Analysis of Peru and Mexico," *American Historical Review*, vol. 93, no. 4 (1988): 898–935; Jason W. Moore, "Silver, Ecology, and the Origin of the Modern World (1450–1640)," in *Rethinking Environmental History: World-System and Global Environmental Change*, edited by Alf Hornborg, John R. McNeill, and Joan Martinez-Alier (Lanham, MD: AltaMira Press, 2007), 123-142.

3. Martin Lynch, *Mining in World History* (London: Reaktion Books, 2002), prologue.

4. Donald J. Hughes, *Environmental Problems of the Greeks and Romans: Ecology in the Ancient Mediterranean* (Baltimore, MD: Johns Hopkins University Press, 2014). The first edition of this work was titled *Pan's Travail: Environmental Problems of the Greeks and Romans* (Baltimore, MD: Johns Hopkins University Press, 1994).

5. John Ulric Nef, *The Conquest of the Material World* (New York: Meridian, 1964), 44.

6. Martin Novak et al., "Origin of Lead in Eight Central European Peat Bogs," *Environmental Science and Technology*, vol. 37, no. 3 (2003): 437–445.

7. Christoph Bartels, "The Administration of Mining in Late Medieval and Early Modern Europe (Fourteenth to Eighteenth Centuries)," in *Copper in the Early Modern Sino-Japanese Trade*, edited by Keiko Nagase-Reimer (Leiden: Brill, 2015), 115–130.

8. Jason W. Moore, "*Ecology and the Rise of Capitalism*," 2 vols., (PhD diss., University of California, Berkeley, 2007), chap. 2.

9. Michel Angel, *Mines et fonderies au xvie siècle d'après le "De re metallica" d'Agricola* (Paris: Les Belles Lettres/Total Édition, 1989); Pamela O. Long, "Of Mining, Smelting and Printing: Agricola's De re metallica," *Technology and Culture*, vol. 44, no. 1 (2003): 97–101; Marie-Claude Deprez-Masson, *Technique, mot et image: Le "De re metallica" d'Agricola* (Turnhout, Belgium: Brepols, 2006); Robert Halleux, "La nature et la formation des métaux selon Agricola et ses contemporains," *Revue d'histoire des sciences*, vol. 27, no. 3 (1974): 211–222.

10. Georgius Agricola, *De re metallica*, translated by Herbert Clark Hoover and Lou Henry Hoover (London: Mining Magazine, 1912). Translated from the first edition of 1556.

11. Bruce T. Moran, *Distilling Knowledge: Alchemy, Chemistry, and the Scientific Revolution* (Cambridge, MA: Harvard University Press, 2005), chap. 3; Charles Webster, *Paracelsus: Medicine, Magic, and Mission at the End of Time* (New Haven, CT: Yale University Press, 2008).

12. Data on silver production fluctuate across a number of different and reliable sources. Our numbers reflect the finished material—that is to say, completely refined silver.

13. Kendall W. Brown, *A History of Mining in Latin America from the Colonial Era to Present* (Albuquerque: University of New Mexico Press, 2012); Richards, *The Unending Frontier*, chap. 10; Peter J. Bakewell, "Mining in Colonial Spanish America," in *The Cambridge History of Latin America*, edited by Leslie Bethell (Cambridge: Cambridge University Press, 1984), 105–152.

14. Brown, *A History of Mining*.

15. Antonio Matilla-Tascón, *Historia de las minas de Almadén, vol. 2, 1646–1799* (Madrid: Ministerio de Hacienda, 1987); Alfredo Menéndez-Navarro, *Un mundo sin sol: La salud de los trabajadores de las minas de Almadén (1750–1900)* (Granada, Spain: University of Granada, 1996); Arthur Preston Whitaker, *The Huancavelica Mercury Mine: A Contribution to the History of the Bourbon Renaissance in the Spanish Empire* (Cambridge, MA: Harvard University Press, 1941); Richards, *The Unending Frontier*, chap. 10.

16. Kendall W. Brown, "Workers' Health and Colonial Mercury Mining at Huancavelica, Peru," *The Americas*, vol. 57, no. 4 (2001): 467–496; Brown, *A History of Mining*, chap. 8.

17. Nicolas A. Robins, *Mercury, Mining, and Empire: The Human and Ecological Cost of Colonial Silver Mining in the Andes* (Bloomington: Indiana

University Press, 2011); Matthew C. LaFevor, "Building a Colonial Resource Monopoly: The Expansion of Sulphur Mining in New Spain (1600–1820)," *Geographical Review*, vol. 102, no. 2 (2012): 202–224.

18. Jerome O. Nriagu, "Mercury Pollution from the Past Mining of Gold and Silver in the Americas," *Science of the Total Environment*, no. 149 (1994): 167–181; Antonio Martínez-Cortizas et al., "Mercury in a Spanish Peat Bog: Archive of Climate Change and Atmospheric Metal Deposition," *Science*, vol. 284, no. 5416 (1999), 939–942.

19. Sven Rydberg, *The Great Copper Mountain: The Stora Story* (Hedemora, Sweden: Gidlunds, 1988).

20. Takehiro Watanabe, "Talking Sulfur Dioxide: Air Pollution and the Politics of Science in Late Meiji Japan," in *Japan at Nature's Edge: The Environmental Context of a Global Power*, edited by Ian Jared Miller, Julia Andeney Thomas, and Brett L. Walker (Honolulu: University of Hawaii Press, 2013), 73–89; Imai Noriko, "Copper in Edo-Period Japan."

21. Lynch, *Mining in World History*, 55, 118–119; Yang Yuda, "Silver Mining in Frontier Zones: Chinese Mining Communities along the Southwestern Borders of the Qing Empire," in *Mining, Monies, and Culture in Early Modern Societies: East Asian and Global Perspectives,* edited by Nanny Kim and Keiko Nagase-Reimer (Leiden: Brill, 2013), 87–114; Guangle Qiu et al., "Methylmercury Accumulation in Rice Grown at Abandoned Mercury Mines in Guizhou, China," *Journal of Agriculture and Food Chemistry*, vol. 56, no. 7 (2008): 2465–2468.

22. Eric Ash, *Power, Knowledge, and Expertise in Elizabethan England* (Baltimore, MD: Johns Hopkins University Press, 2004); Michael J. Braddick, *State Formation in Early Modern England (1550–1770)* (Cambridge: Cambridge University Press, 2000).

23. Lynch, *Mining in World History*, 63–64.

24. Edmund Newell, "Atmospheric Pollution and the British Copper Industry (1690-1920)," *Technology and Culture*, vol. 38, no. 3 (1997): 655–689.

25. J. Morton Briggs, "Pollution in Poullaouen," *Technology and Culture*, vol. 38, no. 3 (1997): 635–654; Anne-François Garçon, "*Les Métaux non ferreux en France aux xviiie et xixe siècles: Ruptures, blocages, évolution au sein des systèmes techniques*" (PhD diss., Université de Paris 1, 1995).

26. E. D. Clarke, *Voyage de Sundsvall en Medelpadie à Drontheim en Norvège* (Paris: Librairie Gide fils, 1822). Our thanks to Geneviève Dufresne for alerting us to this work.

27. Karl Polanyi, *La Grande Transformation: Aux origines politiques et économiques de notre temps* (Paris: Gallimard, 2009 [1944]).

28. Robert Allen, *The British Industrial Revolution in Global Perspective* (Cambridge: Cambridge University Press, 2009); Jan De Vries, *The Industrious Revolution: Consumer Demand and the Household Economy (1650 to the Present)* (Cambridge: Cambridge University Press, 2008); Pomeranz, *Une grande divergence*; Liliane Hilaire-Pérez, *La Pièce et le geste: Artisans, marchands et*

culture technique à Londres au xviiie siècle (Paris: Albin Michel, 2013); Hilaire-Pérez, *L'Invention technique au siècle des Lumières* (Paris: Albin Michel, 2000); Philippe Minard, *La Fortune du colbertisme: État et industrie dans la France des Lumières* (Paris: Fayard, 1998); Jeff Horn, *The Path Not Taken: French Industrialization in the Age of Revolution (1750–1830)* (Cambridge, MA: MIT Press, 2006); Joel Mokyr, *The Enlightened Economy: An Economic History of Britain (1700–1850)* (New Haven, CT: Yale University Press, 2010); Peter M. Jones, *Industrial Enlightenment: Science, Technology and Culture in Birmingham and the West Midlands (1760–1820)* (Manchester, UK: Manchester University Press, 2008); Patrick Verley, *La Révolution industrielle*; Charles F. Sabel and Jonathan Zeitlin, eds., *World of Possibilities: Flexibility and Mass Production in Western Industrialization* (Cambridge: Cambridge University Press, 1999); Jonathan Zeitlin, "Les voies multiples de l'industrialisation," *Le Mouvement social,* no. 153 (1985): 25-34.

29. Sarah B. Pritchard and Thomas Zeller, "The Nature of Industrialization," in *The Illusory Boundary: Environment and Technology in History,* edited by Stephen Cutcliffe and Martin Reuss (Charlottesville: University of Virginia Press, 2010), 69–100; Charles Coulston Gillispie, "The Natural History of Industry," *Isis,* vol. 48, no. 4 (1957): 398–407.

30. Giorgio Riello, *Cotton: The Fabric That Made the Modern World* (Cambridge: Cambridge University Press, 2013).

31. Maxine Berg, *The Age of Manufactures (1700–1820): Industry, Innovation and Work in Britain* (London: Routledge, 2005 [1994]), chap. 9.

32. Arthur Young, *Annals of Agriculture and Other Useful Arts* (London, 1785), 4:168.

33. Francis D. Klingender, "Le sublime et le pittoresque," *Actes de la recherche en sciences sociales,* vol. 75, no. 1 (1988): 2–13.

34. Anna Seward, *The Poetical Works of Anna Seward* (Edinburgh: James Ballantyne, 1810) 2:314–315.

35. Malcolm McKinnon Dick, "Discourses for the New Industrial World: Industrialisation and the Education of the Public in Late Eighteenth-Century Britain," *History of Education,* vol. 37, no. 4 (2008): 567–584.

36. Charles-François Mathis, *In Nature We Trust: Les paysages anglais à l'ère industrielle* (Paris: Presses de l'université Paris-Sorbonne, 2010), 87–116.

37. Mark Crosby, Troy Patenaude, and Angus Whitehead, eds., *Re-envisioning Blake* (Basingstoke, UK: Palgrave Macmillan, 2012).

38. Martin Lynch, *Mining in World History*; Peter Perdue, "Is There a Chinese View of Technology and Nature?," in *The Illusory Boundary: Environment and Technology in History,* edited by Stephen Cutcliffe and Martin Reuss (Charlottesville: University of Virginia Press, 2010), 103; Joseph Needham, *The Development of Iron and Steel Technology in China* (Cambridge, MA: W. Heller & Sons, 1964); Teng T'o, "En Chine du xvie au xviiie siècle: les mines de charbon de Men-t'ou-kou," *Annales ESC,* vol. 22, no. 1 (1967): 54–87.

39. Conrad Totman, *Japan: An Environmental History* (London: I.B. Tauris, 2014), 174.

40. Richard W. Unger, "Energy Sources for the Dutch Golden Age: Peat, Wind and Coal," *Research in Economic History*, vol. 9 (1984): 221–253.

41. Brinley Thomas, "Escaping from Constraints: The Industrial Revolution in a Malthusian Context," *Journal of Interdisciplinary History*, vol. 15, no. 4 (1985): 729–753.

42. John Ulric Nef, *The Rise of the British Coal Industry*, 2 vols. (London: Routledge, 1932); Richards, *The Unending Frontier*, chap. 6; Michael W. Flinn and David Stoker, *The History of the British Coal Industry, vol. 2, 1700–1830: The Industrial Revolution* (Oxford: Clarendon Press, 1993).

43. Astrid Kander, Paolo Malanima and Paul Warde, *Power to the People: Energy in Europe over the Last Five Centuries* (Princeton, NJ: Princeton University Press, 2013); Pomeranz, *Une grande divergence*; Edward A. Wrigley, *Energy and the English Industrial Revolution* (Cambridge: Cambridge University Press, 2010); Wrigley, *The Path to Sustained Growth: England's Transition from an Organic Economy to an Industrial Revolution* (Cambridge: Cambridge University Press, 2016).

44. Michael W. Flinn and David Stoker, *The History of the British Coal Industry*, 69–145.

45. J. Kanefsky and J. Robey, "Steam Engines in Eighteenth-Century Britain: A Quantitative Assessment" *Technology and Culture*, vol. 21, no. 2 (1980): 161–186.

46. Nicholas von Tunzelmann, *Steam Power and British Industrialization to 1860* (Oxford: Oxford University Press, 1978).

47. Marcel Rouff, *Les Mines de charbon en France au xviiie siècle (1744–1791)* (Paris: Rieder, 1922); Denis Woronoff, "Une nouvelle source d'énergie: le charbon en France à l'époque modern," in *Economia e Energia*, (Florence: Le Monnier, 2003), 711–724.

48. Jennifer Tann and M. J. Breckin, "The International Diffusion of the Watt Engine (1775–1825)," *The Economic History Review*, vol. 31, no. 4 (1978): 541–564.

49. David Levine and Keith Wrightson, *The Making of an Industrial Society: Whickham (1560–1765)* (Oxford, Oxford University Press, 1991), 106–134.

50. Cavert, *The Smoke of London*, 93–98.

51. Le Roux, *Le Laboratoire*, 49–50.

52. Richards, *The Unending Frontier*, chap. 5, 185.

53. Reynald Abad, "L'Ancien Régime à la recherche d'une transition énergétique? La France du xviiie siècle face au bois," in *L'Europe en transitions: Énergie, mobilité, communication, XVIIIe–XXIe siècles*, edited by Yves Bouvier and Léonard Laborie (Paris: Nouveau Monde Éditions, 2016), 23–84.

54. Isabelle Parmentier, *Histoire de l'environnement en pays de Charleroi (1730–1830): Pollution et nuisances dans un paysage en voie d'industrialisation*

(Brussels: Académie royale de Belgique, 2008), 137; Paul Benoit and Catherine Verna, eds., *Le Charbon de terre en Europe occidentale avant l'usage industriel de la coke* (Turnhout, Belgium: Brépols, 1999); Le Roux, *Le Laboratoire*; Eloy Martín Corrales, "La contaminación industrial en el litoral catalán durante el siglo XVIII," in *Pollutions industrielles*, edited by Laura Centemeri and Xavier Daumalin, 215–237.

55. Quoted in François Crouzet, "Naissance du paysage industriel," in *Les Sources de l'histoire de l'environnement: Le xixe siècle*, edited by Andrée Corvol (Paris: L'Harmattan, 1999), 61.

56. Quoted in Barrie M. Ratcliffe and W. H. Chaloner, eds., *A French Sociologist Looks at Britain: Gustave d'Eichthal and British Society in 1828* (Manchester, UK: Manchester University Press, 1977), 133–135.

57. Brian W. Clapp, *An Environmental History of Britain* (London: Longman, 1994).

58. André Guillerme, *La Naissance de l'industrie à Paris: Entre sueurs et vapeurs (1780–1830)* (Seyssel, France: Champ Vallon, 2007).

59. Michael Sonenscher, *The Hatters of Eighteenth-Century France* (Berkeley: University of California Press, 1987).

60. André Guillerme, "Le mercure dans Paris: usages et nuisances (1780–1830)," *Histoire urbaine*, vol. 18, no. 1 (2007): 77–95.

61. Liliane Mottu-Weber, "Inventeurs genevois aux prises avec la maladie des doreuses et doreurs en horlogerie (fin xviiie–début xixe siècle)" in *Artisans, industrie: Nouvelles révolutions du Moyen Âge à nos jours*, edited by Liliane Hilaire-Pérez et al. (Lyon, France: ENS Éditions, 2004), 283–296.

62. Thomas Le Roux, "Santés ouvrières et développement des arts et manufactures au xviiie siècle en France," in *Economic and Biological Interactions in Pre-industrial Europe from the 13th to the 18th Centuries*, edited by Simonetta Cavaciocchi (Florence: Firenze University Press, 2010), 573–585.

63. Jean Delumeau, *L'Alun de Rome (xve–xixe siècle)* (Paris: École pratique des hautes études, 1962).

64. Charles Singer, *The Earliest Chemical Industry: An Essay in the Historical Relations of Economics and Technology Illustrated from the Alum Trade* (London: Folio Society, 1948).

65. John Svidén and Mats Eklund, "From Resource Scarcity to Pollution Problem: The Production and Environmental Impact of a Swedish Alum Works (1723–1877)," *Environment and History*, vol. 2, no. 1 (1996): 39–61.

66. Mats Eklund et al., "Reconstruction of Historical Cadmium and Lead Emissions from a Swedish Alumworks (1726–1840)," *The Science of the Total Environment*, vol. 170, no. 1–2 (1995): 21–30.

67. Maurice Picon, "Des aluns naturels aux aluns artificiels et aux aluns de synthèse: Matières premières, gisements et proceeds," in *L'Alun de Méditerranée*, edited by Philippe Borgard, Jean-Pierre Brun, and Maurice (Naples: Publications du Centre Jean-Bérard, 2005), 13–38.

68. Serge Chassagne, *Oberkampf, un entrepreneur capitaliste au siècle des Lumières* (Paris: Aubier Montaigne, 1980), 157–160; Chassagne, *Le Coton et ses patrons (France, 1760–1840)* (Paris: EHESS, 1991).

69. Bernadette Bensaude-Vincent and Isabelle Stengers, *Histoire de la chimie* (Paris: La Découverte, 1993).

70. Le Roux, *Le Laboratoire*, chap. 3.

71. John Graham Smith, *The Origins and Early Development of the Heavy Chemical Industry in France* (Oxford: Clarendon Press, 1979).

72. Xavier Daumalin, "La pollution des soudières marseillaises au début du xixe siècle. Réflexions autour d'une étude de cas" in *Letté Débordements industriels*, 75; Jean-Baptiste Fressoz, *L'Apocalypse Joyeuse: Une histoire du risque technologique* (Paris: Le Seuil, 2012).

73. Smith, *The Origins and Early Development.*

74. René Davy, *L'Apothicaire Baumé (1728–1804): Les origines de la droguerie pharmaceutique et de l'industrie du sel ammoniac en France* (Cahors, France: Couesant, 1955).

75. Leslie Tomory, *Progressive Enlightenment: The Origin of the Gas Lighting Industry (1780–1820)* (Cambridge, MA: MIT Press, 2012); Jean-Pierre Williot, *Naissance d'un service public, le gaz à Paris* (Paris: Rive Droite/Institut d'histoire de l'industrie, 1999).

76. David E. Nye, *American Technological Sublime* (Cambridge, MA: MIT Press, 1994), chap. 5: "The Factory: From the Pastoral Mill to the Industrial Sublime," 109–142.

77. Anna Seward, *The Poetical Works of Anna Seward* (Edinburgh: James Ballantyne, 1810) 2:318.

78. Cavert, *The Smoke of London*, 3–5.

79. David Barnett, *London, Hub of the Industrial Revolution: A Revisionary History (1775–1825)* (London: I.B. Tauris, 1998); Leonard D. Schwarz, *London in the Age of Industrialisation: Entrepreneurs, Labour Force and Living Conditions (1700–1850)* (Cambridge: Cambridge University Press, 1992).

80. Brimblecombe, *The Big Smoke.*

81. Thomas Salmon, *Modern History, or The Present State of All Nations,* 5 vols. (Dublin: 1739), 4:489.

82. Cavert, *The Smoke of London*, notably 72 and 193–194, 213–231.

83. Cavert, *The Smoke of London*, 32–39; Peter Brimblecombe, "London Air Pollution (1500–1900)," *Atmospheric Environment*, vol. 11, no. 12 (1967): 1157–1162.

84. William Frend, "Is It Impossible to Free the Atmosphere of London in a Very Considerable Degree, from the Smoke and Deleterious Vapours with Which It Is Hourly Impregnated?" *The Pamphleteer*, vol. 15, no. 29 (1819): 62–65.

85. Catherine Bowler and Peter Brimblecombe, "Control of Air Pollution in Manchester Prior to Public Health Act (1875)," *Environment and History*, vol.

6, no. 1 (2000): 71–98; Stephen Mosley, *The Chimney of the World: A History of Smoke Pollution in Victorian and Edwardian Manchester* (Cambridge: White Horse Press, 2001).

86. Alexis de Tocqueville, *Journeys to England and Ireland* (New Haven, CT: Yale University Press, 1958) 104–107.

87. Julien Vincent, "La réforme sociale à l'heure du thé: La porcelain anglaise, l'Empire britannique et la santé des ouvrières dans le Staffordshire (1864–1914)," *Revue d'histoire moderne et contemporaine*, vol. 56, no. 1 (2009): 29–60.

88. Parmentier, *Histoire de l'environnement*.

89. Claude Lachaise, *Topographie médicale de Paris* (Paris: 1822); Alexandre Parent-Duchâtelet and Charles Pavet de Courteille, *Recherches et considerations sur la rivière de Bièvre ou des Gobelins* (Paris: 1822).

Chapter 3

1. On the relationship between the chemical industry and the first industrialization, however, let us mention the old but still important work of William A. Campbell, *The Chemical Industry* (Harlow, UK: Longman, 1971); Archibald Clow and Nan L. Clow, *The Chemical Revolution: A Contribution to Social Technology* (London: Batchworth Press, 1952); Albert E. Musson, ed., *Science, Technology, and Economic Growth in the Eighteenth Century* (London: Routledge, 2008 [1972]).

2. Ernst Homburg, "Quimica e industria (1500–2000)," *Anales de Quimica*, vol. 105, no. 1 (2009): 58–66. Production intensified tenfold again between 1830 and 1870.

3. Theodore J. Kreps, *The Economics of the Sulfuric Acid Industry* (Stanford, CA: Stanford University Press, 1938).

4. Peter J. T. Morris, "Chemical Industries before 1850," in *The Oxford Encyclopedia of Economic History*, vol. 1 (Oxford: Oxford University Press, 2003), 394–398.

5. Cited in Le Roux, *Le Laboratoire*, 111–116; Oldham, *The Mansfield Manuscripts*, 886–893.

6. Corbin, *Le Miasme et la jonquille*.

7. Thomas Le Roux, "Du bienfait des acides: Guyton de Morveau et le grand basculement de l'expertise sanitaire et environnementale (1773–1809)," *Annales historiques de la Révolution française*, no. 383 (2016): 153–175.

8. Louis-Bernard Guyton de Morveau, *Encyclopédie méthodique. Chimie, pharmacie et métallurgie*, vol. 1 (Paris: Panckoucke, 1786), 27.

9. Antonio García Belmar and José Ramón Bertomeu-Sánchez, "L'Espagne fumigée: Consensus et silences autour des fumigations d'acides minéraux en Espagne (1770–1804)," *Annales historiques de la Révolution française*, no. 383 (2016): 177–202; Elena Serrano, "Miasmas, Politics, and Material History: The Voyages of Guyton's Disinfection Apparatus through Spain

(1790–1805)," in *Compound Histories: Materials, Chemical Governance and Production (1760–1840)*, edited by Lissa Roberts and Simon Werrett (Leiden: Brill, 2017).

10. Lars Oberg, "De mineralsura rökningarna: En episod ur desinfektions-medlens historia," *Lychnos*, 1965–1966, 159–180; John Johnstone, *An Account of the Discovery of the Power of Mineral Acid Vapours to Destroy Contagion* (London: J. Mawman, 1803).

11. Bernadette Bensaude-Vincent, *Lavoisier: Mémoires d'une révolution* (Paris: Flammarion, 1993); Jan Golinski, *Science as Public Culture: Chemistry and Enlightenment in Britain (1760–1820)* (Cambridge: Cambridge University Press, 1992).

12. Georges Vigarello, "L'hygiène des Lumières," in *Les Hygiénistes. Enjeux, modèles et pratiques*, edited by Patrice Bourdelais (Paris: Belin, 2001), 26–40.

13. R. O'Reilly, *Essai sur le blanchiment* (Paris: Bureau des Annales des arts et manufactures, 1801), 99.

14. Hasok Chang and Catherine Jackson, eds., *An Element of Controversy: The Life of Chlorine in Science, Medicine, Technology and War* (London: BSHS, 2007). See, especially, chapter 6.

15. Le Roux, *Le Laboratoire*, 85–92 and 339–342.

16. Le Roux, *Le Laboratoire*, 127–131.

17. John Perkins, ed., "Sites of Chemistry in the Eighteenth Century," special issue of *Ambix*, vol. 60, no. 2 (2013): 18.

18. Ursula Klein and Emma Spary, eds., *Materials and Expertise in Early Modern Europe: Between Market and Laboratory* (Chicago: University of Chicago Press, 2010); Christelle Rabier, ed., *Fields of Expertise: A Comparative History of Expert Procedures in Paris and London (1600 to Present)* (Cambridge: Cambridge University Press, 2007).

19. Louis-Bernard Guyton de Morveau, *Traité des moyens de désinfecter l'air, de prévenir la contagion et d'en arrêter le progrès* (1805 [first edition 1801]), 311–312.

20. Arthur Donovan, *Antoine Lavoisier: Science, Administration, and Revolution* (Oxford: Blackwell, 1993); Gérard Jorland, *Une société à soigner. Hygiène et salubrité publiques en France au xixe siècle* (Paris: Gallimard, 2010), 19–26.

21. William Coleman, "L'hygiène et l'État selon Montyon," *Dix-huitième siècle*, no. 9 (1977): 101–108.

22. Christian Laval, *Jeremy Bentham, les artifices du capitalisme* (Paris: Presses universitaires de France, 2003); William Cavert, *The Smoke of London*, chap. 6.

23. Bernard Delmas, Thierry Demals, and Philippe Steiner, eds., *La Diffusion internationale de la physiocratie (xviiie–xixe siècle)* (Grenoble, France: Presses universitaires de Grenoble, 1995).

24. Quoted in Philippe Minard, "L'inspection des manufactures et la réglementation industrielle à la fin du xviiie siècle," in *Naissance des libertés économiques*

(1791–fin xixe siècle), edited by Alain Plessis (Paris: Institut d'histoire de l'industrie, 1993), 49–60.

25. Paolo Napoli, *Naissance de la police moderne: Pouvoir, normes, société* (Paris: La Découverte, 2003), 67–68 and 83.

26. Liliane Pérez, "Technology, Curiosity and Utility in France and England in the Eighteenth Century," in *Science and Spectacle in the European Enlightenment*, edited by Bernadette Bensaude-Vincent and Christine Blondel (Aldershot, UK: Ashgate, 2008), 25–42; William Clark, Jan Golinski, and Simon Schaffer, eds., *The Sciences in Enlightened Europe* (Chicago: University of Chicago Press, 1999); William Sewell, "Visions of Labour: Illustrations of the Mechanical Arts Before, In, and After Diderot's *Encyclopédie*," in *Work in France: Representations, Meaning, Organization and Practice*, edited by Steven Kaplan and Cynthia J. Koepp (Ithaca, NY: Cornell University Press, 1986), 258–286.

27. Michel Foucault, *Surveiller et punir* (Paris: Gallimard, 1975); Foucault, *Naissance de la biopolitique: Cours au Collège de France (1978–1979)* (Paris: Gallimard/Seuil, 2004).

28. Ken Adler, *Engineering the Revolution: Arms and Enlightenment in France (1763–1815)* (Princeton, NJ: Princeton University Press, 1999).

29. Kendall W. Brown, *A History of Mining in Latin America*, 34–35; Mark Chambers, "*River of Gray Gold: Cultural and Material Changes in the Land of Ores, Country of Minerals (1719–1839)*" (PhD diss., Stony Brook University, 2012); Hjalmar Fors, *The Limits of Matter: Chemistry, Mining and Enlightenment* (Chicago: University of Chicago Press, 2014); Pierre-Claude Reynard, "Public Order and Privilege."

30. Guillaume Garner, *État, économie, territoire en Allemagne. L'espace dans le caméralisme et l'économie politique (1740–1820)* (Paris: EHESS, 2005); Pascale Laborier, Frédéric Audren, Paolo Napoli, and Jakob Vogel, eds., *Les Sciences camérales. Activités pratiques et histoire des dispositifs publics* (Paris: Presses universitaires de France, 2011); Philippe Minard, "Économie de marché et État en France: Mythes et légendes du colbertisme," *L'Économie politique*, vol. 37, no. 1 (2008): 77–94.

31. Christian Hick, "'Arracher les armes des mains des enfants.' La doctrine de la police médicale chez Johann Peter Frank et sa fortune littéraire en France," in *Les Hygiénistes*, edited by Patrice Bourdelais, 41–59; Virginie Tournay, "Le concept de police médicale: D'une aspiration militante à la production d'une objectivité administrative," *Politix*, vol. 77, no. 1 (2007): 173–199.

32. Andrew Harris, "Policing and Public Order in the City of London," *London Journal*, vol. 28 (2003): 1–20; F. M. Dodsworth, "The Idea of Police in Eighteenth-Century England: Discipline, Reformation, Superintendance (c. 1780–1880)," *Journal of the History of Ideas*, vol. 69, no 4. (2008): 583–605.

33. Thomas Le Roux, "Les effondrements de carrières de Paris: La grande réforme des années 1770," *French Historical Studies*, vol. 36, no. 2 (2013): 205–237; Le Roux, *Le Laboratoire*, 85–101.

34. Mark Jenner, "Monopoly, Markets and Public Health: Pollution and Commerce in the History of London Water (1780–1830)," in *Medicine and the Market in Pre-Modern England and Its Colonies (c. 1450–1850)*, edited by Mark Jenner and Patrick Wallis (Basingstoke, UK: Palgrave Macmillan, 2007), 216–237; Frédéric Graber, "La qualité de l'eau à Paris (1760–1820)," *Entreprises et histoire*, no. 50 (2008): 119–133.

35. Guillaume Leyte, "Les évocations, entre régulation juridique et arbitrage politique," in "Cassations et évocations. Le Conseil du roi et les parlements au xviiie siècle," special issue of *Histoire, économie et société*, vol. 29, no. 3 (2010): 37–43.

36. Le Roux, *Le Laboratoire*, 116–141.

37. Le Roux, *Le Laboratoire*, 116–141.

38. Oldham, *The Mansfield Manuscripts*, 1:184, 2:882–925.

39. Parmentier, *Histoire de l'environnement*, 138; Eloy Martín Corrales, "La contaminación industrial."

40. David Armitage and Sanjay Subrahmantan, eds., *The Age of Revolutions in Global Context (c. 1760–1840)* (Basingstoke, UK: Palgrave Macmillan, 2009); Bayly, *La Naissance du monde moderne*.

41. Jehan de Malafosse, "Un obstacle à la protection de la nature: le droit révolutionnaire," *Dix-huitième siècle*, vol. 9, no. 1 (1977): 91–100; Andrée Corvol, ed., *La Nature en Révolution (1750–1800)* (Paris: Groupe d'histoire des forêts françaises/L'Harmattan, 1993); Christian Dugas de La Boissonny, "La législation révolutionnaire," in *Nature, environnement et paysage, l'héritage du xviiie siècle: Guide de recherches archivistiques et bibliographiques*, edited by Andrée Corvol and Isabelle Richefort (Paris: L'Harmattan, 1995), 59–72; Reynald Abad, *La Conjuration contre les carpes: Enquête sur l'ori- gine du décret de dessèchement des étangs du 14 frimaire an II* (Paris: Fayard, 2006). For a more positive reevaluation of the environmental impact of the French Revolution, see Peter McPhee, *Revolution and Environment in Southern France: Peasants, Lords and Murder in the Corbières (1780–1830)* (Oxford: Oxford University Press, 1999).

42. Mark Fiege, *The Republic of Nature: An Environmental History of the United States* (Seattle: University of Washington Press, 2012), chap. 1 and 2; Charles A. Miller, *Jefferson and Nature: An Interpretation* (Baltimore, MD: Johns Hopkins University Press, 1988).

43. Richard W. Judd, *The Untilled Garden: Natural History and the Spirit of Conservation in America (1790–1840)* (New York: Cambridge University Press, 2009); Richard Slotkin, *The Fatal Environment: The Myth of the Frontier in the Age of Industrialization (1800–1890)* (New York: Atheneum, 1985), 107–119.

44. François-Xavier Guerra, *Modernidad e independencias* (Madrid: Editorial MAPFRE), 1992.

45. Plessis, ed., *Naissance des libertés économiques*; Steven Kaplan and Philippe Minard, eds., *La Fin des corporations* (Paris: Belin, 2003); Le Roux, *Le Laboratoire*, 168–173.

46. Quoted in Thomas Le Roux, "Régime des droits *vs* utilité publique—Justice, police et administration: faire face à l'industrialisation (France–Grande-Bretagne, 1750–1850)," in *Justice et police, le nœud gordien (1750–1850)* edited by Marco Cicchini, Vincent Denis, Vincent Milliot, and Michel Porret (Geneva: Georg/L'Équinoxe, 2017).

47. Robert Knight, *Britain against Napoleon: The Organization of Victory (1793–1815)* (London: Penguin, 2014).

48. Jean-François Belhoste and Denis Woronoff, "Ateliers et manufactures: une réévaluation nécessaire," in *À Paris sous la Révolution. Nouvelles approches de la ville*, edited by Françoise Monnier (Paris: Publications de la Sorbonne, 2008), 79–91; Patrice Bret, *L'État, l'armée, la science: L'invention de la recherche publique en France (1763–1830)* (Rennes: Presses universitaires de Rennes, 2002); Charles C. Gillispie, *Science and Polity in France: The Revolutionary and Napoleonic Years* (Princeton, NJ: Princeton University Press, 2004); Mary Ashburn Miller, *A Natural History of Revolution: Violence and Nature in the French Revolution* (Ithaca, New York: Cornell University Press, 2011); Thomas Le Roux, "Accidents industriels et régulation des risques: l'explosion de la poudrerie de Grenelle en 1794," *Revue d'histoire moderne et contemporaine*, vol. 58, no. 3 (2011): 34–62.

49. Le Roux, *Le Laboratoire*, 195–212.

50. Horn, *The Path Not Taken*; Igor Moullier, "Le Ministère de l'Intérieur sous le Consulat et le Premier Empire (1799–1814): Gouverner la France après le 18 Brumaire," (PhD diss., Université de Lille 3, 2004).

51. "Rapport ... sur la question de savoir si les manufactures qui exhalent une odeur désagréable peuvent être nuisibles à la santé" (17 December 1804), *Procès-verbaux des séances de l'Académie des sciences, tenues depuis la fondation de l'Institut jusqu'au mois d'août 1835* (Hendaye: Observatoire d'Abbadia, 1910–1922), 3:165–168.

52. Catherine Bowler and Peter Brimblecombe, "Control of Air Pollution."

53. Le Roux, *Le Laboratoire*, chap. 5 and 6.

54. "Rapport sur les manufactures de produits chimiques qui peuvent être dangereuses" (30 October 1809), *Procès-verbaux des séances de l'Académie des sciences*, 4:268–273.

55. The 1804 and 1809 reports—and the 1810 law—were revisited by Massard-Guilbaud, *Histoire de la pollution industrielle*, 34–45; Le Roux, *Le Laboratoire*, 255–261, 274–283; Fressoz, *L'Apocalypse joyeuse*, 150–165.

56. Le Roux, *Le Laboratoire*, 246–249, 315.

57. Ann Fowler La Berge, *Mission and Method: The Early-Nineteenth-Century French Public Health Movement* (Cambridge: Cambridge University Press, 2002 [1992]).

58. Patrice Bret, "Des essais de la Monnaie à la recherche et à la certification des métaux: un laboratoire modèle au service de la guerre et de l'industrie (1775–1830)," *Annales historiques de la Révolution française*, vol. 320, no. 2 (2000): 137–148.

59. Jean-Pierre Darcet, *Collection de mémoires relatifs à l'assainissement des ateliers, des édifices publics et des habitations particulières*, 1843, 21–22.

60. Julien Vincent, "Bernardino Ramazzini, historien des maladies humaines et médecin de la société civile: La carrière franco-britannique du *De morbis artificum diatriba* (1777–1855)," in *La Société civile: Savoirs, enjeux et acteurs en France et en Grande-Bretagne (1780–1914)*, edited by Christophe Charle and Julien Vincent (Rennes: Presses universitaires de Rennes, 2011), 169–202; Thomas Le Roux, "L'effacement du corps de l'ouvrier : La santé au travail lors de la première industrialisation de Paris (1770–1840)," *Le Mouvement social*, no. 234 (2011): 103–119.

61. Merel Klein, "Risques industriels à une période de transferts et de transition: la gestion des manufactures et des ateliers dangereux à Amsterdam (1810–1830)," in *Risques industriels. Savoirs, régulations, politiques d'assistance (fin xviie-début xxe siècle)*, edited by Thomas Le Roux (Rennes: Presses universitaires de Rennes, 2016), 257–278.

62. Mary Anderson and Eric Ashby, *The Politics of Clean Air* (Oxford: Clarendon Press, 1981); Bill Luckin, "Country, Town and Metropolis: The Formation of an Air Pollution Problem in London (1800–1870)," in *Energie und Stadt in Europa*, edited by Dieter Schott (Stuttgart: Steiner, 1997), 77–91; Masahiko Akatsu, "The Problem of Air Pollution during the Industrial Revolution: A Reconsideration of the Enactment of the Smoke Nuisance Abatement Act of 1821," in *Economic History of Energy and Environment*, edited by S. Sugiyama (Tokyo: Springer, 2015), 85–110.

63. David Zylberberg, "Plants and Fossils: Household Fuel Consumption in Hampshire and the West Riding of Yorkshire (1750–1830)," (PhD diss., University of York, 2014), chap. 6.

64. Grégoire Bigot, *L'Autorité judiciaire et le contentieux de l'administration: Vicissitudes d'une ambition (1800–1872)* (Paris: LGDJ, 1999).

65. Thomas Le Roux, "Déclinaisons du 'conflit': Autour des atteintes environnementales de l'affinage des métaux précieux (Paris, années 1820)," in *Débordements industriels*, edited by Thomas Le Roux and Michel Letté, 179–198.

66. Fressoz, *L'Apocalypse joyeuse*, 178–188; Daumalin, "La pollution des soudières marseillaises au début du xixe siècle."

67. Adolphe Trébuchet, *Code administratif des établissements dangereux, insalubres ou incommodes* (Paris: 1832), 110–119.

68. Le Roux, *Le Laboratoire*, 473–477. Quotation is from page 287.

69. Jean-Baptiste Fressoz, "L'émergence de la norme technique de sécurité en France vers 1820," *Le Mouvement social*, no. 249 (2014): 73–89; Fressoz, "The Gas Lighting Controversy: Technological Risk, Expertise and Regulation in Nineteenth-Century Paris and London," *Journal of Urban History*, vol. 33, no. 5 (2007): 729–755.

70. Thomas Percival, *Medical Ethics* (Oxford: Parker, 1849 [1803]), 118.

71. Jean-Antoine Chaptal, *De l'industrie française* (Paris: Renouard, 1819), 2:75–76.

72. Le Roux, *Le Laboratoire*, chap. 7 and 9.

73. François Jarrige, *Au temps des "tueuses de bras": Les bris de machines à l'aube de l'ère industrielle* (Rennes: Presses universitaires de Rennes, 2009).

74. Alexandre Parent-Duchâtelet, *Hygiène publique* (Paris: Baillière, 1836), 323.

75. François-Emmanuel Fodéré, *Traité de médecine légale et d'hygiène publique* (Paris: Mame, 1813), 6:318.

76. Antonio García Belmar and José Ramón Bertomeu-Sánchez, "L'Espagne fumigée."

77. Gilles Denis, "Dégâts sur les plantes."

Part 2

1. Pomeranz, *Une grande divergence*; Allen, *The British Industrial Revolution*.

2. Emmanuel Fureix and François Jarrige, *La Modernité désenchantée: Relire l'histoire du xixe siècle français* (Paris: La Découverte, 2015).

3. David S. Landes, *L'Europe technicienne ou le Prométhée libéré: Révolution technique & libre essor industriel en Europe occidentale, de 1750 à nos jours* (Paris: Gallimard, 1975 [1969]).

4. Fernand Braudel and Ernest Labrousse, eds., *Histoire économique & sociale de la France* (Paris: Presses universitaires de France, 1976).

5. Denis Woronoff, *Histoire de l'industrie en France, du xvie siècle à nos jours* (Paris: Le Seuil, 1998 [1994]), 149; Geneviève Massard-Guilbaud, *Histoire de la pollution industrielle*, 8.

Chapter 4

1. Victor Hugo, "Voyage en Belgique," *Œuvres complètes. Voyages* (Paris: Robert Laffont, 2002 [1837]), 611.

2. Pascal Grousset, "De l'épuisement probable des mines de houille en Angleterre," *L'Économiste belge*, no. 22 (31 October 1868), 261–262.

3. Thorsheim, *Inventing Pollution*.

4. Paul Bairoch, "Niveaux de développement économique de 1810 à 1910," *Annales ESC*, vol. 20, no. 6 (1965): 1091–1117.

5. Bill Luckin, "Pollution in the City," in *The Cambridge Urban History of Britain*, vol. 3, *1840–1950*, edited by Martin Daunton (Cambridge: Cambridge University Press, 2000), 207–228.

6. Jean-Luc Pinol, *Le Monde des villes au xixe siècle* (Paris: Hachette, 1991); Friedrich Lenger, *European Cities in the Modern Era (1850–1914)* (Leiden: Brill, 2012); Jean-Luc Pinol and François Walter, eds., *Histoire de l'Europe urbaine*, vol. 4, *La Ville contemporaine* (Paris: Le Seuil, 2012).

7. Jean-Pierre Goubert, *La Conquête de l'eau: L'avènement de la santé à l'âge industriel* (Paris: Robert Laffont, 1986).

8. Ian MacLachlan, "A Bloody Offal Nuisance: The Persistence of Private Slaughter-Houses in Nineteenth-Century London," *Urban History*, vol. 34, no. 2 (2007): 228–254.

9. Clay McShane and Joel A. Tarr, *The Horse in the City: Living Machines in the Nineteenth Century* (Baltimore, MD: Johns Hopkins University Press, 2007), 16; Daniel Roche, *La Culture équestre de l'Occident (xvie-xixe siècle), vol. 1, Le Cheval moteur* (Paris: Fayard, 2008), 65; Joel Tarr, *The Search for the Ultimate Sink: Urban Pollution in Historical Perspective* (Akron, OH: University of Akron Press, 1996), 323–333.

10. Jean-Pierre Williot, "Odeurs, fumées & écoulement putrides: Les pollutions de la première génération d'usines à gaz à Paris (1820–1860)," in *Le Démon moderne*, edited by Christoph Bernhardt and Geneviève Massard-Guilbaud, 273–282.

11. E. Landrin, "Utilisation des eaux d'égout à Gennevilliers," *La Nature: Revue des sciences & de leurs applications aux arts & à l'industrie*, no. 23 (8 November 1873): 353–357.

12. Jacques Léonard, *Archives du corps. La santé au xixe siècle* (Rennes: Ouest France, 1986); Pierre Darmon, *L'Homme & les microbes*.

13. Hanley, "Urban Sanitation in Preindustrial Japan."

14. Pratik Chakrabarti, "Purifying the River: Pollution and Purity of Water in Colonial Calcutta," *Studies in History (New Delhi)*, vol. 31, no. 2 (2015): 178–205.

15. Sabine Barles, *L'Invention des déchets urbains (France, 1790–1970)* (Seyssel, France: Champ Vallon, 2005).

16. Guillaume Carnino, "L'environnement & la science: Acclimater la population de Gennevilliers aux débordements des eaux usées parisiennes (1870–1880)," in *Débordements industriels*, edited by Thomas Le Roux and Michel Letté, 199–224; Ananda Kohlbrenner, "From Fertiliser to Waste, Land to River: A History of Excrement in Brussels," *Brussels Studies*, no. 78 (23 June 2014); Marion W. Gray, "Urban Sewage and Green Meadows: Berlin's Expansion to the South (1870–1920)," *Central European History*, vol. 47, no. 2 (2014): 275–306; Sylvia Gierlinger, Gertrud Haidvogl, Simone Gingrich, and Fridolin Krausmann, "Feeding and Cleaning the City: The Role of the Urban Waterscape in Provision and Disposal in Vienna during the Industrial Transformation," *Water History*, vol. 5, no. 2 (2013): 219–239; Arn Keeling, "Urban Waste Sinks as a Natural Resource: The Case of the Fraser River," *Urban History Review/Revue d'histoire urbaine*, vol. 34, no. 1 (2005): 58–70; Christopher Hamlin, *What Becomes of Pollution? Adversary Science and the Controversy on the Self-Purification of Rivers in Britain (1850–1900)* (London: Garland Publishing, 1987).

17. David S. Barnes, *The Great Stink of Paris and the Nineteenth-Century Struggle against Filth and Germs* (Baltimore, MD: Johns Hopkins University Press, 2006).

18. Émile Raspail, *Des odeurs de Paris* (Paris: Larousse, 1880).

19. Stéphane Frioux, *Les Batailles de l'hygiène: Villes & environnement, de Pasteur aux Trente Glorieuses* (Paris: Presses universitaires de France, 2013); Timothy Cooper, "Modernity and the Politics of Waste in Britain," in *Nature's End: History and the Environment*, edited by Paul Warde and Sverker Sörlin (Basingstoke, UK: Palgrave Macmillan, 2009), 247–272.

20. Thomas Le Roux, *Les Paris de l'industrie (1750–1920)* (Grane, France: Créaphis, 2013), 56–57; Alain Faure, "Autorités publiques & implantation industrielle en agglomération parisienne (1860–1914)," in "Région parisienne, approches d'une notion (1860–1980)," *Les Cahiers de l'IHTP*, no. 12, edited by Danièle Voldman (1989), 93–104.

21. André Guillerme, Anne-Cécile Lefort, and Gérard Jigaudon, *Dangereux, insalubres & incommodes: paysages industriels en banlieue parisienne (xixe-xxe siècle)* (Seyssel, France: Champ Vallon, 2004), 118.

22. Richard Harris, *Unplanned Suburbs: Toronto's American Tragedy (1900 to 1950)* (Baltimore, MD: Johns Hopkins University Press, 1996); Robert D. Lewis, ed., *Manufacturing Suburbs: Building Work and Home on the Metropolitan Fringe* (Philadelphia, PA: Temple University Press, 2004).

23. Andrew Hurley, "Creating Ecological Wastelands: Oil Pollution in New York City (1870–1900)," *Journal of Urban History*, vol. 20, no. 3 (1994): 340–364.

24. Sabine Barles, *La Ville délétère: Médecins & ingénieurs dans l'espace urbain (xviiie-xixe siècle)* (Seyssel, France: Champ Vallon, 1999), 334.

25. René Leboutte, *Vie & mort des bassins industriels en Europe (1750–2000)* (Paris: L'Harmattan, 1997).

26. Louis C. Hunter, *Waterpower in the Century of the Steam Engine* (Charlottesville: University Press of Virginia, 1979).

27. Kenneth Pomeranz, *La Force de l'empire: Révolution industrielle & écologie, ou Pourquoi l'Angleterre a fait mieux que la Chine* (Maison-Alfort, France: Ere, 2009), 106.

28. Peter Thorsheim, "The Paradox of Smokeless Fuels: Gas, Coke and the Environment in Britain (1813–1949)," *Environment and History*, vol. 8, no. 4 (2002): 381–401.

29. Mosley, *The Chimney of the World*.

30. Kander et al., *Power to the People*.

31. Charles Dickens, *Les Temps difficiles (Hard Times)* (Paris: Librairie Hachette, 1859 [1854]), 75.

32. Clapp, *An Environmental History of Britain*, 68; McNeill, *Du nouveau sous le soleil*, 96.

33. Olivier Raveux, *Marseille, ville des métaux & de la vapeur au xixe siècle* (Paris: CNRS Éditions, 1998), 84–86.

34. Mark Cioc, "The Impact of the Coal Age on the German Environment: A Review of the Historical Literature," *Environment and History*, vol. 4, no. 1 (1998): 105–124.

35. Franz-Josef Brüggemeier, "The Ruhr Basin (1850–1980): A Case of Large-Scale Environmental Pollution," in *The Silent Countdown: Essays in European Environmental History*, edited by Peter Brimblecombe and Christian Pfister (Berlin: Springer Verlag, 1990), 210–227; Brüggemeier, "A Nature Fit for Industry: The Environmental History of the Ruhr Basin (1840–1990)," *Environmental History Review*, vol. 19, no. 1 (1994): 35–54.

36. Edward J. Cocks and Bernhardt Walters, *A History of the Zinc Smelting Industry in Britain* (London: Harrap, 1968), 7.

37. Anne-Françoise Garçon, *Mine & métal (1780–1880): Les non-ferreux & l'industrialisation* (Rennes, France: Presses universitaires de Rennes, 1998).

38. Arnaud Péters, "'L'affaire de Saint-Léonard' & l'abandon du berceau de l'industrie du zinc (1809–1880)," in *Débordements industriels*, edited by Thomas Le Roux and Michel Letté, 77–98.

39. Bill Luckin, "Demographic, Social and Cultural Parameters of Environmental Crisis: The Great London Smoke Fogs in the Late 19th and Early 20th Centuries," in *Le Démon moderne*, edited by Christoph Bernhardt and Geneviève Massard-Guibaud, 219–238; Brimblecombe, *The Big Smoke*, 108–135; Thorsheim, *Inventing Pollution*, 14–30.

40. Anthony E. Dingle, "'The Monster Nuisance of All': Landowners, Alkali Manufacturers, and Air Pollution (1828–1864)," *The Economic History Review*, vol. 35, no. 4 (1982): 529–548.

41. Antoine Poggiale, *Rapport à M. le préfet de police sur l'insalubrité des eaux de la Bièvre*, 1875, 21.

42. Jarrige, "Quand les eaux de rouissage débordaient dans la cité."

43. Anthony S. Wohl, *Endangered Lives: Public Health in Victorian Britain* (London: Dent & Sons, 1983); Bill Lückin, *Pollution and Control: A Social History of the Thames in the Nineteenth Century* (Bristol, UK: Adam Hilger, 1986).

44. Massard-Guilbaud, *Histoire de la pollution industrielle*, 202.

45. Stéphane Frioux, ed., "Fléau, ressource, exutoire: Visions & usages des rivières urbaines (xviiie-xxie siècle)," special issue of *Géocarrefour*, vol. 85, no. 3 (2010).

46. Brüggemeier, "A Nature Fit for Industry."

47. Romain Garcier, "Une étude de cas: la pollution de la Fensch," in *Géographie du droit. Épistémologie, développements & perspectives*, edited by Patrick Forest (Québec: Presses de l'université Laval, 2009), 219–236.

48. Denys B. Barton, *A History of Tin Mining and Smelting in Cornwall* (Exeter, UK: Cornwall, 1989 [1967]); Bryan Earl, *The Cornish Arsenic Industry* (Camborne, UK: Penhellick Publications, 1996).

49. Edmund Newell, "The Irremediable Evil: British Copper Smelters' Collusion and the Cornish Mining Industry (1725–1865)," in *From Family Firms to Corporate Capitalism*, edited by Kristine Bruland and Patrick Karl O'Brien (Oxford: Oxford University Press, 1998), 170–198; Newell, "'Copperopolis': The Rise and Fall of the Copper Industry in the Swansea District (1826–1921)," *Business History*, vol. 32, no. 3 (1990): 75–97.

50. Duane A. Smith, *Mining America: The Industry and the Environment (1800–1980)* (Lawrence: University Press of Kansas, 1987); Charles K. Hyde, *Copper for America: The United States Copper Industry from Colonial Times to 1990* (Tucson: University of Arizona Press, 1998); David Stiller, *Wounding the West: Montana, Mining and the Environment* (Lincoln: University of Nebraska Press, 2000).

51. Robert Stolz, *Bad Water: Nature, Pollution and Politics in Japan (1870–1950)* (Durham, NC: Duke University Press, 2014); Kichiro Shoji and Masuro Sugai, "The Ashio Copper Mine Pollution Case: The Origins of Environmental Destruction," in *Industrial Pollution in Japan*, edited by Jun Ui (Tokyo: United Nations University Press, 1992), chap. 1; Brett Walker, *Toxic Archipelago: A History of Industrial Disease in Japan* (Seattle: University of Washington Press, 2010).

52. Martin V. Melosi, *Effluent America: Cities, Industry, Energy, and the Environment* (Pittsburgh, PA: Pittsburgh University Press, 2001), 53; Matthieu Auzanneau, *Or noir. La grande histoire du pétrole* (Paris: La Découverte, 2015), 58.

53. Myrna Santiago, *The Ecology of Oil: Environment, Labor and the Mexican Revolution (1900–1938)* (Cambridge: Cambridge University Press, 2006), 125–126.

54. Yves-Henri Nouailhat, *Évolution économique des États-Unis, du milieu du xixe siècle à 1914* (Paris: CEDES-CDU, 1982).

55. Quoted in McNeill, *Du nouveau sous le soleil*, 97–98.

56. Christine Meisner Rosen, "Businessmen against Pollution in Late Nineteenth-Century Chicago," *The Business History Review*, vol. 69, no. 3 (1995): 351–397.

57. Joel A. Tarr, ed., *Devastation and Renewal: An Environmental History of Pittsburgh and Its Region* (Pittsburgh, PA: Pittsburgh University Press, 2003), 3; Joel A. Tarr and Karen Clay, "Pittsburgh as an Energy Capital: Perspective on Coal and Natural Gas Transitions and the Environment," in *Energy Capitals: Local Impact, Global Influence*, edited by Joseph A. Pratt, Martin V. Melosi, and Kathleen A. Brosnan (Pittsburgh, PA: Pittsburgh University Press, 2014), 13.

58. Gérard Chastagnaret, *L'Espagne, puissance minière dans l'Europe du xixe siècle* (Madrid: Casa de Velásquez, 2000); Chastagnaret, "Mourir pour un air pur? Le massacre de Río Tinto du 4 février 1888," in *Pollutions industrielles*, edited by Laura Centemeri and Xavier Daumalin, 45–67.

59. Charles E. Harvey, *The Rio Tinto Company: An Economic History of a Leading International Mining Concern (1873–1954)* (Cornwall, UK: Alison Hodge, 1981), 89–90.

60. Michio Hashimoto, "History of Air Pollution Control in Japan," in *How to Conquer Air Pollution: A Japanese Experience*, edited by Hajime Nishimura (Amsterdam: Elsevier, 1989), chap. 1.

61. Lynch, *Mining in World History.*

62. William J. Lines, *Taming the Great South Land: A History of the Conquest of Nature in Australia* (Berkeley: University of California Press, 1991).

63. Claude Markovits, "Bombay as a Business Centre in the Colonial Period: A Comparison with Calcutta," in *Bombay, Metaphor for Modern India*, edited by Sujata Patel and Alice Thorner (New Delhi: Oxford University Press, 2003), 28–29.

64. Shashi Bhushan Upadhyay, *Existence, Identity and Mobilization: The Cotton Millworkers of Bombay (1890–1919)* (New Delhi: Manohar Publishers, 2004), 28.

65. David Arnold, "Pollution, Toxicity and Public Health in Metropolitan India (1850–1939)," *Journal of Historical Geography*, vol. 42 (2013): 124–133; Arnold, *Toxic Histories*, 194.

66. Shawn William Miller, *An Environmental History of Latin America* (Cambridge: Cambridge University Press, 2007).

67. Donald Quataert, "Premières fumées d'usines," in *Salonique (1850–1918): La "ville des Juifs" et le réveil des Balkans*, edited by Gilles Veinstein (Paris: Autrement, 1992), 177–194; Quataert, *Ottoman Manufacturing in the Age of Industrial Revolution* (Cambridge: Cambridge University Press, 1993).

68. Brent H. Usher and D. Vermeulen, "The Impacts of Coal and Gold Mining on the Associated Water Resources in South Africa," in *Groundwater Pollution in Africa*, edited by Yongxin Xu and Brent H. Usher (London: Taylor & Francis, 2006), 301–314.

69. Bensaude-Vincent and Stengers, *Histoire de la chimie*; Maurice Daumas, "La montée de la grande industrie chimique," in *Histoire générale des techniques*, edited by M. Daumas (Paris: Presses universitaires de France, 1996 [1968]), 3:617–645; Sacha Tomic, *Aux origines de la chimie organique: Méthodes et pratiques des pharmaciens et des chimistes (1785–1835)* (Rennes, France: Presses universitaires de Rennes, 2010).

70. François Caron, *La Dynamique de l'innovation": Changement technique & changement social (xvie-xxe siècle)* (Paris: Gallimard, 2010), 216.

71. Bernadette Bensaude-Vincent, *Matière à penser: Essais d'histoire & de philosophie de la chimie* (Nanterre, France: Presses universitaires de Paris Ouest, 2008), 175–191; M. Boas Hall, "La croissance de l'industrie chimique en Grande-Bretagne au xixᵉ siècle," *Revue d'histoire des sciences*, vol. 26, no. 1 (1973): 49–68; Ludwig Fritz Haber, *The Chemical Industry during the Nineteenth Century: A Study of the Economic Aspect of Applied Chemistry in Europe and North America* (London: Clarendon Press, 1958).

72. Ulrike Fell, ed., *Chimie & industrie en Europe: L'apport des sociétés savantes industrielles, du xixe siècle à nos jours* (Paris: Éd. des Archives contemporaines, 2001); Anita Kildebæk Nielsen and Soňa Štrbáňová, eds., *Creating Networks in Chemistry: The Founding and Early History of Chemical Societies in Europe* (London: Royal Society of Chemistry Publishing, 2008).

73. Robert Fox, "The Savant Confronts His Peers: Scientific Societies in France (1815–1914)," in *The Organization of Science and Technology in France (1808–1914)*, edited by Robert Fox and George Weisz (Cambridge: Cambridge University Press, 1980), 241–282.

74. Werner Abelshauser et al., *German Industry and Global Enterprise. BASF: The History of a Company* (Cambridge: Cambridge University Press, 2004).

75. Benjamin Ross and Steven Amter, *The Polluters: The Making of Our Chemically Altered Environment* (Oxford: Oxford University Press, 2010), 18–19.

76. Ernst Homburg, "Pollution and the Dutch Chemical Industry: The Turning Point of the 1850s," in *The Chemical Industry in Europe (1850–1914): Industrial Growth, Pollution and Professionalization*, edited by Ernst Homburg, Anthony S. Travis, and Harm G. Schröter (Dordrecht, Netherlands: Kluwer Academic Publishers, 1998), 165–181; Sacha Tomic, *Comment la chimie a transformé le monde* (Paris: Le Square éditeur, 2013).

77. Mark Cioc, *The Rhine: An Eco-Biography (1815–2000)* (Seattle: University of Washington Press, 2002).

78. Alain Beltran and Pascal Griset, *Histoire des techniques (xixe-xxe siècle)* (Paris: Armand Colin, 1990), 75.

79. Anselme Payen, "Les industries chimiques au xixe siècle," *Revue des Deux Mondes*, vol. 63 (1866): 958–983.

80. Kenneth Warren, *Chemical Foundations: The Alkali Industry in Britain to 1926* (Oxford: Clarendon Press, 1980).

81. Dingle, "'The Monster Nuisance of All.'"

82. Julien Maréchal, *La Guerre aux cheminées: Pollutions, peurs & conflits autour de la grande industrie chimique (Belgique, 1810–1880)* (Namur, France: Presses universitaires de Namur, 2016), 337.

83. Kenneth Bertrams, Nicolas Coupain, and Ernst Homburg, *Solvay: History of a Multinational Family Firm* (Cambridge: Cambridge University Press, 2013); Kenneth Bertrams, *Une entreprise au cœur de l'histoire: Solvay* (Cambridge: Cambridge University Press, 2013), 67.

84. Romain Garcier, "La Pollution industrielle de la Moselle française: Naissance, développement & gestion d'un problème environnemental (1850–2000)" (PhD diss., Université de Lyon 2, 2005), 264.

85. Catherine Paquot, *Henri Saint-Claire Deville: Chimie, recherche & industrie* (Paris: Vuibert, 2005); Thierry Renaux, "L'Aluminium au xixe siècle" (PhD diss., EHESS Paris, 2017).

86. Muriel Le Roux, "Les industries chimiques & l'environnement: Le cas des industries de l'aluminium de 1854 à nos jours," in *Les Sources de l'histoire de l'environnement*, vol. 3, *Le xxe siècle*, edited by Andrée Corvol (Paris: L'Harmattan/Direction des Archives de France, 2003), 171–195.

87. Olivier Chatterji, "Les débuts de l'aluminium en Maurienne: Conflits & mobilisations contre les 'émanations délétères' (1895–1914)," *Revue française d'histoire économique*, vol. 4–5, no. 2 (2015): 214–226.

88. Tomic, *Comment la chimie a transformé le monde*.

89. W. M. Mathew, "Peru and the British Guano Market (1840–1870)," *The Economic History Review*, vol. 23, no. 1 (1970): 112–128; Pierre Vayssière, *Un siècle de capitalisme minier au Chili (1830–1930)* (Paris: CNRS Éditions,

1980); Robert G. Greenhill and Rory R. Miller, "The Peruvian Government and the Nitrate Trade (1873–1879)," *Journal of Latin American Studies*, vol. 5, no. 1 (1973): 107–131; Brett Clark and John Bellamy Foster, "Ecological Imperialism and the Global Metabolic Rift: Unequal Exchange and the Guano/Nitrates Trade," *International Journal of Comparative Sociology*, vol. 50, no. 3–4 (2009): 311–334; Jean Boulaine, "Quatre siècles de fertilisation," *Étude & gestion des sols*, vol. 2, no. 3–4 (1995): 201–208, 219–226; Stephen Mosley, *The Environment in World History* (London: Routledge, 2010), 73–74.

90. Nathalie Jas, *Au carrefour de la chimie & de l'agriculture: Les sciences agronomiques en France & en Allemagne (1840–1914)* (Paris: Éd. des Archives contemporaines, 2001); William H. Brock, *Justus von Liebig: The Chemical Gatekeeper* (Cambridge: Cambridge University Press, 1997).

91. Pierre-Paul Dehérain, *Traité de chimie agrícola: Développement des végétaux, terre arable, amendements et engrais* (Paris: Masson, 1892), vi.

92. Sacha Tomic, "La 'science des engrais' et le monde agricole en France au xixe siècle," *Journal for the History of Environment and Society*, vol. 2 (2017): 63–93.

93. Journal *L'Indépendant*, April 1858.

94. Nathalie Jas, "Déqualifier le paysan, introniser l'agronome (France, 1840–1914)," *Écologie et politique*, vol. 31, no. 2 (2005): 45–55.

95. Georges Dureau, *Traité de la culture de la betterave à sucre* (Paris: Journal des fabricants de sucre, 1886), 224.

96. George V. Dyke, *John Lawes of Rothamsted: Pioneer of Science, Farming and Industry* (Harpenden, UK: Hoos Press, 1993).

97. Bruce M. S. Campbell and Mark Overton, eds., *Land, Labour, and Livestock: Historical Studies in European Agricultural Productivity* (Manchester, UK: Manchester University Press, 1991); Peter Mathias and John Davis, eds., *Agriculture and Industrialization: From the Eighteenth Century to the Present Day* (Oxford: Wiley-Blackwell, 1996).

98. Jean-Pierre Daviet, *Un destin international: La Compagnie de Saint-Gobain de 1830 à 1939* (Paris: Éd. des Archives contemporaines, 1988), 293–294, 326.

99. Xavier Daumalin, *Du sel au pétrole. L'industrie chimique de Marseille-Berre au xixe siècle* (Marseille: Tacussel, 2003), 121.

100. *La Culture intensive illustrée: Organe mensuel de la Société d'encouragement pour développer l'emploi des engrais chimiques en France*, August 1901–1914 [I–XIV].

101. Richard C. Sheridan, "Chemical Fertilizers in Southern Agriculture," *Agricultural History*, vol. 53, no. 1 (1979): 308–318; Jean Boulaine, "Histoire de la fertilisation phosphatée (1762–1914)," *Étude et gestion des sols*, vol. 13, no. 2 (2006): 129–138.

102. Vaclav Smil, *Enriching the Earth: Fritz Haber, Carl Bosch and the Transformation of World Food Production* (Cambridge, MA: MIT Press, 2001).

103. Centre national des ressources textuelles & lexicales: <http://www.cnrtl.fr/etymologie/>.

104. Roger Pouget, *Histoire de la lutte contre le phylloxéra de la vigne en France (1868–1895)* (Paris: Institut national de la recherche agronomique, 2015).

Chapter 5

1. Laurence Lestel, "La production de céruse en France au xixe siècle: Évolution d'une industrie dangereuse," *Techniques et culture*, vol. 38 (2001): 35–66; Anne-Cécile Lefort and Laurence Lestel, eds., "La céruse: Usages et effets (xe-xxe siècle)," special issues of *Documents pour l'histoire des techniques*, no. 12 (2003).

2. Judith Rainhorn, *Poison legal: Une histoire sociale, politique et sanitaire de la céruse et du saturnisme professionnel (xixe-xxe siècle)* (Paris: Presses de Sciences Po, 2017).

3. Rabier, *Fields of Expertise*. Rabier addresses the tension between economic development and pollution, paying particular attention to a more contemporary period of history, cf. chap. 10 & 11.

4. Bourdelais, ed., *Les Hygiénistes*; Vigarello, *Le Sain et le malsain*; Sophie Chauveau, Stéphane Frioux, and Patrick Fournier, *Hygiène et santé en Europe: De la fin du xviiie siècle aux lendemains de la Première Guerre mondiale* (Paris: SEDES, 2011).

5. La Berge, *Mission and Method*.

6. Thomas Le Roux, "Hygiénisme," in *Dictionnaire encyclopédique de l'État*, edited by Pascal Mbongo, François Hervouët, and Carlo Santulli (Paris: Berger-Levrault, 2014), 517–521.

7. Christopher Hamlin, *Public Health and Social Justice in the Age of Chadwick (Britain, 1800–1854)* (Cambridge: Cambridge University Press, 1998).

8. Alain Corbin, *The Foul and the Fragrant: Odor and the French Social Imagination* (New York: Berg Publishers, 1986), 131–132; Thomas Le Roux, "Risques et maladies du travail: Le Conseil de salubrité de Paris aux sources de l'ambiguïté hygiéniste au xixe siècle," in *La Santé au travail: Entre savoirs et pouvoirs (xixe–xxe siècle)* edited by Anne-Sophie Bruno, Eric Geerkens, Nicolas Hatzfeld, and Catherine Omnès (Rennes, France: Presses universitaires de Rennes, 2011), 45–63.

9. Jacques Léonard, *La Médecine entre les savoirs et les pouvoirs* (Paris: Aubier, 1981).

10. Massard-Guilbaud, *Histoire de la pollution industrielle*.

11. *Dictionnaire de l'industrie manufacturière, commerciale et agricole*, vol. 3 (Paris: Baillière, 1835), 165.

12. Ernst Homburg and Johan H. De Vlieger, "A Victory of Practice over Science: The Unsuccessful Modernization of the Dutch White Lead Industry (1780–1865)," *History and Technology*, vol. 13, no. 1 (1996): 33–52.

13. *L'Atelier*, February & November 1845.

14. J. Coulier, *Question de la céruse et du blanc de zinc, envisagée sous les rapports de l'hygiène et des intérêts publics* (Paris: Baillière, 1852), 3.

15. Rainhorn, *Poison légal;* Lestel, "La production de ceruse."

16. William Coleman, *Death Is a Social Disease: Public Health and Political Economy in Early Industrial France* (Madison: University of Wisconsin Press, 1982); Jean-Baptiste Fressoz, "Circonvenir les *circumfusa*: La chimie, l'hygiénisme et la libéralisation des 'choses environnantes' (France, 1750–1850)," *Revue d'histoire moderne et contemporaine*, vol. 56, no 4. (2009): 39–76.

17. Louis-René Villermé, *Tableau de l'état physique et moral des ouvriers employés dans les manufactures de coton, de laine et de soie* (Paris: 1840), chap. 7.

18. Maréchal, *La Guerre aux cheminées*, 104, 113.

19. James Condamin, *Histoire de Saint-Chamond et de la seigneurie de Jarez depuis les temps les plus reculés jusqu'à nos jours* (Paris: Picard, 1890), 649.

20. Mosley, *The Chimney of the World*, 63 and *passim*.

21. David Stradling, *Smokestacks and Progressives: Environmentalists, Engineers, and Air Quality in America (1881–1951)* (Baltimore, MD: Johns Hopkins University Press, 1999), 47.

22. Jules Arnould and André-Justin Martin, *Rapports sur la protection des cours d'eau et des nappes souterraines contre la pollution des résidus industriels* (Paris: Bibliothèque des Annales économiques, 1889); Jean-Pierrre Goubert, "L'eau et l'expertise sanitaire dans la France du xixe siècle: Le rôle de l'Académie de médecine et des congrès internationaux d'hygiène," *Sciences sociales et santé*, vol. 3, no. 2 (1985): 75–102.

23. Mathis, *In Nature We Trust*, 181, 328.

24. Armand Gautier, "Les fumées de Paris: Influence exercée par les produits de combustion sur l'atmosphère de la ville," *Revue d'hygiène et de police sanitaire*, no. 23 (1901); Octave Du Mesnil, "L'hygiène à Paris: Les fumées des machines à vapeur," *Annales d'hygiène publique et de médecine légale*, no. 24 (1890): 534–540.

25. Lion Murard and Patrick Zylberman, *L'Hygiène dans la République: La santé publique ou l'utopie contrariée (1870–1918)* (Paris: Fayard, 1986); Vincent Viet, "L'hygiène en l'État. La collection numérique des travaux du Conseil consultatif d'hygiène publique de France (1872–1910)," *Revue française des affaires sociales*, vol. 1, no. 1–2 (2014): 255–278.

26. John Duffy, *The Sanitarians: A History of American Public Health* (Urbana: University of Illinois Press, 1992).

27. Stradling, *Smokestacks and Progressives*, 50–51.

28. Claire Salomon-Bayet, ed., *Pasteur et la révolution pasteurienne* (Paris: Payot, 1986); Darmon, *L'Homme et les microbes;* Bruno Latour, *Pasteur: Guerre et paix des microbes* (Paris: La Découverte, 2001 [1984]).

29. Ilana Löwy, "Cultures de bactériologie en France (1880–1900): La paillasse et la politique," *Gesnerus*, vol. 67, no. 2 (2010): 188–216.

30. Léon Poincaré, "La contamination des cours d'eau par les soudières," *Annales d'hygiène publique et de médecine légale*, vol. 3, no. 9 (1883): 216–222.

31. Wrigley, *Energy and the English Industrial Revolution*; Verley, *L'Échelle du monde*, 96.

32. René Passet, *Les Grandes Représentations du monde et de l'économie à travers l'histoire: De l'univers magique au tourbillon créateur* (Paris: Les Liens qui libèrent, 2010); Dominique Méda, *La Mystique de la croissance* (Paris: Flammarion, 2013), 81.

33. Maxine Berg, *The Machinery Question and the Making of Political Economy (1815–1848)* (Cambridge: Cambridge University Press, 1980); Alf Hornborg, *The Power of the Machine: Global Inequalities of Economy, Technology and Environment* (Walnut Creek, CA: AltaMira Press, 2001).

34. Quoted in Christophe Charle, *Discordance des temps: Une brève histoire de la modernité (1830–1930)* (Paris: Armand Colin, 2011), 68.

35. Andreas Malm, *Fossil Capital: The Rise of Steam-Power in the British Cotton Industry (c. 1825–1848), and the Roots of Global Warming* (London: Verso Books, 2015); François Jarrige, ed., *Dompter Prométhée. Technologie et socialisme à l'âge romantique* (Besançon, France: Presses universitaires de Franche-Comté, 2016).

36. Andrew Ure, *The Philosophy of Manufactures or, An Exposition of the Scientific, Moral, and Commercial Economy of the Factory System of Great Britain* (London: Charles Knight, 1835), 12, 45.

37. William Cook-Taylor, *Notes of a Tour in the Manufacturing Districts of Lancashire* (London, 1842), 22.

38. Stephen Mosley, "Public Perceptions of Smoke Pollution in Victorian Manchester," in *Smoke and Mirrors: The Politics and Culture of Air Pollution*, edited by E. Melanie DuPuis (New York: New York University Press, 2004), 51.

39. Angus Bethune Reach, *Labour and the Poor in England and Wales (1849–1851)*, vol. 1, *Lancashire, Yorkshire, Cheshire* (London: Frank Cass, 1983), 5; several such witness accounts are collected in Mosley, *The Chimney of the World*, 70; *and* Mathis, *In Nature We Trust*, 175; on Reach, see Harold L. Platt, *Shock Cities: The Environmental Transformation and Reform of Manchester and Chicago* (Chicago: University of Chicago Press, 2005), 25.

40. Eugène Lebel, *Étangs, canaux et usines: La pollution des eaux* (Péronne: A. Doal, 1907), 146.

41. Massard-Guilbaud, *Histoire de la pollution industrielle*, 76–77.

42. Terry Shinn, "Des corps d'État au secteur industriel: Genèse de la profession d'ingénieur, 1750–1920," *Revue française de sociologie*, vol. 19, no. 1 (1978): 39–71.

43. Caroline Moriceau, *Les Douleurs de l'industrie: L'hygiénisme industriel en France* (Paris: Éditions de l'EHESS, 2009); Massard-Guilbaud, *Histoire de la*

pollution industrielle; Bernard Kalaora and Chloé Vlassopoulos, *Pour une sociologie de l'environnement: Environnement, société et politique* (Seyssel, France: Champ Vallon, 2014); Chloé Vlassopoulos, "Protection de l'environnement ou protection du pollueur? L'emprise des industriels sur la politique antipollution," in *Au cœur des combats juridiques*, edited by Emmanuel Dockès (Paris: Dalloz, 2007), 473–485.

44. Stradling, *Smokestacks and Progressives*, 86–87.

45. Bart Van der Herten, Michel Oris, and Jan Roegiers, eds., *La Belgique industrielle en 1850: Deux cents images d'un monde nouveau* (Brussels: Crédit communal, 1995).

46. Thomas Le Roux and Nicolas Pierrot, "Représenter le travail et l'industrie à Paris (1750–1900)," *Histoire de l'art*, no. 74 (2014): 43–54.

47. Gaston Bonnefont, *Souvenirs d'un vieil ingénieur au Creusot* (Paris: F. Juven, 1905), 18.

48. Mathis, *In Nature We Trust*, 163–164.

49. Henri Braconnot and François Simonin, "Notes sur les émanations des industries chimiques," *Annales d'hygiène publique et de médecine légale*, no. 40 (1848): 128–136.

50. Maréchal, *La Guerre aux cheminées*, 83–84.

51. Christopher Hamlin, *A Science of Impurity: Water Analysis in Nineteenth-Century Britain* (Berkeley: University of California Press, 1990).

52. Laurence Lestel and Michel Meybeck, "La mesure de la qualité chimique de l'eau (1850–1970)," *La Houille blanche*, no. 3 (2009): 25–30; Laurence Lestel, "Experts and Water Quality in Paris in 1870," in *Resources of the City: Contributions to an Environmental History of Modern Europe*, edited by Dieter Schott, Bill Lückin, and Geneviève Massard-Guilbaud (Aldershot, UK: Ashgate, 2005), 203–214; Garcier, *La Pollution industrielle de la Moselle française*, chap. 3; Lückin, *Pollution and Control*.

53. *Annuaire des eaux de la France pour 1851–1854* (Paris: Imprimerie nationale, 1854), xxiii.

54. Quoted in Goubert, *La Conquête de l'eau*, 45.

55. Poincaré, "La contamination des cours d'eau par les soudières."

56. Arnold, *Toxic Histories*, 98–115.

57. Arne Andersen, "Pollution and the Chemical Industry: The Case of the German Dye Industry," in *The Chemical Industry in Europe (1850–1914)*, edited by Ernst Homburg, Anthony S. Travis, and Harm G. Schröter, 183–205; Jürgen Büschenfeld, *Flüsse und Kloaken: Umweltfragen im Zeitalter der Industrialisierung (1870–1918)* (Stuttgart: Klett-Cotta, 1997); Garcier, *La Pollution industrielle de la Moselle française*, 71–72.

58. Jean-Marc Drouin, *Réinventer la nature, l'écologie et son histoire* (Paris: Desclée de Brouwer, 1991); Jean-Paul Deléage, *Histoire de l'écologie: Une science de l'homme et de la nature* (Paris: La Découverte, 1991); Donald Worster, *Les Pionniers de l'écologie: Une histoire des idées écologiques* (Paris: Sang de la

Terre, 1992 [1977]); Pascal Acot, ed., *The European Origins of Scientific Ecology (1800–1901)* (Amsterdam: Gordon & Breach, 1998).

59. Quoted in Patrick Matagne, *Aux origines de l'écologie: Les naturalistes en France de 1800 à 1914* (Paris: CTHS, 1999), 117.

60. Tom Williamson, *An Environmental History of Wildlife in England (1650–1950)* (London: Bloomsbury, 2013), 162.

61. Svante Arrhenius, *L'Évolution des mondes* (Paris: Librairie polytechnique Ch. Béranger, 1910 [1907]).

62. Louis de Launay, "Les ressources en combustibles du monde," *La Nature*, no. 2127 (28 February 1914): 238.

63. Diana K. Davis, *Les Mythes environnementaux de la colonisation française au Maghreb* (Seyssel, France: Champ Vallon, 2012).

64. Arnold, *Toxic Histories*.

65. Victor Hugo, *The Rhine* (New York: Wiley & Putnam, 1845), 47–48.

66. John Stuart Mill, "Walking Tour of Yorkshire and the Lake District (July–August 1831)," in *The Collected Works of John Stuart Mill, Newspaper Writings (December 1822–July 1831)*, vol. 22 (Toronto: University of Toronto, 1986), 503–504.

67. Flora Tristan, *London Journal: A Survey of London Life in the 1830s*, translated by Dennis Palmer and Giselle Pincetl (London: George Prior Publishers, 1980), 7; Serge Audier, *La Société écologique et ses ennemis: Pour une histoire alternative de l'émancipation* (Paris: La Découverte, 2017).

68. Klingender, *Art and the Industrial Revolution*.

69. Jonathan Ribner, "La poétique de la pollution," in *Turner, Whistler, Monet* exposition catalog, edited by Katharine Lochnan (Grand Palais, 11 October 2004–17 January 2005) (Paris: Réunion des musées nationaux, 2004).

70. James H. Rubin, *Impressionism and the Modern Landscape: Productivity, Technology and Urbanization from Manet to Van Gogh* (Berkeley: University of California Press, 2008); Le Roux and Pierrot, "Représenter le travail et l'industrie à Paris."

71. Ernest Jones, "The Factory Town," *The Northern Star and National Trades' Journal* (13 February 1847), 3.

72. Malm, *Fossil Capital*, 505–506, quotation 525.

73. *L'Atelier*, no. 9 (June 1846): 323–324.

74. Eugène Huzar, *La Fin du monde par la science* (Paris: E. Dentu, 1855).

75. François-Vincent Raspail, *Appel urgent au concours des hommes éclairés de toutes les professions contre les empoisonnements industriels ou autres qui compromettent de plus en plus la santé publique et l'avenir des générations* (Paris/Bruxelles: 1863), 51–52; Thomas Le Roux, "Contre les poisons industriels: La voix dissonante de Raspail," in *Une imagination républicaine, François-Vincent Raspail (1794–1878)*, edited by Ludovic Frobert (Besançon, France: Presses universitaires de Franche-Comté, 2017), 103–127.

76. Stradling, *Smokestacks and Progressives*.

77. Peter C. Gould, *Early Green Politics: Back to Nature, Back to the Land, and Socialism in Britain (1880–1900)* (New York: St. Martin's Press, 1988).

78. Mathis, *In Nature We Trust*, 369–371.

79. The Conservative politician R. Asheton Cross during a difficult parliamentary debate on toxic gases in 1881, quoted in Mathis, *In Nature We Trust*, 338.

80. Louis Bertrand, *L'Invasion: Roman contemporain* (Paris: Fasquelle, 1907), quoted in A. Mata, *L'Industrie des produits chimiques et ses travailleurs* (Paris: Gaston Doin, 1925), xv.

81. François Jarrige, *Technocritiques: Du refus des machines à la contestation des technosciences* (Paris: La Découverte, 2014), 172–176.

82. Kaj Noschis, *Monte Verità, Ascona, le génie du lieu* (Lausanne, Switzerland: Presses polytechniques et universitaires romandes, 2011).

83. Text reproduced in *Naturiens, végétariens, végétaliens et crudivégétaliens dans le mouvement anarchiste français (1895–1938)*, supplement to no. 9, series IV, *Invariance* (July 1993): 91.

84. Arnaud Baubérot, *Histoire du naturisme. Le mythe du retour à la nature* (Rennes, France: Presses universitaires de Rennes, 2004).

85. Massard-Guilbaud, *Histoire de la pollution industrielle*, chap. 2.

86. Quoted in Massard-Guilbaud, *Histoire de la pollution industrielle*, 78. The quotation comes from Douai in 1857.

87. Archives départementales de la Côte-d'Or, 14 S A 46: 1864–1865.

88. Quoted in Massard-Guilbaud, *Histoire de la pollution industrielle*, 86.

89. Archives de l'Académie François-Bourdon (Le Creusot), SS O 107–01: plaintes concernant les fours à coke Coppée en 1908.

90. Quoted in Arnaud Péters, "L'essor de l'industrie du zinc et la prise en compte de l'environnement," in *La Recherche en histoire de l'environnement: Belgique, Luxembourg, Congo, Rwanda, Burundi*, edited by Isabelle Parmentier (Namur: Presses universitaires de Namur, 2011), 159–174, 161.

91. Anthony E. Dingle, "'The Monster Nuisance of All'"; Steven Mosley, "Public Perceptions of Smoke Pollution in Victorian Manchester," in *Technologies of Landscape: From Reaping to Recycling*, edited by David E. Nye (Amherst: University of Massachusetts Press, 1999), 161–186.

92. Christine Meisner Rosen, "Noisome, Noxious, and Offensive Vapors: Fumes and Stenches in American Towns and Cities (1840–1865)," *Historical Geography*, no. 25 (1997): 49–82.

93. Frank Uekötter, *The Greenest Nation? A New History of German Environmentalism* (Cambridge, MA: MIT Press, 2014), 38.

94. Stéphane Frioux and Jean-François Malange, "'L'eau pure pour tous!' Mobilisations sociales contre la pollution des eaux douces françaises (1908-années 1960)," *Histoire et sociétés*, no. 27 (December 2008); Wanda Balcers and Chloé Deligne, "Les sociétés de pêche à la ligne, 'consciences' de la

pollution des rivières en Belgique (1880–1940)," in *La Recherche en histoire de l'environnement*, edited by Isabelle Parmentier, 175–186.

95. Quoted in Balcers and Deligne, "Les sociétés de pêche à la ligne," 181.

96. Chloé Deligne and Wanda Balcers, "Environmental Protest Movements against Industrial Waste in Belgium (1850–1914)," in *Environmental and Social Justice in the City: Historical Perspectives*, edited by Geneviève Massard-Guilbaud and Richard Rodger (Cambridge: White Horse Press, 2011).

97. Mosley, *The Chimney of the World*.

98. Brüggemeier, "A Nature Fit for Industry," 45.

99. Quoted in Jean-Claude Devinck, "La lutte contre les poisons industriels et l'élaboration de la loi sur les maladies professionnelles," *Sciences sociales et santé*, vol. 28, no. 2 (2010): 65–93, quotation is from page 71; Madeleine Rebérioux, "Mouvement syndical et santé en France, 1880–1914," *Prévenir*, vol. 18, no. 1 (1989): 15–30.

100. Quoted in Devinck, "La lutte contre les poisons industriels."

101. Henry Napias, "Les revendications ouvrières au point de vue de l'hygiène," *Revue d'hygiène et de police sanitaire*, no. 12 (1890), quotations from pages 691 and 698.

102. Bonnie Gordon, "Ouvrières et maladies professionnelles sous la Troisième République: La victoire des allumettiers français sur la nécrose phosphorée de la mâchoire," *Le Mouvement social*, no. 164 (1993): 77–93.

103. Léon Bonneff and Maurice Bonneff, *Les Métiers qui tuent: Enquête auprès des syndicats ouvriers sur les maladies professionnelles* (Paris: Bibliographie sociale, 1905); Bonneff and Bonneff, *La Vie tragique des travailleurs. Enquêtes sur les conditions économiques et morales des ouvriers et ouvrières de l'industrie* (Paris: J. Rouff, 1908).

104. Devinck, "La lutte contre les poisons industriels," 89.

105. Maréchal, *La Guerre aux cheminées*, chap. 1.

106. Xavier Daumalin and Isabelle Laffont-Schwob, eds., *Les Calanques industrielles de Marseille et leurs pollutions: Une histoire au présent* (Aix-en-Provence, France: REF.2C Éditions, 2016), 171–177.

107. Temma Kaplan, "De l'émeute à la grève de masse: Conscience de classe et communauté ouvrière en Andalousie au xixᵉ siècle," *Le Mouvement social*, no. 107 (1979): 15–50; María Dolores Ferrero Blanco, *Capitalismo minero y resistencia rural en el suroeste andaluz (Río Tinto, 1873–1900)* (Huelva, Spain: Universidad de Huelva, 1998); Chastagnaret, "Mourir pour un air pur?"; Ximo Guillem-Llobat et al., "Medioambiente y salud en los espacios industriales y urbanos de la España contemporánea (1881–1923)," in *Pollutions industrielles*, edited by Laura Centimeri and Xavier Daumalin, 239–258.

108. Dominique Bourg and Augustin Fagnière, *La Pensée écologique: Une anthologie* (Paris: Presses universitaires de France, 2014), 108–111; Pierre-François Souyri, *Moderne sans être occidental: Aux origines du Japon d'aujourd'hui* (Paris: Gallimard, 2016), 344–363: "Contre la mine mortifère";

Cyrian Pitteloud, "Modernité et luttes sociales: La mine d'Ashio à la fin du xixe siècle" in *Japon pluriel 11: Le Japon au début du xxie siècle: Dynamiques et mutations: Actes du 11e colloque de la Société française des études japonaises*, edited by David-Antoine Malinas and Julien Martine (Paris: Philippe Picquier, 2017), 91–98.

109. Kazuo Nimura, *The Ashio Riot of 1907: A Social History of Mining in Japan* (Durham, NC: Duke University Press, 1997).

Chapter 6

1. René Musset, "La métallurgie du cuivre," *Annales de géographie*, vol. 44, no. 250 (1935): 439–440; Christopher J. Schmitz, "The Rise of Big Business in the World Copper Industry (1870–1930)," *The Economic History Review*, vol. 39, no. 3 (1986): 392–410.

2. Edmund Newell, "'Copperopolis': The Rise and Fall of the Copper Industry in the Swansea District (1826–1921)," *Business History*, vol. 32, no. 3 (1990): 75–97; Newell, "Atmospheric Pollution."

3. Paul H. Rubin, *Business Firms and the Common Law: The Evolution of Efficient Rules* (New York: Praeger, 1983).

4. Christine M. Rosen, "Differing Perceptions of the Value of Pollution Abatement across Time and Place: Balancing Doctrine in Pollution Nuisance Law (1840–1906)," *Law and History Review*, vol. 11, no. 2 (1993): 303–381.

5. Christine M. Rosen, "'Knowing' Industrial Pollution: Nuisance Law and the Power of Tradition in a Time of Rapid Economic Change (1840–1864)," *Environmental History*, vol. 8, no. 4 (2003): 565–597.

6. Leslie Rosenthal, *The River Pollution Dilemma in Victorian England: Nuisance Law versus Economic Efficiency* (London: Ashgate, 2014), chap. 4: "Nuisance Law and Nuisance Economics."

7. Joel Franklin Brenner, "Nuisance Law and the Industrial Revolution," *Journal of Legal Studies*, vol. 3, no. 2 (1974): 403–333; John S. MacLaren, "Nuisance Law and the Industrial Revolution: Some Lessons from Social History," *Oxford Journal of Legal Studies*, vol. 3, no. 2 (1983): 155–221.

8. Ronald Coase, "The Problem of Social Cost," *Journal of Law and Economics* (October 1960): 1–44; Richard A. Posner, *Economic Analysis of Law*, 3rd ed. (Boston: Little, Brown and Company, 1986).

9. Martin V. Melosi, "Hazardous Waste and Environmental Liability: An Historical Perspective," *Houston Law Review*, no. 25 (July 1988): 741–779.

10. Morton J. Horwitz, *The Transformation of American Law (1780–1860)* (Cambridge, MA: Harvard University Press, 1977); Jouni Paavola, "Water Quality as Property: Industrial Water Pollution and Common Law in the Nineteenth-Century United States," *Environment and History*, no. 8 (2002): 295–318.

11. William J. Novak, *People's Welfare: Law and Regulation in Nineteenth-Century America* (Chapel Hill: University of North Carolina Press, 1996); Peter

Karsten, *Heart versus Head: Judge-Made Law in Regulation in Nineteenth-Century America* (Chapel Hill: University of North Carolina Press, 1997).

12. Martin V. Melosi, *Pollution and Reform in American Cities (1870–1930)* (Austin: University of Texas Press, 1980); Melosi, *Garbage in the Cities: Refuse, Reform and the Environment (1880–2000)* (Pittsburgh, PA: University of Pittsburgh Press, 2005); Tarr, *The Search for the Ultimate Sink*; Mosley, The Chimney of the World.

13. Mary Anderson and Eric Ashby, "Studies in the Politics of Environmental Protection: The Historical Roots of the British *Clean Air Act* (1956): The Awakening of Public Opinion over Industrial Smoke (1843–1853)," *Interdisciplinary Science Review*, vol. 1, no. 4 (1976): 279–290.

14. Bowler and Brimblecombe, "Control of Air Pollution."

15. Wohl, *Endangered Lives*, 220–222.

16. Richard Hawes, "The Municipal Reform of Smoke Pollution in Liverpool (1853–1866)," *Environment and History*, vol. 4, no. 1 (1998): 75–90.

17. Anderson and Ashby, *The Politics of Clean Air*; Wohl, *Endangered Lives*, 214–215 and 220–222.

18. Anderson and Ashby, *The Politics of Clean Air*, 16–18.

19. Catherine Mills, *Regulating Health and Safety in the British Mining Industries (1800–1914)* (Surrey, UK: Routledge, 2010); Peter Bartrip and Sandra Burman, *The Wounded Soldiers of Industry: Industrial Compensation Policy (1833–1897)* (Oxford: Oxford University Press, 1983).

20. Dingle, "'The Monster Nuisance of All'"; Richard Hawes, "The Control of Alkali Pollution in St. Helens (1862–1890)," *Environment and History*, vol. 1, no. 2 (1995): 159–171; Peter Reed, "The Alkali Inspectorate (1874–1906): Pressure for Wider and Tighter Pollution Regulation," *Ambix*, vol. 59, no. 2, 2012, 131–151.

21. Peter Reed, *Acid Rain and the Rise of the Environmental Chemist in Nineteenth-Century Britain: The Life and Work of Robert Angus Smith* (London: Ashgate, 2014).

22. Ben Pontin, "Tort Law and Victorian Government Growth: The Historiographical Significance of Tort in the Shadow of Chemical Pollution and Factory Safety Regulation," *Oxford Journal of Legal Studies*, vol. 18, no. 4 (1998): 661–680.

23. Mathis, *In Nature We Trust*, 316.

24. Reed, "The Alkali Inspectorate"; Wohl, *Endangered Lives*, 216–232; Noga Morag-Levine, *Chasing the Wind: Regulating Air Pollution in the Common Law State* (Princeton, NJ: Princeton University Press, 2003).

25. Rosenthal, *The River Pollution Dilemma in Victorian England*, 22–23.

26. Wohl, *Endangered Lives*, 233–255.

27. Morag-Levine, *Chasing the Wind*.

28. Frank Uekoetter, *The Age of Smoke: Environmental Policy in Germany and the United States (1880–1970)* (Pittsburgh, PA: University of Pittsburgh Press, 2009), chap. 2.

29. Stradling, *Smokestacks and Progressives*, 61–75; Melosi, *Effluent America*, 66.

30. Uekoetter, *The Age of Smoke*, chap. 2.

31. Désiré Dalloz, *Répertoire méthodique et alphabétique de législation, de doctrine et de jurisprudence* (Paris: Bureau de la jurisprudence générale, 1845–1869), 31:4.

32. Massard-Guilbaud, *Histoire de la pollution industrielle*; Gabriel Ullmann, *Les Installations classes: Deux siècles de législation et de nomenclature, vol. 1, Le Décret fondateur du 15 octobre 1810 et la loi du 19 décembre 1917: La protection progressive des droits des tiers* (Paris: Cogiterra, 2016).

33. Maréchal, *La Guerre aux cheminées*, 196–200; Arnaud Péters, *La Vieille-Montagne (1806–1873): Innovations et mutations dans l'industrie du zinc* (Liège, France: Éd. de la province de Liège, 2016); Maréchal, "L'affaire de Saint-Léonard."

34. Christophe Verbruggen, "Nineteenth Century Reactions to Industrial Pollution in Ghent, the Manchester of the Continent: The Case of the Chemical Industry," in *Le Démon moderne*, edited by Christoph Bernhardt and Geneviève Massard-Guilbaud, 377–391.

35. Herman Diederiks and Charles Jeurgens, "Environmental Policy in 19th-Century Leyden," in *The Silent Countdown*, edited by Peter Brimblecombe and Christian Pfister, 167–181; Homburg, "Pollution and the Dutch Chemical Industry."

36. Cioc, "The Impact of the Coal Age on the German Environment."

37. Uekoetter, *The Age of Smoke*, 43–66.

38. Franz-Josef Brüggemeier, "A Nature Fit for Industry," see especially 48; Andersen, "Pollution and the Chemical Industry."

39. Simone Neri Serneri, "Dealing with Industrial Pollution in Italy (1880–1940)," in *Nature and History in Modern Italy*, edited by Marco Armiero and Marcus Hall (Athens: Ohio University Press, 2010), 161–179; Serneri, "Industrial Pollution and Urbanization: Ancient and New Industrial Areas in Early 20th Century Italy," in *Environmental Problems in European Cities in the 19th and 20th Century*, edited by Christoph Bernhardt (Munich: Waxmann, 2004), 165–182.

40. Ian Inkster, *Japanese Industrialisation: Historical and Cultural Perspectives* (London: Routledge, 2001).

41. Stolz, *Bad Water*, see especially 37–100.

42. Arnold, *Toxic Histories*, 176–208; Awadhendra Sharan, *In the City, Out of Place: Nuisance, Pollution and Dwelling in Delhi (c. 1850–2000)* (New Delhi: Oxford University Press, 2014); Colin McFarlane, "Governing the Contaminated City: Infrastructure and Sanitation in Colonial and Post-Colonial Bombay," *International Journal of Urban and Regional Research*, vol. 32 (2008): 415–435; Michael R. Anderson, "The Conquest of Smoke: Legislation and Pollution in Colonial Calcutta," in *Nature, Culture, Imperialism: Essays on the*

Environmental History of South Asia, edited by David Arnold and Ramachandra Guha (New Delhi: Oxford University Press, 1995), 293–335.

43. Pontin, "Tort Law and Victorian Government Growth."

44. Laurence Lestel et al., "La transaction comme régulation des déversements industriels en rivière: Le cas de la Seine-et-Marne au xxᵉ siècle," in *Débordements industriels*, edited by Thomas Le Roux and Michel Letté, 225–245.

45. Jean-Baptiste Fressoz, "Payer pour polluer: L'industrie chimique et la compensation des dommages environnementaux (1800–1850)," *Histoire et mesure*, vol. 27, no. 1 (2013): 145–186; Ariane Debourdeau and Christelle Gramaglia, "La fabrication d'un héritage encombrant: Les pollutions métallurgiques de Viviez (Aveyron)," in *Débordements industriels*, edited by Thomas Le Roux and Michel Letté, 335–360.

46. William Scott Prudham, "Commodification," in *A Companion to Environmental Geography*, edited by Noel Castree et al. (Oxford: Wiley-Blackwell, 2009), 123–142.

47. Jeffrey K. Stine and Joel A. Tarr, "At the Intersection of Histories: Technology and the Environment," *Technology and Culture*, vol. 39, no. 4 (1998): 601–640.

48. For the French example, see the classic work by Charles de Freycinet, *Traité d'assainissement industriel comprenant les principaux procédés employés dans les centres manufacturiers de l'Europe occidentale* (Paris: Dunod, 1870).

49. Anselme Payen, "Sur le procédé de rouissage à l'eau chaude de Schenck," *Le Technologiste, ou Archives des progrès de l'industrie française et étrangère* (Paris: Roret, 1852), 13:291–297.

50. Thorsheim, "The Paradox of Smokeless Fuels."

51. Thomas Le Roux, "Les fourneaux fumivores, progrès technologique, recul écologique (France/Grande-Bretagne, 1780–1860)," in *Innovations et transferts de technologie en Europe du Nord-Ouest aux xixe et xxe siècles*, edited by Jean-François Eck and Pierre Tilly (Brussels: Peter Lang, 2011), 139–161.

52. Carlos Flick, "The Movement for Smoke Abatement in 19th-Century Britain," *Technology and Culture*, vol. 21, no. 1 (1980): 29–50.

53. *Bulletin de la Société d'encouragement pour l'industrie nationale*, no. 54 (1855): 272.

54. *Bulletin de la Société d'encouragement pour l'industrie nationale*, no. 93 (1894): 380.

55. Frank Uekoetter, "Solving Air Pollution Problems Once and For All: The Potential and the Limits of Technological Fixes," in *The Technological Fix: How People Use Technology to Create and Solve Problems*, edited by Lisa Rosner (New York: Routledge, 2004), 155–174.

56. Mosley, *The Chimney of the World*.

57. William S. Jevons, *The Coal Question: An Inquiry concerning the Progress of the Nation, and the Probable Exhaustion of Our Coal-Mines* (London: Macmillan, 1865).

58. Laurence Lestel, "Pollution atmosphérique en milieu urbain : De sa régulation à sa surveillance," *Vertigo. La revue électronique en sciences de l'environnement*, no. 15 (February 2013); Frank Uekoetter, "The Strange Career of the Ringelmann Smoke Chart," *Environmental Monitoring and Assessment*, no. 106 (2005), 11–26.

59. Haber, *The Chemical Industry during the Nineteenth Century*.

60. Massard-Guilbaud, *Histoire de la pollution industrielle*, 248–253.

61. Newell, "Atmospheric Pollution," 673–678.

62. Peter Reed, "Acid Towers and the Control of Chemical Pollution (1823–1876)," *Transactions of the Newcomen Society*, no. 78 (2008): 99–126; Reed, *Entrepreneurial Venture in Chemistry: The Muspratts of Liverpool (1793–1934)* (London: Ashgate, 2015); Sarah Wilmot, "Pollution and Public Concern: The Response of the Chemical Industry in Britain to Emerging Environmental Issues (1860–1901)," in *The Chemical Industry in Europe (1850–1914)*, edited by Ernest Homburg, Anthony S. Travis, and Harm G. Schröter, 121–149.

63. Georg Lunge, *A Theoretical and Practical Treatise on the Manufacture of Sulphuric Acid and Alkali with the Collateral Branches* (London: John Van Voorst, 1880), 2:252.

64. Jean-Baptiste Fressoz, "La main invisible a-t-elle le pouce vert? Le libéralisme et la naissance de l'écologie industrielle au xixe siècle," *Technique et culture*, no. 65–66 (2016): 328–343.

65. Erland Mårald, "Everything Circulates: Agricultural Chemistry and Recycling Theories in the Second Half of the Nineteenth Century," *Environment and History*, vol. 8, no. 1 (2002): 65–84; Laurent Herment and Thomas Le Roux, "Recycling: The Industrial City and Its Surrounding Countryside, 1750–1940," *Journal of the History of Environment and Society*, no. 2 (2017): 1–24.

66. Victor Hugo, *Les Miserables*, translated by Christine Donougher (London: Penguin Classics, 2015), 1126.

67. Danna Simmons, "Waste Not, Want Not: Excrement and Economy in Nineteenth Century," *Representations*, vol. 96, no. 1 (2006): 73–79.

68. Christopher Hamlin, "Recycling as a Goal of Sewage Treatment in Mid-Victorian Britain," in *The History and Sociology of Technology*, edited by Donald Hoke (Milwaukee, WI: Milwaukee Public Museum, 1982), 299–304; Hamlin, *What Becomes of Pollution?*

69. Christopher Hamlin, "Environmental Sensibility in Edinburgh (1839–1840): The Fetid Irrigation Controversy," *Journal of Urban History*, vol. 20, no. 3 (1994): 311–339; Sabine Barles, "Experts contre experts: Les champs d'épandage de la ville de Paris dans les années 1870," *Histoire urbaine*, vol. 14, no. 3 (2005): 65–80; Gray, "Urban Sewage"; Carnino, "L'environnement et la science."

70. Harold W. Brace, *History of Seed Crushing in Great Britain* (London: Land Books, 1960).

71. John Martin, "The Origins and Impacts of Chemical Fertilizers (1840–1940): The Science and Practice," oral presentation at research symposium, "L'agriculture, une solution pour recycler les déchets urbains et industriels?," Centre de recherches historiques (CNRS/EHESS), Paris, 10 April 2015.

72. Freycinet, *Traité d'assainissement industriel*, 368; Freycinet, *Rapport sur l'assainissement des fabriques ou des procédés d'industries insalubres en Angleterre* (Paris: Dunod, 1864); Patrick Fournier, "Charles de Freycinet, théoricien et acteur de l'assainissement à l'âge de l'hygiénisme," in "Autour de Charles de Freycinet: Sciences, techniques et politique," special issue of *Bulletin de la SABIX (Société des amis de la bibliothèque et de l'histoire de l'École polytechnique)*, edited by Fabien Conord and Jean-Claude Caron, no. 58 (2016): 19–29.

73. André Thépot, "Frédéric Kuhlmann, industriel et notable du Nord (1803–1881)," *Revue du Nord*, vol. 265 (1985): 527–554; Massard-Guilbaud, *Histoire de la pollution industrielle*, 244.

74. Reed, "The Alkali Inspectorate (1874–1906)."

75. Alexandre Parent-Duchâtelet, "De l'influence que peuvent avoir sur la santé les émanations provenant de la fonte et des préparations diverses que l'on fait subir au bitume asphaltique," *Annales d'hygiène publique et de médecine légale*, vol. 14 (1835): 65–87.

76. Georg Lunge, *Traité de la distillation du goudron de houille et du traitement de l'eau ammoniacale* (Paris: Savy, 1885 [1882]), 10.

77. Bensaude-Vincent and Stengers, *Histoire de la chimie*; Ernst Homburg, "Pollution and the Dutch Chemical Industry"; Andersen, "Pollution and the Chemical Industry."

78. John Clark, "'The Incineration of Refuse is Beautiful': Torquay and the Introduction of Municipal Refuse Destructors," *Urban History*, vol. 34, no. 2 (2007): 255–277; Timothy Cooper, "Modernity and the Politics of Waste in Britain"; Cooper, "Peter Lund Simmonds and the Political Ecology of Waste Utilization in Victorian Britain," *Technology and Culture*, vol. 52, no. 1 (2011): 21–44.

79. Walter Francis Goodrich, *The Economic Disposal of Towns' Refuse* (London: P. S. King & Son, 1901); Goodrich, *Modern Destructor Practice* (London: Charles Griffin, 1912).

80. Melosi, *Garbage in the Cities*; Melosi, "The British Destructor: Transfer of a Waste Destruction Technology," in *Technology and the Rise of the Networked City in Europe and America*, edited by Joel A. Tarr and Gabriel Dupuy (Philadelphia, PA: Temple University Press, 1988).

81. Frioux, *Les Batailles de l'hygiène*, 51; Frioux, "Amélioration de l'environnement urbain et transferts de technologie entre la France et ses voisins nord-européens (années 1880 années 1910)," in *Innovations et transferts de technologie*, edited by Jean-François Eck and Pierre Tilly, 235–249.

82. Freycinet, *Traité d'assainissement industriel*, 172, 177, 329.

83. Quoted in Mosley, *The Chimney of the World*, 25.

84. Mosley, *The Chimney of the World*, 23–25.

85. Louis-Augustin de Buzonnière, *Fourneau à double foyer pour la combustion de la fumée de la houille* (Orléans, France: impr. de Pagnerre, 1851), 4–5.

86. Arne Andersen, *Historische Technikfolgenabschätzung am Beispiel des Metallhüttenwesens und der Chemieindustrie (1850–1933)* (Stuttgart: Franz Steiner Verlag, 1996).

87. Freycinet, *Traité d'assainissement industriel*, 330.

88. François Walter, *Les Suisses et l'environnement: Une histoire du rapport à la nature du xviiie siècle à nos jours* (Geneva: Zoé, 1990), chap. 3.

89. William Cronon, *Nature's Metropolis: Chicago and the Great West* (New York: Norton, 1991); McNeill, *Du nouveau sous le soleil*, 193–195.

90. Antoine-Fortuné Marion, *Esquisse d'une topographie zoologique du golfe de Marseille* (Marseille: Cayer, 1883).

91. Faget, "Une cité sous les cendres."

92. Freycinet, *Traité d'assainissement industriel*, 340.

93. Hurley, "Creating Ecological Wastelands."

94. Hurley, "Creating Ecological Wastelands."

95. Dieter Schott, "The Formation of an Urban Industrial Policy to Counter Pollution in German Cities (1890–1914)," in *Le Démon moderne*, edited by Christoph Bernhardt and Geneviève Massard-Guilbaud, 311–332.

96. Wohl, *Endangered Lives*, 214–215.

97. Barry Doyle, "Managing and Contesting Industrial Pollution in Middlesbrough (1880–1940)," *Northern History*, vol. 47, no. 1 (2010): 135–154; Minoru Yasumoto, *The Rise of a Victorian Ironopolis: Middlesbrough and Regional Industrialization* (Woodbridge, UK: Boydell & Brewer, 2011).

98. Thorsheim, "The Paradox of Smokeless Fuels."

99. Fressoz, "Payer pour polluer," 177–178.

100. Lynch, *Mining in World History*, 120–214.

101. Michael R. Anderson, "The Conquest of Smoke: Legislation and Pollution in Colonial Calcutta," in *Nature, Culture, Imperialism*, edited by David Arnold and Ramachandra Guha, 293–335; Michael Mann, "Delhi's Belly: On the Management of Water, Sewage and Excreta in a Changing Urban Environment during the Nineteenth Century," *Studies in History*, vol. 23, no. 1 (2007): 1–31; Awadhendra Sharan, "From Source to Sink: 'Official' and 'Improved' Water in Delhi (1868–1956)," *Indian Economic and Social History Review*, vol. 48 (2011): 425–462; Janine Wilhelm, *Environment and Pollution in Colonial India: Sewerage Technologies along the Sacred Ganges* (London: Routledge, 2016); Chakrabarti, "Purifying the River."

102. Fabien Bartolotti, "Mobilités d'entrepreneurs et circulations des techniques: Les chantiers portuaires de Dussaud frères d'un rivage à l'autre (1848–1869)," *Revue d'histoire du xixe siècle*, no. 51 (2015): 171–185.

Part III

1. Arnulf Grübler, *Technology and Global Change* (Cambridge: Cambridge University Press, 2003), 197; Fridolin Krausmann, Simone Gingrich, and Reza Nourbakhch-Sabet, "The Metabolic Transition in Japan: A Material Flow Account for the Period From 1878 to 2005," *Journal of Industrial Ecology*, vol. 15, no. 6 (2011): 877–892.

2. François Fourquet, *Les Comptes de la puissance: Histoire de la comptabilité nationale et du plan* (Paris: Éd. Recherches, 1980).

3. Linda Nash, "Un siècle toxique. L'émergence de la 'santé environnementale,'" in *Histoire des sciences et des savoirs*, vol. 3, *Le Siècle des technosciences*, edited by Christophe Bonneuil and Dominique Pestre (Paris: Le Seuil, 2015), 145–165; Nash, *Inescapable Ecologies: A History of Environment, Disease, and Knowledge* (Berkeley: University of California Press, 2006).

4. McNeill, *Du nouveau sous le soleil*, 58–59.

5. Fabrice Nicolino, *Un empoisonnement universel: Comment les produits chimiques ont envahi la planète* (Paris: Les Liens qui libèrent, 2014); William Cronon, "Foreword: The Pain of a Poisoned World," in *Toxic Archipelago: A History of Industrial Disease in Japan*, edited by Brett L. Walker (Seattle: University of Washington Press, 2010), ix–xii; Rachel Carson, *Silent Spring* (London: Hamish Hamilton, 1963), 143.

6. Julia K. Steinberger, Fridolin Krausmann, and Nina Eisenmenger, "Global Patterns of Materials Use: A Socioeconomic and Geophysical Analysis," *Ecological Economics*, no. 69 (2010): 1148–1158.

7. Christopher Sellers and Joseph Melling, eds., *Dangerous Trade: Histories of Industrial Hazards across a Global World* (Philadelphia, PA: Temple University Press, 2012).

Chapter 7

1. Brian Allen Drake, ed., *The Blue, the Gray, and the Green: Toward an Environmental History of the Civil War* (Athens: University of Georgia Press, 2015).

2. Mimoun Charqi, *Armes chimiques de destruction massive sur le Rif: Histoire, effets, droits, préjudices et réparations* (Rabat, Morocco: Amazigh, 2014); Robert M. Neer, *Napalm: An American Biography* (Cambridge, MA: Belknap Press, 2013); David Zierler, *The Invention of Ecocide: Agent Orange, Vietnam, and the Scientists Who Changed the Way We Think about the Environment* (Athens: University of Georgia Press, 2011).

3. Bonneuil and Fressoz, *L'Événement anthropocène*, 141–171.

4. "L'industrie du coke et des benzols," *La Nature*, no. 2205 (1 January 1916): 6–10.

5. François Guedj, ed., *Le xxe siècle des guerres* (Paris: Éd. de l'Atelier, 2004), 23.

6. Rafael Reuveny, Andreea S. Mihalache-O'Keef, and Quan Li, "The Effect of Warfare on the Environment," *Journal of Peace Research*, vol. 47, no. 6,

(November 2010): 749–761; Richard Tucker and Edmund Russell, eds., *Natural Enemy, Natural Ally: Toward an Environmental History of War* (Corvallis: Oregon State University Press, 2004); Jay E. Austin and Carl E. Bruch, eds., *The Environmental Consequences of War: Legal, Economic, and Scientific Perspectives* (Cambridge: Cambridge University Press, 2007); M. Gutmann, "The Nature of Total War: Grasping the Global Environmental Dimensions of World War II," *History Compass*, vol. 13, no. 5 (2015): 251–261; Jacob Darwin Hamblin, "Environmental Dimensions of World War II," in *A Companion to World War II*, edited by Thomas W. Zeiler (Malden, UK: Wiley-Blackwell, 2013), 698–716.

7. John R. McNeill, "Woods and Warfare in World History," *Environmental History*, vol. 9, no. 3 (2004): 388–410; McNeill, *Du nouveau sous le soleil*, 454–455.

8. Peter Coates, Tim Cole, Marianna Dudley, and Chris Pearson, "Defending Nation, Defending Nature? Militarized Landscapes and Military Environmentalism in Britain, France and the US," *Environmental History*, vol. 16, no. 3 (2011): 456–491.

9. Kenneth Mouré, "'Les canons avant le beurre': Consommation et marchés civils en temps de guerre," in *1937–1947: La guerre-monde*, edited by Alya Aglan and Robert Frank, vol. 2 (Paris: Gallimard, 2015 [1973]).

10. Simo Laakkonen, "War: An Ecological Alternative to Peace? Indirect Impacts of World War II on the Finnish Environment," in *Natural Enemy, Natural Ally*, edited by Richard Tucker and Edmund Russell, 175–194; Chris Pearson, *Scarred Landscapes: War and Nature in Vichy France* (Basingstoke, UK: Palgrave, Macmillan, 2008).

11. Stradling, *Smokestacks and Progressives*, 141; Peter Thorsheim, *Waste into Weapons: Recycling in Britain during the Second World War* (Cambridge: Cambridge University Press, 2015).

12. William Tsutsui, "Landscapes in the Dark Valley: Toward an Environmental History of Wartime Japan," *Environmental History*, vol. 8, no. 2 (2003): 294–311.

13. Kenneth Mouré, "'Les canons avant le beurre.'"

14. Asit Biswas, "Scientific Assessment of the Long-Term Environmental Consequences of War," in *The Environmental Consequences of War*, edited by Jay E. Austin and Carl E. Bruch, 303–315.

15. Richard Tucker, *Insatiable Appetite: The United States and the Ecological Degradation of the Tropical World* (Berkeley: University of California Press, 2000).

16. Rémy Porte, *La Mobilisation industrielle, "premier front" de la Grande Guerre?* (Saint-Cloud, France: Éd. 14–18, 2006); Dominique Barjot, ed., *Deux guerres totales: 1914–1918, 1939–1945: La mobilisation de la nation* (Paris: Economica, 2012); Matthew Evenden, *Allied Power: Mobilizing Hydro-Electricity during Canada's Second World War* (Toronto: University of Toronto Press, 2015).

17. Pap Ndiaye, "La société américaine et la 'bonne guerre,'" in *1937–1947: la guerre-monde*, edited by Alya Aglan and Robert Frank, 2:1469.

18. Quoted in Auzanneau, *Or noir*, 107; Timothy C. Winegard, *The First World Oil War* (Toronto: University of Toronto Press, 2016).

19. Adeline Blaszkiewicz-Maison, *Albert Thomas: le socialisme en guerre (1914–1918)* (Rennes, France: Presses universitaires de Rennes, 2016).

20. Stradling, *Smokestacks and Progressives*, chap. 7: "War Meant Smoke," 148–149, 151–152.

21. Frank Uekoetter, "Total War: Administering Germany's Environment in Two World Wars," in *War and the Environment: Military Destruction in the Modern Age*, edited by Charles Closmann (Austin: University of Texas Press, 2009), 92–111.

22. Uekoetter, "Total War"; Uekoetter, "Polycentrism in Full Swing: Air Pollution Control in Nazi Germany," in *How Green Were the Nazis? Nature, Environment, and Nation in the Third Reich*, edited by Franz-Josef Brüggemeier, Mark Cioc, and Thomas Zeller (Athens: Ohio University Press, 2006), 101–128.

23. Alon Tal, *Pollution in a Promised Land: An Environmental History of Israel* (Berkeley: University of California Press, 2002), 68.

24. David Edgerton, *Britain's War Machine: Weapons, Resources and Experts in the Second World War* (London: Penguin Books, 2012), 14.

25. Umesh Chandra Jha, *Armed Conflict and Environmental Damage* (New Delhi: Vij Book, 2014), 161–162, 183.

26. Marianna Dudley, *An Environmental History of the UK Defence Estate (1945 to the Present)* (London: Continuum, 2012); Chris Pearson, *Mobilizing Nature: The Environmental History of War and Militarization in Modern France* (Manchester, UK: Manchester University Press, 2012).

27. Jeffrey Sasha Davis, Jessica S. Hayes-Conroy, and Victoria M. Jones, "Military Pollution and Natural Purity: Seeing Nature and Knowing Contamination in Vieques, Puerto Rico," *GeoJournal*, vol. 69, no. 3 (2007): 165–179.

28. Carla Goffi and Ria Verjauw, "La Sardaigne, poubelle de l'OTAN et du complexe militaro-industriel," on website: "La Plume à gratter," http://www.laplumeagratter.fr/2012/02/06/la-sardaigne-poubelle-de-lotan-et-du-complexe-militaro-industriel/ (last accessed 23 March 2017).

29. Michael Renner, "Assessing the Military's War on the Environment," in *State of the World 1991*, edited by Lester Brown (New York: Norton, 1991); Joseph Hupy, "The Environmental Footprint of War," *Environment and History*, vol. 14, no. 3 (2008): 405–421; John R. McNeill and David S. Painter, "The Global Environmental Footprint of the US Military (1789–2003)," in *War and the Environment*, edited by Charles Closmann, 10–31; Bonneuil and Fressoz, *L'Événement anthropocène*, 143.

30. Paul Josephson, "Industrial Deserts: Industry, Science and the Destruction of Nature in the Soviet Union," *Slavonic and East European Review*, vol. 85, no. 2 (2007): 294–321.

31. Paul Josephson, "War on Nature as Part of the Cold War: The Strategic and Ideological Roots of Environmental Degradation in the USSR," in *Environmental Histories of the Cold War*, edited by John R. McNeill and Corinna R. Unger (New York: Cambridge University Press, 2010), 21–49; Paul Josephson, Nicolai Dronin, Ruben Mnatsakanian, Aleh Cherp, Dmitry Efremenko, and Vladislav Larin, *An Environmental History of Russia* (Cambridge: Cambridge University Press, 2013).

32. Tsutsui, "Landscapes in the Dark Valley."

33. Paul R. Josephson, *Industrialized Nature: Brute Force Technology and the Transformation of the Natural World* (Washington, DC: Island Press, 2002).

34. Peter Galison and Bruce Hevly, eds., *Big Science: The Growth of Large-Scale Research* (Stanford, CA: Stanford University Press, 1992); Amy Dahan and Dominique Pestre, eds., *Les Sciences pour la guerre (1940–1960)* (Paris: EHESS, 2004); Anne Rasmussen, "Sciences et guerres," in *Histoire des sciences et des savoirs, vol. 3, Le Siècle des technosciences*; edited by Christophe Bonneuil and Dominique Pestre, 46–65.

35. Thomas Hugues, *American Genesis: A Century of Invention and Technological Enthusiasm (1870–1970)* (Chicago: University of Chicago Press, 2004 [1989]).

36. Edgerton, *Britain's War Machine*.

37. Sebastian Grevsmühl, *La Terre vue d'en haut: L'invention de l'environnement global* (Paris: Le Seuil, 2015); Sarah Bridger, *Scientists at War: The Ethics of Cold War Weapons Research* (Cambridge, MA: Harvard University Press, 2015).

38. Silvan S. Schweber, "Big Science in Context: Cornell and MIT," in *Big Science*, edited by Peter Galison and Bruce Hevly, 149–183.

39. Alain Gras, *Grandeur et dépendance: Sociologie des macro-systèmes techniques* (Paris: Presses universitaires de France, 1993), 161–165.

40. Thomas Hippler, *Le Gouvernement du ciel: Histoire globale des bombardements aériens* (Paris: Les Prairies ordinaires, 2014); John Buckley, *Air Power in the Age of Total War* (Bloomington: Indiana University Press, 1999).

41. Edgerton, *Britain's War Machine*, 185.

42. Hermione Giffard, *Making Jet Engines in World War II: Britain, Germany, and the United States* (Chicago: University of Chicago Press, 2016).

43. Guido De Luigi, Edgar Meyer, and Andrea Saba, "Industrie, pollution et politique: la 'zone noire' de la Societa Italiana del l'Alumino dans la province de Trente (1928–1938)," in *Industrialisation et société en Europe occidentale, de la fin du xixe siècle à nos jours: L'âge de l'aluminium*, edited by Ivan Grinberg and Florence Hachez-Leroy (Paris: Armand Colin, 1997), 314–323.

44. Matthew Evenden, "Aluminium, Commodity Chains and the Environmental History of the Second World War," *Environmental History*, vol. 16, no. 1 (2011): 69–93.

45. Edmund Russell, *War and Nature: Fighting Humans and Insects with Chemicals from World War I to Silent Spring* (Cambridge: Cambridge

University Press, 2001); Ross and Amter, *The Polluters*, 18; Nicolino, *Un empoisonnement universel*, part 2: "Le temps des assassins."

46. Quoted in Russell, *War and Nature*, 18.

47. Pap Ndiaye, *Du nylon et des bombes. DuPont de Nemours, le marché et l'État américain (1900–1970)* (Paris: Belin, 2001).

48. Nathalie Jas, "Gouverner les substances chimiques dangereuses dans les espaces internationaux," in *Le Gouvernement des technosciences: Gouverner le progrès et ses dégâts depuis 1945*, edited by Dominique Pestre (Paris: La Découverte, 2014), 31–63.

49. Russell, *War and Nature*, 53–94; Ross and Amter, *The Polluters*, 45–57.

50. James E. McWilliams, *American Pests: The Losing War on Insects from Colonial Times to DDT* (New York: Columbia University Press, 2008).

51. Russell, *War and Nature*, 165–228.

52. George L. Mosse, *De la Grande Guerre au totalitarisme: La brutalisation des sociétés européennes* (Paris: Hachette, 2015 [1990]).

53. Jean-Pierre Guéno, *Paroles de poilus: Lettres et carnets du front (1914–1918)* (Paris: Librio/Flammarion, 1998), introduction to chap. 4, p. 97.

54. Pearson, *Mobilizing Nature*, chap. 4–6; Dorothee Brantz, "Environments of Death: Trench Warfare on the Western Front (1914–1918)," in *War and the Environment*, edited by Charles E. Closmann, 68–91; Jean-Paul Amat, "Guerre et milieux naturels: les forêts meurtries dans l'est de la France 70 ans après Verdun," *Espace géographique*, vol. 16, no. 3 (1987) 217–233; Jeffrey Sasha Davis, "Military Natures: Militarism and the Environment," *GeoJournal*, vol. 69, no. 3, 2007, 131–134; Jean-Yves Puyo, "Les conséquences de la Première Guerre mondiale pour les forêts et les forestiers français," *Revue forestière française*, vol. 56, no. 6, 2004, 573–584; Hugh D. Clou, *After the Ruins: Restoring the Countryside of Northern France after the Great War* (Exeter, UK: Exeter University Press, 1996); Pierre-Alain Tallier, "La reconstitution du patrimoine forestier belge après 1918," in *Forêt et guerre*, edited by Andrée Corvol and Jean-Paul Amat (Paris: L'Harmattan, 1994), 215–225.

55. Paul Arnould, Micheline Hotyat, and Laurent Simon, *Les Forêts d'Europe* (Paris: Nathan, 1997), 114.

56. *La Nature*, no. 2206 (8 January 2016) : 30.

57. Tobias Bausinger, Éric Bonnaire, and Johannes Preuß, "Contribution à l'étude des conséquences écologiques de la Première Guerre mondiale dans les forêts dévastées de la Zone rouge," in *Des milieux aux territoires forestiers*, edited by Marc Galochet and Éric Glon (Arras, France: Artois Presses Université, 2010), 157–167; Benoît Hopquin, "Le poison de la guerre coule toujours à Verdun," *Le Monde*, 20 January 2014.

58. Henri Barbusse, *Under Fire*, translated by Fitzwater Wray. http://www.gutenberg.org/files/4380/4380-h/4380-h.htm (last accessed 13 November 2018).

59. Barbusse, *Under Fire*.

60. Olivier Lepick, *La Grande Guerre chimique (1914–1918)* (Paris: Presses universitaires de France, 1998).

61. Jean-Marie Moine, "Un mythe aéronautique et urbain dans la France de l'entre-deux-guerres: le péril aérochimique," *Revue historique des armées*, no. 256 (2009): 94–119.

62. Judith A. Bennett, "War, Emergency and the Environment: Fiji, 1939–1946," *Environment and History*, vol. 7, no. 3 (2001): 255–287; Bennett, *Natives and Exotics: World War II and Environment in the Southern Pacific* (Honolulu: University of Hawai'i Press, 2009).

63. Jörg Friedrich, *The Fire: The Bombing of Germany (1940–1945)* (New York: Columbia University Press, 2007); Randall Hansen, *Foudre et devastation: Les bombardements alliés sur l'Allemagne (1942–1945)* (Québec: Presses universitaires de Laval, 2012).

64. Greg Bankoff, "A Curtain of Silence: The Fate of Asia's Fauna in the Cold War," in *Environmental Histories of the Cold War*, edited by John R. McNeill and Corinna R. Unger, 203–226.

65. Barry Weisberg, *Ecocide in Indochina: The Ecology of War* (San Francisco: Canfield Press, 1969).

66. Claude-Marie Vadrot, *Guerres et environnement: Panorama des paysages et des écosystèmes bouleversés* (Paris: Delachaux & Niestlé, 2005); Thao Tran, Jean-Paul Amat, and Françoise Pirot, "Guerre et défoliation dans le Sud Viêt-Nam (1961–1971)," *Histoire et mesure*, vol. 22, no. 1 (2007): 71–107.

67. Hopquin, "Le poison de la guerre coule toujours à Verdun"; Olivier Saint-Hilaire, "Déchets de guerre," *Médiapart*, 16 May 2014, https://www.mediapart.fr/studio/portfolios/dechets-de-guerre (last accessed 13 November 2018).

68. "Cimetières de sous-marins et zones d'essais nucléaires: les séquelles de la guerre froide," in "Environnement et pollution en Russie et en Asie centrale: l'héritage soviétique" (Paris: La Documentation française, 2007), http://www.ladocumentationfrancaise.fr/dossiers/heritage-sovietique/index.shtml (last accessed 13 November 2018).

69. Richard Rhodes, *The Making of the Atomic Bomb* (New York: Simon & Schuster, 1986); Walter E. Grunden, Mark Walker, and Masakatsu Yamazaki, "Wartime Nuclear Weapons Research in Germany and Japan," in "Politics and Science in Wartime," special issue of *Osiris*, vol. 20 (2005): 107–130.

70. Tsutsui, "Landscapes in the Dark Valley."

71. Toshihiro Higuchi, "Atmospheric Nuclear Weapons Testing and the Debate on Risk Knowledge in Cold War America (1945–1963)," in *Environmental Histories of the Cold War*, edited by John R. McNeill and Corinna R. Unger, 301–322.

72. Jacques Villain, *Le Livre noir du nucléaire militaire* (Paris: Fayard, 2014).

73. "Environnement et pollution en Russie et en Asie centrale"; McNeill, *Du nouveau sous le soleil*, 452.

74. *Les Incidences environnementales et sanitaires des essais nucléaires effectués par la France entre 1960 et 1996 et les éléments de comparaison avec les*

essais des autres puissances nucléaires, French parliamentary report no. 207 (2001–2002); Bruno Barrillot, *Essais nucléaires français: L'héritage empoisonné* (Lyon, France: Observatoire des armements, 2012); Barillot, *Les Essais nucléaires français (1960–1996): Conséquences sur l'environnement et la santé* (Lyon, France: Centre de documentation et de recherche sur la paix et les conflits, 1996).

75. Michele S. Gerber, *On the Home Front: The Cold War Legacy of the Hanford Nuclear Site* (Lincoln: University of Nebraska Press, 1992); McNeill, *Du nouveau sous le soleil,* 450–451.

76. Jacob Darwin Hamblin, *Poison in the Well: Radioactive Waste in the Oceans at the Dawn of the Nuclear Age* (New Brunswick, NJ: Rutgers University Press, 2008); Hamblin, "Gods and Devils in the Details: Marine Pollution, Radioactive Waste, and an Environmental Regime (circa 1972)," *Diplomatic History,* vol. 32, no. 4 (2008): 539–560; Hamblin, "Hallowed Lords of the Sea: Scientific Authority and Radioactive Waste in the United States, Britain, and France," *Osiris,* vol. 21, no. 1 (2006): 209–228.

77. Jean-Pierre Queneudec, "Le rejet à la mer de déchets radioactifs," *Annuaire français de droit international,* vol. 11, no. 1 (1965): 750–782.

78. *Report on the Health Consequences to the American Population from Nuclear Weapons Tests Conducted by the United States and Other Nations,* online: http:// www.cdc.gov/nceh/radiation/fallout/ (last accessed 13 November 2018).

79. Hicham-Stéphane Afeissa, *La Fin du monde et de l'humanité. Essai de généalogie du discours écologique* (Paris: Presses universitaires de France, 2014).

80. Worster, *Les Pionniers de l'écologie.*

81. Quoted in Yannick Mahrane and Christophe Bonneuil, "Gouverner la biosphère: De l'environnement de la guerre froide à l'environnement néolibéral," in *Le Gouvernement des technosciences,* edited by Dominique Pestre, 134–135.

82. Michael Egan, *Barry Commoner and the Science of Survival: The Remaking of American Environmentalism* (Cambridge, MA: MIT Press, 2007); Jacob Darwin Hamblin, *Arming Mother Nature: The Birth of Catastrophic Environmentalism* (Oxford: Oxford University Press, 2013).

83. Austin and Bruch, eds., *The Environmental Consequences of War.*

Chapter 8

1. Bruce Podobnik, *Global Energy Shifts: Fostering Sustainability in a Turbulent Age* (New Delhi: Teri Press, 2006); Kander, Malanima, and Warde, *Power to the People,* 348; Jean-Claude Debeir, Jean-Paul Deléage, and Daniel Hémery, *Une histoire de l'énergie. Les servitudes de la puissance* (Paris: Flammarion, 2013 [1986]); Alfred W. Crosby, *Children of the Sun: A History of Humanity's Unappeasable Appetite for Energy* (New York: Norton, 2006).

2. Vaclav Smil, *Energy in World History* (Boulder: Westview Press, 1994), 205–207; Smil, *Energy Myths and Realities: Bringing Science to the Energy Policy Debate* (Washington, DC: AEI Press, 2010); Joel Darmstadter, *Energy in the World Economy* (Baltimore: Johns Hopkins University Press, 1971).

3. McNeill, *Du nouveau sous le soleil*, 42.

4. Bouda Etemad, *La Production mondiale d'énergie primaire commerciale* (Paris: Unesco, 1993), online: http://unesdoc.unesco.org/images/0009/000965 /096534fb.pdf (last accessed 15 November 2018).

5. Sam H. Schurr, ed., *Energy, Economic Growth, and the Environment* (New York: RFF Press, 2013); Andrew Nikiforuk, *L'Énergie des esclaves: Le pétrole et la nouvelle servitude* (Montréal: Écosociété, 2015).

6. Jean-Marie Martin-Amouroux, *Charbon, les métamorphoses d'une industrie* (Paris: Technip, 2008); Régine Perron, *Le Marché du charbon, un enjeu entre l'Europe et les États-Unis (de 1945 à 1958)* (Paris: Publications de la Sorbonne, 1996).

7. Vaclav Smil, *China's Past, China's Future: Energy, Food, Environment* (London: Routledge, 2004), 11–12.

8. Odette Hardy-Hémery, "Rationalisation technique et rationalisation du travail à la compagnie des mines d'Anzin (1927–1938)," *Le Mouvement social*, no. 72 (1970): 3–48; Thomas G. Andrews, *Killing for Coal: America's Deadliest War on Labor* (Cambridge, MA: Harvard University Press, 2008); Diana Cooper-Richet, *Le Peuple de la nuit: Mines et mineurs en France (xixe–xxe siècle)* (Paris: Perrin, 2002); Rolande Trempé, *Les Trois Batailles du charbon (1936–1947)* (Paris: La Découverte, 1989).

9. Arthur McIvor and Ronald Johnston, *Miners' Lung: A History of Dust Disease in British Coal Mining* (Aldershot, UK: Ashgate, 2007); David Rosner and Gerald Markowitz, *Deadly Dust: Silicosis and the On-Going Struggle to Protect Workers' Health* (Ann Arbor: University of Michigan Press, 2006 [1991]).

10. Paul-André Rosental, "La silicose comme maladie professionnelle transnationale," *Revue française des affaires sociales*, no. 2 (2008): 255–277; Paul-André Rosental and Jean-Claude Devinck, "Statistique et mort industrielle: La fabrication du nombre de victimes de la silicose dans les houillères en France de 1946 à nos jours," *Vingtième siècle: Revue d'histoire*, vol. 95, no. 3 (2007): 75–91.

11. Eric Geerkens, "Quand la silicose n'était pas une maladie professionnelle: Genèse de la réparation des pathologies respiratoires des mineurs en Belgique (1927–1940)," *Revue d'histoire moderne et contemporaine*, vol. 56, no. 1 (2009): 127–141; Judith Rainhorn, ed., *Santé et travail à la mine (xixe–xxie siècle)* (Villeneuve-d'Ascq: Presses universitaires du Septentrion, 2014); Stefania Barca, "Bread and Poison: The Story of Labor Environmentalism in Italy (1968–1998)," in *Dangerous Trade*, edited by Christopher Sellers and Joseph Melling, 126–139.

12. Georges Franju, *Les Poussières*, documentary film, France, 1954 (22 min).

13. Rosner and Markowitz, *Deadly Dust*; Chad Montrie, *To Save the Land and People: A History of Opposition to Surface Coal Mining in Appalachia* (Chapel Hill, NC: University of North Carolina Press, 2003).

14. Smil, *China's Past, China's Future*, 11–12, 17.

15. Wang Bing, *À l'ouest des rails* (2003) and *L'Argent du charbon* (2008), documentary films.

16. Philip Roth, *I Married a Communist* (New York, Vintage Books, 1998), 225–226.

17. Josephson et al., *An Environmental History of Russia*, 83; Jean-Marie Martin-Amouroux, "Le développement énergétique de l'Union soviétique de 1917 à 1950," *Encyclopédie de l'énergie*, October 2015, online: https://www.encyclopedie-energie.org/le-developpement-energetique-de-lunion-sovietique-de-1917-a-1950/ (last accessed 15 November 2018).

18. Alexis Zimmer, "'Le brouillard mortel de la vallée de la Meuse' (décembre 1930): Naturalisation de la catastrophe," in *Débordements industriels*, edited by Thomas Le Roux and Michel Letté, 115–131, quotation from page 125; Zimmer, *Brouillards toxiques: Vallée de la Meuse, 1930, contre-enquête* (Brussels: Zones sensibles, 2016).

19. Tarr, ed., *Devastation and Renewal*; Lynne Snyder, "The Death-Dealing Smog over Donora, Pennsylvania: Industrial Air Pollution, Public Health Policy, and the Politics of Expertise (1948–1949)," *Environmental History Review*, vol. 18, no. 1 (1994): 117–139; Ross and Amter, *The Polluters*, 86–97.

20. Sabine Barles and Eunhye Kim, "The Energy Consumption of Paris and Its Supply Areas from the Eighteenth Century to the Present," *Regional Environmental Change*, vol. 12, no. 2 (2002): 295–310.

21. "Le gaz carbonique, son action sur l'organisme, ses dangers," *L'Humanité* (11 October 1926): 4.

22. Stephen Mosley, "'A Network of Trust': Measuring and Monitoring Air Pollution in British Cities (1912–1960)," *Environment and History*, vol. 15, no. 3 (2009): 273–302; Florian Charvolin, Stéphane Frioux, Léa Kamoun, François Mélard, and Isabelle Roussel, *Un air familier? Sociohistoire des pollutions atmosphériques* (Paris: Presses de l'École des mines, 2015).

23. Peter Thorsheim, "Interpreting the London Fog Disaster of 1952," in *Smoke and Mirrors*, edited by E. Melanie DuPuis, 154–169; Thorsheim, *Inventing Pollution*; Brimblecombe, *The Big Smoke*, chap. 8.

24. McNeill, *Du nouveau sous le soleil*, 113.

25. Frédéric Montandon and André Picot, *Écotoxicochimie appliquée aux hydrocarbures* (Paris: Technique et documentation, 2013).

26. Daniel Yergin, *The Prize: The Epic Quest for Oil, Money, and Power* (New York: Free Press, 2008); Nikiforuk, *L'Énergie des esclaves*; Auzanneau, *Or noir*; Alain Gras, *Oil: Petite anthropologie de l'or noir* (Paris: Éd. B2, 2015).

27. Timothy Mitchell, *Carbon Democracy: Le pouvoir politique à l'ère du pétrole* (Paris: La Découverte, 2013).

28. Serge Mallet, *La Nouvelle Classe ouvrière* (Paris: Le Seuil, 1963); Marion Fontaine, *Fin d'un monde ouvrier (Liévin, 1974)* (Paris: EHESS, 2014), 47–77.

29. McNeill, *Du nouveau sous le soleil*, 394.

30. Elsa Devienne, "Des plages dans la ville: Une histoire sociale et environnementale du littoral de Los Angeles (1920–1972)" (PhD diss., EHESS, 2014).

31. Upton Sinclair, *Pétrole!* (Paris: Stock, 2012 [1927]).

32. Auzanneau, *Or noir*, 122; "Oil in American History," special issue of the *Journal of American History*, vol. 99, no. 1 (2012).

33. Hugh S. Gorman, *Redefining Efficiency: Pollution Concerns, Regulatory Mechanisms, and Technological Change in the US Petroleum Industry* (Akron, OH: University of Akron Press, 2001).

34. Martin V. Melosi and Joseph A. Pratt, eds., *Energy Metropolis: An Environmental History of Houston and the Gulf Coasts* (Pittsburgh, PA: University of Pittsburgh Press, 2007), 3; Kathleen A. Brosnan, Martin V. Melosi, and Joseph A. Pratt, eds., *Energy Capitals: Local Impact, Global Influence* (Pittsburgh, PA: Pittsburgh University Press, 2014).

35. Joseph A. Pratt, "A Mixed Blessing: Energy, Economic Growth, and Houston's Environment," in *Energy Metropolis*, edited by Martin V. Melosi and Joseph A. Pratt, 21–52.

36. Miller, *An Environmental History of Latin America*, 156–157; Christopher Sellers, "Petropolis and Environmental Protest in Cross-National Perspective: Beaumont-Port Arthur, Texas, versus Minatitlan-Coatzacoalcos, Veracruz," *Journal of American History*, vol. 99, no. 1 (2012): 111–123.

37. Santiago, *The Ecology of Oil*.

38. Didier Ramousse, "L'industrie pétrolière au Venezuela: Rupture, conflits et gestion des espaces côtiers," in *Les Littoraux latino-américains. Terres à découvrir*, edited by Violette Brustlein-Waniez and Alain Musset (Paris: Éd. de l'IHEAL, 1998), 152–177.

39. Marcel Amphoux, "Une nouvelle industrie française: le raffinage du pétrole," *Annales de géographie*, vol. 44, no. 251 (1935): 509–533.

40. André Nouschi, *La France et le pétrole* (Paris: Picard, 2001); Samir Saul, "Politique nationale du pétrole, sociétés nationales et 'pétrole franc,'" *Revue historique*, no. 638 (2006): 355–388; Saul, *Intérêts économiques français et décolonisation de l'Afrique du Nord (1945–1962)* (Geneva: Droz, 2016).

41. Archives départementales de Gironde, 5 M 210, Pollution des eaux de la Garonne et de la Gironde (1935–1937).

42. *L'Humanité*, 15 January 1936, 1.

43. Xavier Daumalin, *Le Patronat marseillais et la deuxième industrialisation (1880–1930)* (Aix-en-Provence, France: Presses universitaires de Provence, 2014).

44. Xavier Daumalin and Christelle Gramaglia, "'Ni partir, ni mourir, mais vivre ici': Jalons pour une sociohistoire des mobilisations contre les pollutions dans la zone industrialo-portuaire de Berre/Fos-sur-Mer," in *Santé et environnement*, edited by Valérie Chansigaud, forthcoming.

45. Bernard Paillard (in collaboration with Claude Fischler), *La Damnation de Fos* (Paris: Le Seuil, 1981).

46. Pamphlet "30 ans de concertation, SPPPI PACA," edited in 2001, online: https://www.spppi-paca.org/_depot_sppi/_depot_arko/articles/20/livrecomplet -30ans-spppipaca_doc.pdf (last accessed 15 November 2018); the pollutions on the site inspired several documentary films, including Jacques Windenberger: *Tumeurs et silences. Fos-étang de Berre: pollutions industrielles et cancers* (2013).

47. Jean Chapelle and Sonia Ketchian, *URSS, second producteur de pétrole du monde* (Paris: Technip, 1963); Robert W. Campbell, *The Economics of Soviet Oil and Gas* (Baltimore, MD: Johns Hopkins University Press, 1968).

48. Josephson et al., *An Environmental History of Russia*, 230; Rachel E. Neville, "Two Black Golds: Petroleum Extraction and Environmental Protection in the Caspian Sea," *The Journal of Public and International Affairs*, vol. 12 (2001): 109–123.

49. Hocine Malti, *Histoire secrète du pétrole algérien* (Paris: La Découverte, 2010).

50. Douglas Yates, "Port Gentil: From Forestry Capital to Energy Capital," in *Energy Capitals*, edited by Kathleen A. Brosnan, Martin V. Melosi, and Joseph A. Pratt, 159–178.

51. Toyin Falola and Matthew M. Heaton, *A History of Nigeria* (Cambridge: Cambridge University Press, 2008), 182; Benoît Paraut, *Le Pétrole au Nigeria, un instrument au service de quel développement? Pillage, crise identitaire et résistance dans le delta du Niger* (Paris: L'Harmattan, 2009); Anna Zalik, "The Niger Delta: Petro-Violence and Partnership Development," *Review of African Political Economy*, vol. 101, no. 4 (2004): 401–424.

52. Alain R. Bertrand, *Transport maritime et pollution accidentelle par le pétrole. Faits et chiffres (1951–1999)* (Paris: Technip, 2000); John Sheail, "Torrey Canyon: The Political Dimension," *Journal of Contemporary History*, vol. 42, no. 3 (2007): 485–504.

53. Emile Guillaumin, *The Life of a Simple Man*, translated by Margaret Holden (London: Selwyn & Blount, 1919), 269–270.

54. Pierre Thiesset, ed., *Écraseurs! Les méfaits de l'automobile* (Vierzon: Le Pas de côté, 2015), n. 15.

55. Centre national de ressources textuelles et lexicales.

56. Steve Bernardin, "La Fabrique privée d'un problème public: La sécurité routière entre industriels et assureurs aux États-Unis (années 1920 à 2000)" (PhD diss., Université de Paris 1, 2015), part 1.

57. Jean-Pierre Bardou, Jean-Jacques Chanaron, Patrick Fridenson, and James M. Laux, *La Révolution automobile* (Paris: Albin Michel, 1977).

58. Lewis H. Siegelbaum, *Cars for Comrades: The Life of the Soviet Automobile* (Ithaca, NY: Cornell University Press, 2008).

59. Michael Walsh, "Global Trends in Motor Vehicle Use and Emissions," *Annual Review of Energy and the Environment*, vol. 15 (1990): 217–243.

60. Frank Uekötter, "The Merits of the Precautionary Principle: Controlling Automobile Exhausts in Germany and the United States before 1945," in *Smoke and Mirrors*, edited by E. Melanie DuPuis, 19–153.

61. Scott Dewey, "Working for the Environment: Organized Labor and the Origins of Environmentalism in the United States (1948–1970)," *Environmental History*, vol. 3, no. 1 (1998): 45–63; Andrew Van Alstyne, "The United Auto-Workers and the Emergence of Labor Environmentalism," *Working USA: The Journal of Labor and Society*, vol. 18 (2015): 613–627; Chad Montrie, *Making a Living: Work and Environment in the United States* (Chapel Hill: University of North Carolina Press, 2008), chap. 5.

62. Chloé Vlassopoulos, "Car Pollution: Agenda Denial and Agenda Setting in Early Twentieth Century France and Greece," in *History and Sustainability*, edited by Marco Armiero et al. (Florence: Universita degli studi di Firenze, 2005), 252–257.

63. McNeill, *Du nouveau sous le soleil*, 114–118; Ross and Amter, *The Polluters*, 73–85.

64. Rajiv K. Sinha, "Automobile Pollution in India and Its Human Impact," *The Environmentalist*, vol. 13, no. 2 (1993): 111–115.

65. Daniel Boullet, *Entreprises et environnement en France de 1960 à 1990: Les chemins d'une prise de conscience* (Geneva: Droz, 2006), 220.

66. Quoted by Uekoetter, *The Age of Smoke*, 215.

67. "Vers l'automobile non polluante," *Pétrole-Progrès*, Spring 1971, nos. 94–95, 7.

68. Louis Tsagué, *La Pollution due au transport urbain et aéroportuaire: Caractéristiques et méthode de réduction* (Paris: L'Harmattan, 2009).

69. Joel Tarr, "Transforming an Energy System: The Evolution of the Manufactured Gas Industry and the Transition to Natural Gas in the United States (1807–1954)," in *The Governance of Large Technical Systems*, edited by Olivier Coutard (London: Routledge, 1999), 19–37.

70. Martin V. Melosi, *Coping with Abundance: Energy and Environment in Industrial America* (Philadelphia, PA: Temple University Press, 1985), 262.

71. Podobnik, *Global Energy Shifts*, 102–103.

72. Serge Paquier and Jean-Pierre Williot, eds., *L'Industrie du gaz en Europe aux xixe et xxe siècles: L'innovation entre marchés privés et collectivités publiques* (Brussels: Peter Lang, 2005); Alain Beltran and Jean-Pierre Williot, *Les Routes du gaz: Histoire du transport de gaz naturel en France* (Paris: Le Cherche Midi, 2012).

73. Vaclav Smil, *Natural Gas: Fuel for the 21st Century* (Chichester: John Wiley, 2015), 120.

74. Christophe Defeuilley, "Le gaz naturel en Europe: Entre libéralisation des marchés et géopolitique," *Flux*, vol. 75, no. 1 (2009): 99–111.

75. Jean-Pierre Digard, Bernard Hourcade, and Yann Richard, *L'Iran au xxe siècle* (Paris: Fayard, 1996), chap. 8: "Une économie noyée dans le pétrole."

76. Louis Bonnaure reports appeared in *Le Dauphiné libéré*, 15 November 1966.

77. "Lacq, capitale du gaz," *Syndicalisme Hebdo*, May 1961, quoted in Renaud Bécot, "Syndicalisme et environnement en France de 1944 aux années quatre-vingt" (PhD diss., EHESS, 2015), 108.

78. Christophe Briand, "Les enjeux environnementaux du complexe industriel de Lacq (1957–2005)," *Flux*, vol. 63–64, no. 1 (2006): 20–31.

79. Bernard Charbonneau, *Tristes campagnes* (Vierzon, France: Le Pas de côté, 2013 [Denoël, 1973]), 88; Mauriac is quoted in Gérard Fayolle, *L'Aquitaine au temps de François Mauriac (1885–1970)* (Paris: Hachette, 2004), 33–34.

80. Alain Beltran, "Du luxe au cœur du système: Électricité et société dans la région parisienne (1800–1939)," *Annales ESC*, vol. 44, no. 5 (1989): 1113–1136; Alain Beltran and Patrice A. Carré, *La Fée et la servante: La société française face à l'électricité (xixe–xxe siècle)* (Paris: Belin, 1991).

81. Thomas Hughes, *Networks of Power: Electrification in Western Society* (Baltimore, MD: Johns Hopkins University Press, 1983); Gras, *Grandeur et dépendance.*

82. David E. Nye, *Electrifying America: Social Meanings of a New Technology* (Cambridge, MA: MIT Press, 1990); Maurice Lévy-Leboyer and Henri Morsel, eds., *Histoire de l'électricité en France*, vol. 2, *1919–1946* (Paris: Fayard, 1994); Arnaud Berthonnet, "L'électrification rurale: Ou le développement de la 'fée électricité' au cœur des campagnes françaises dans le premier xxe siècle," *Histoire et sociétés rurales*, vol. 19, no. 1 (2003): 193–219; Vincent Lagendijk, *Electrifying Europe: The Power of Europe in the Construction of Electricity Networks* (Amsterdam: Aksant, 2008).

83. Josephson et al., *An Environmental History of Russia*, 162.

84. Timothy J. LeCain, *Mass Destruction: The Men and Giant Mines That Wired America and Scarred the Planet* (New Brunswick, NJ: Rutgers University Press, 2009).

85. Cioc, "The Impact of the Coal Age on the German Environment."

86. Judith Shapiro, *Mao's War against Nature: Politics and the Environment in Revolutionary China* (Cambridge: Cambridge University Press, 2001); Kenneth Pomeranz, "Les eaux de l'Himalaya: Barrages géants et risques environnementaux en Asie contemporaine," *Revue d'histoire moderne et contemporaine*, vol. 62, no. 1 (2015): 7–47.

87. McNeill, *Du nouveau sous le soleil*, 218–251.

88. Josephson, *Industrialized Nature*, chap. 1.

89. Patrick McCully, *Silenced Rivers: The Ecology and Politics of Large Dams* (London: Zed Books, 1996).

90. Robert W. Adler, *Restoring Colorado River Ecosystems: A Troubled Sense of Immensity* (Washington, DC: Island Press, 2007); Edward Goldsmith and

Nicholas Hildyard eds., *The Social and Environmental Effects of Large Dams*, vol. 1, *Overview* (Camelford, UK: Wadebridge Ecological Centre, 1984); Jacques Leslie, *La Guerre des barrages: Développement forcé, populations sacrifiées, environnement dévasté* (Paris: Buchet-Chastel, 2008); M. Bakre, J. Bethemont, R. Commère, and A. Vant, *L'Égypte et le haut-barrage d'Assouan. De l'impact à la valorisation* (Saint-Étienne, France: Presses de l'université de Saint-Étienne, 1980).

91. Alain Pelosato, *Au fil du Rhône: Histoires d'écologie (1971–1991)* (Paris: Messidor, 1991); Sara B. Pritchard, *Confluence: The Nature of Technology and the Remaking of the Rhône* (Cambridge, MA: Harvard University Press, 2011), 208.

92. David Pace, "Old Wine, New Bottles: Atomic Energy and Ideology of Science in Post-War France," *French Historical Studies*, vol. 17, no. 1 (1991): 38–61; Gabriel Hecht, *Le Rayonnement de la France* (Paris: La Découverte, 1998); Robert Belot, *L'Atome et la France: Aux origines de la technoscience française* (Paris: Odile Jacob, 2015).

93. Soraya Boudia, "Naissance, extinction et rebonds d'une controverse scientifique: les dangers de la radioactivité pendant la guerre froide," *Mil neuf cent: Revue d'histoire intellectuelle*, no. 25 (2007): 157–170; Boudia, "Les problèmes de santé publique de longue durée: Les effets des faibles doses de radioactivité," in *La Définition des problèmes de santé publique*, edited by Claude Gilbert and Emmanuel Henry (Paris: La Découverte, 2009), 35–53.

94. Rachel Rothschild, "Environmental Awareness in the Atomic Age: Radioecologists and Nuclear Technology," *Historical Studies in the Natural Sciences*, vol. 43, no. 4 (2013): 492–530.

95. Sezin Topçu, "Atome, gloire et désenchantement: Résister à la France atomique avant 1968," and Gabrielle Hecht, "L'empire nucléaire: Les silences des 'Trente Glorieuses,'" in *Une autre histoire des "Trente Glorieuses": Modernisation, contestations et pollutions dans la France d'après guerre*, edited by Christophe Bonneuil, Céline Pessis, and Sezin Topçu (Paris: La Découverte, 2013), 189–209 and 159–178.

96. Quoted in Bécot, "Syndicalisme et environnement en France," 83.

97. Françoise Zonabend, *La Presqu'île au nucléaire: Three Mile Island, Tchernobyl, Fukushima ... et après?* (Paris: Odile Jacob, 2014).

98. Quoted in Philippe Brunet, *La Nature dans tous ses états: Uranium, nucléaire et radioactivité en Limousin* (Limoges, France: Presses universitaires du Limousin, 2004), 167.

99. Gabrielle Hecht, *Uranium africain: Une histoire globale* (Paris: Le Seuil, 2016).

100. These numbers come from the Club of Rome report, *The Limits to Growth*.

101. Philippe Lhoste, Michel Havard, and Éric Vall, *La Traction animale* (Gembloux, Belgium: Presses agronomiques de Gembloux/Wageningen, CTA, 2010); Debeir, Deléage, and Hémery, *Une histoire de l'énergie*, 289; David Edgerton,

Quoi de neuf? Du rôle des techniques dans l'histoire globale (Paris: Le Seuil, 2013), 65.

Chapter 9

1. Oiwa Keibo and Ogata Masato, *Rowing the Eternal Sea: The Story of a Minamata Fisherman* (Lanham, MD: Rowman & Littlefield, 2001 [1996]); Walker, *Toxic Archipelago*; Paul Jobin, "La maladie de Minamata et le conflit pour la reconnaissance," *Ebisu*, vol. 31, no. 1 (2003): 27–56.

2. Bonneuil and Fressoz, *L'Événement anthropocène*, 31; John McNeill and Peter Engelke, *The Great Acceleration: An Environmental History of the Anthropocene since 1945* (Cambridge, MA: Belknap Press of Harvard University Press, 2014).

3. Manuel Charpy, *Le Théâtre des objets: Culture matérielle et identité bourgeoise au xixe siècle* (Paris: Flammarion, 2017); Frank Trentmann, *Empire of Things: How We Became a World of Consumers, from the Fifteenth Century to the Twenty-First* (London: Allen Lane, 2016).

4. Robert Skidelsky, *John Maynard Keynes (1883–1946), Economist, Philosopher, Statesman* (London: Macmillan, 2003).

5. Lizabeth Cohen, *A Consumers' Republic: The Politics of Mass Consumption in Postwar America* (New York: A. A. Knopf, 2003).

6. Marshall I. Goldman, *The Spoils of Progress: Environmental Pollution in the Soviet Union* (Cambridge, MA: MIT Press, 1972).

7. Jackson Lears, *Fables of Abundance: A Cultural History of Advertising in America* (New York: Basic Book, 1994); Stuart Ewen, *Captains of Consciousness: Advertising and the Social Roots of the Consumer Culture* (New York: Basic Books, 2001 [1976]).

8. Kathleen G. Donohue, *Freedom from Want: American Liberalism and the Idea of the Consumer* (Baltimore, MD: Johns Hopkins University Press, 2003); Gary Cross, *An All-Consuming Century: Why Commercialism Won in Modern America* (New York: Columbia University Press, 2000); Emily S. Rosenberg, "'Le "modèle américain' de la consommation de masse," *Cahiers d'histoire: Revue d'histoire critique*, no. 108 (2009): 111–142.

9. Sheldon Garon and Patricia L. Maclachlan, eds., *The Ambivalent Consumer: Questioning Consumption in East Asia and the West* (Ithaca, NY: Cornell University Press, 2006); S. Jonathan Wiesen, *Creating the Nazi Marketplace: Commerce and Consumption in the Third Reich* (New York: Cambridge University Press, 2011).

10. Jean-Claude Daumas, "L'invention des usines à vendre : Carrefour et la révolution de l'hypermarché," *Réseaux*, vol. 135–136, no. 1 (2006): 59–91.

11. Roy S. Thomson, "Chrome Tanning in the Nineteenth Century," *Journal of the Society of Leather Technologists and Chemists*, vol. 69 (1985): 93–98; *Leather Processing and Tanning Technology, Handbook* (New Delhi: NIIR Project Consultancy Service, 2011); Cédric Perrin, "Le développement durable

en perspective historique: L'exemple des tanneries," *L'Homme et la société*, vol. 193–194, no. 3 (2014): 37–56.

12. Subodh Kumar Rastogi, Amit Pandey, and Sachin Tripathi, "Occupational Health Risks among the Workers Employed in Leather Tanneries at Kanpur," *Indian Journal of Occupational and Environmental Medicine*, vol. 12, no. 3 (2008): 132–135; United Nations Industrial Development Organization, *Chrome Management in the Tanyard*, Regional Programme for Pollution Control in the Tanning Industry in South-East Asia, 2000.

13. Quynh Delaunay, *Histoire de la machine à laver: Un objet technique dans la société française* (Rennes, France: Presses universitaires de Rennes, 1994); Kristin Ross, *Rouler plus vite, laver plus blanc: Modernisation de la France et décolonisation au tournant des années 1960* (Paris: Flammarion, 2006 [1995]).

14. Adrian Forty, *Objects of Desire: Design and Society from Wedgewood to IBM* (New York: Pantheon Books, 1986); Claire Leymonerie, *Le Temps des objets: Une histoire du design industriel en France (1945–1980)* (Saint-Étienne: Cité du design, 2016).

15. Susan Strasser, *Waste and Want: A Social History of Trash* (New York: Metropolitan Books, 2000); Denis Woronoff, *Histoire de l'emballage en France (du xviiie siècle à nos jours)* (Valenciennes, France: Presses universitaires de Valenciennes, 2015).

16. Bernard London, *L'Obsolescence planifiée: Pour en finir avec la Grande Dépression*, followed by Serge Latouche, *Bernard London, ou la Planification de l'obsolescence à des fins sociales* (Paris, Éd. B2, 2013).

17. Giles Slade, *Made to Break: Technology and Obsolescence in America* (Cambridge, MA: Harvard University Press, 2006); Ndiaye, *Du nylon et des bombes*.

18. McNeill, *Du nouveau sous le soleil*, 382–383; Melosi, *Garbage in the Cities*; Timothy Cooper, "Modernity and the Politics of Waste in Britain," in *Nature's End*, edited by Paul Warde and Sverker Sörlin, 247–272; Cooper, "War on Waste? The Politics of Waste and Recycling in Post-War Britain, 1950–1975," *Capitalism Nature Socialism*, vol. 20, no. 4 (2009): 53–72.

19. Renaud Bécot, "L'invention syndicale de l'environnement dans la France des années 1960," *Vingtième Siècle: Revue d'histoire*, vol. 113, no. 1 (2012): 169–178.

20. Cross, *An All-Consuming Century*, 159–160.

21. Aurélien Boutaud and Natacha Gondran, *L'Empreinte écologique* (Paris: La Découverte, 2009); Frédéric-Paul Piguet, Isabelle Blanc, Tourane Corbière-Nicollier, and Suren Erkman, "L'empreinte écologique: un indicateur ambigu," *Futuribles*, no. 334 (2007): 5–24.

22. Tony Allen, *Virtual Water: Tackling the Threat to Our Planet's Most Precious Resource* (London: I.B. Tauris, 2011).

23. Jacques Theys, "Quelques données quantitatives sur le développement des problèmes d'environnement en France entre 1945 et 1975," in *Matière et énergie dans les écosystèmes et les systèmes socioéconomiques*, edited by Philippe

Mirenowicz (Paris: GERMES, 1980), 371–424; Christophe Bonneuil and Sté-phane Frioux, "Les 'Trente Ravageuses'? L'impact environnemental et sanitaire des décennies de haute croissance," in *Une autre histoire des "Trente Glo-rieuses,"* edited by Christophe Bonneuil, Céline Pessis, and Sezin Topçu, 41–59.

24. Brenda Vale and Robert Vale, eds., *Living within a Fair Share Ecological Footprint* (London: Routledge, 2013).

25. Jean Lebacqz, "Les industries extractives," in *Centenaire de l'indépendance de la Belgique. Exposition internationale de Liège 1930. Rapport général du Commissariat général du gouvernement* (Liège: 1931), 353. We are grateful to Alexis Zimmer for bringing this document to our attention.

26. Jas, "Gouverner les substances chimiques dangereuses," 37; Fred Aftalion, *A History of the International Chemical Industry* (Philadelphia: University of Pennsylvania Press, 1991); Bensaude-Vincent and Stengers, *Histoire de la chimie.*

27. Jeffrey L. Meikle, *American Plastic: A Cultural History* (New Brunswick, NJ: Rutgers University Press, 1995); Suzan Freinkel, *Plastic: A Toxic Love Story* (New York: Houghton Mifflin Harcourt, 2011); Jennifer Gabrys, Gay Hawkins, and Mike Michael, eds., *Accumulation: The Material Politics of Plastic* (London: Routledge, 2013); Christian Marais, *L'Âge du plastique: Découvertes et utilisations* (Paris: L'Harmattan, 2005).

28. Roland Barthes, "Plastic," in *Mythologies*, translated by Annette Lavers (London: Paladin, 1973 [1957]), 97–99. Quotation is from page 98–99.

29. Nicolino, *Un empoisonnement universel*, chap. 8.

30. A. Pinton, "La soie artificielle à Lyon," *Les Études rhodaniennes*, vol. 6, no. 3 (1930): 229–250; Hervé Joly, *Les Gillet de Lyon: Fortunes d'une grande dynastie industrielle (1838–2015)* (Geneva: Droz, 2015).

31. Paul D. Blanc, *Fake Silk: The Lethal History of Viscose Rayon* (New Haven, CT: Yale University Press, 2016); Blanc, "Rayon, Carbon Disulfide, and the Emergence of the Multinational Corporation in Occupational Disease," in *Dangerous Trade*, edited by Christopher Sellers and Joseph Melling, 73–84.

32. Andrea Wackerman, "When Consumer Citizens Spoke Up: West Germany's Early Dealings with Plastic Waste," *Contemporary European History*, vol. 22, no. 3 (2013): 477–498.

33. Bensaude-Vincent and Stengers, *Histoire de la chimie*, 258–259.

34. René Musset, "Les progrès de la pétrochimie en France," *Annales de géog-raphie*, vol. 71, no. 386 (1962): 441–444; Claude Mercier, *L'Industrie pétrochi-mique et ses possibilités d'implantation dans les pays en voie de développement* (Paris: Technip, 1966).

35. Brian C. Black, *Crude Reality: Petroleum in World History* (Lanham, MD: Rowman & Littlefield, 2012).

36. Emmanuel Martinais, "L'emprise du risque sur les espaces industriels," in *Habiter les territoires à risques*, edited by Valérie November, Marion Penelas, and Pascal Viot (Lausanne, Switzerland: Presses polytechniques et universitaires

romandes, 2011), 101–119; Gwenola Le Naour, "Feyzin (1959–1971): Composer avec les débordements de l'industrie dans le sud lyonnais," in *Débordements industriels*, edited by Thomas Le Roux and Michel Letté, 99–114; François Duchêne and Léa Marchand, *Lyon, vallée de la chimie: Traversée d'un paysage industriel* (Lyon, France: Libel, 2016).

37. Quoted in Bensaude-Vincent and Stengers, *Histoire de la chimie*, 257.

38. Quoted in Wackerman, "When Consumer Citizens Spoke Up," 487.

39. Quoted in Meikle, *American Plastic*, 230.

40. Baptiste Monsaingeon, "Plastiques: ce continent qui cache nos déchets," in "Où va l'homo détritus?," special issue of *Mouvements*, no. 87 (2016): 48–58.

41. Aurélien Féron, "Le Problème PCB (polychlorobiphényles) des années 1960 à 2010: Enquête sociohistorique sur une pollution visible, massive et durable" (PhD diss., EHESS, 2018).

42. K. Breivik, A. Sweetman, J. M. Pacyna, and K. C. Jones, "Towards a Global Historical Emission Inventory for Selected PCB Congeners: A Mass Balance Approach," *Science of the Total Environment*, vol. 290, no. 1–3 (2002): 81–224.

43. Émile Guillaumin, *The Life of a Simple Man*, translated by Margaret Crosland (Hanover, NH: University Press of New England, 1983), 194.

44. Roland Barthes, "The New Citroën," in *Mythologies*, translated by Annette Lavers (London: Paladin, 1973 [1957]), 88–90. Quotation is from page 88.

45. Yoann Demoli, "Carbone et tôle froissée: L'espace social des modèles de voitures," *Revue française de sociologie*, vol. 56, no. 2 (2015): 223–260.

46. Jacques Payen, ed., *Analyse historique de l'évolution des transports en commun dans la région parisienne* (Paris: Centre de documentation de l'histoire des techniques, 1977).

47. Richard Bergeron, *Le Livre noir de l'automobile: Exploration du rapport malsain de l'homme contemporain à l'automobile* (Montréal: Hypothèse, 2005).

48. Christopher W. Wells, *Car Country: An Environmental History* (Seattle: University of Washington Press, 2012).

49. Thomas Zeller, *Driving Germany: The Landscape of the German Autobahn (1930–1970)* (New York: Berghahn Books, 2007).

50. Mathieu Flonneau, "City Infrastructures and City Dwellers: Accommodating the Automobile in Twentieth-Century Paris," *The Journal of Transport History*, vol. 27, no. 1, 2006, 93–114.

51. Mike Davis, *City of Quartz: Los Angeles, capitale du futur* (Paris: La Découverte, 1999 [1990]); Adam Rome, *The Bulldozer in the Countryside: Suburban Sprawl and the Rise of American Environmentalism* (Cambridge: Cambridge University Press, 2001).

52. Steven Watts, *The People's Tycoon: Henry Ford and the American Century* (New York: Vintage Books, 2005).

53. George Galster, *Driving Detroit: The Quest for Respect in the Motor City* (Philadelphia: University of Pennsylvania Press, 2012).

54. Tom McCarthy, *Auto Mania: Cars, Consumers, and the Environment* (New Haven, CT: Yale University Press, 2007), 65.

55. McCarthy, *Auto Mania: Cars, Consumers, and the Environment*, 65.

56. Paul Jobin, *Maladies industrielles et renouveau syndical au Japon* (Paris: EHESS, 2006).

57. Peter Hayes, *Industry and Ideology: IG Farben in the Nazi Era* (Cambridge: Cambridge University Press, 1987).

58. Peter Freund and George Martin, *The Ecology of the Automobile* (Montréal: Black Rose Books, 1993).

59. Stephen L. Harp, *A World History of Rubber: Empire, Industry and the Everyday* (Chichester, UK: John Wiley & Sons, 2016).

60. Quentin R. Skrabec, *Rubber: An American Industrial History* (Jefferson, NC: McFarland, 2014).

61. "Le diesel n'est pas le seul responsable de la pollution automobile," *Le Monde*, 19 January 2015; "Les océans pollués par des particules invisibles de plastique," *Le Monde*, 22 February 2017.

62. "Retour sur terre pour des milliers de pneus usagés sortis de la Méditerranée," *Le Monde*, 13 May 2015.

63. William Kovarik, "Ethyl-Leaded Gasoline: How a Classic Occupational Disease Became an International Public Health Disaster," *International Journal of Occupational and Environmental Health*, vol. 11, no. 4 (2005): 384–397; Alan Loeb, "Birth of the Kettering Doctrine: Fordism, Sloanim and the Discovery of Tetraethyl Lead," *Business and Economic History*, vol. 24, no. 1 (1995): 72–87; quotation is from Ross and Amter, *The Polluters*, 34.

64. Video from the INA website: "La ville et l'automobile," 6 October 1973, 14 min 27 s: http://www.ina.fr/video/CAF93027915 (last accessed 15 November 2018).

65. Jacques Lob and José Bielsa, *Les Mange-Bitume* (Paris: Dargaud, 1974).

66. Michel Freyssenet, *Les Ventes d'automobiles neuves dans le monde, par continent et principaux marchés (1898–2016)*, inquiry document 2007: http://freyssenet.com/?q=node/364 (last accessed 15 November 2018).

67. Éric Drezet, "Épuisement des ressources naturelles," *EcoInfo*, 11 March 2014, electronic article: http://ecoinfo.cnrs.fr/?p=11055 (last accessed 15 November 2018).

68. Fred G. Bell and Laurance J. Donnelly, *Mining and Its Impact on the Environment* (Boca Raton, FL: CRC Press, 2006); Leonard J. Arrington and Gary B. Hansen, *The Richest Hole on Earth: A History of the Bingham Copper Mine* (Logan: Utah State University Press, 1963).

69. Stiller, *Wounding the West*; Smith, *Mining America*; LeCain, *Mass Destruction*.

70. Numerous such examples can be found in Philippe Bihouix and Benoît de Guillebon, *Quel futur pour les métaux? Raréfaction des métaux: un nouveau défi pour la société* (Les Ulis, France: EDP Sciences, 2010).

71. Mark E. Schlesinger, Matthew J. King, Kathryn C. Sole, and William G. Davenport, *Extractive Metallurgy of Copper* (Amsterdam: Elsevier, 2011), 232.

72. *Anaconda Reduction Works* (Anaconda, MT: Anaconda Copper Mining Company, 1920).

73. Alyson Warhurst, ed., *Mining and the Environment: Case Studies from the Americas* (Ottawa: International Development Research Centre, 1999).

74. Lynch, *Mining in World History.*

75. Yann Bencivengo, *Nickel: La naissance de l'industrie calédonienne* (Tours, France: Presses universitaires François-Rabelais, 2014); Ian McNeil, ed., *An Encyclopaedia of the History of Technology* (London: Routledge, 1990), 96–100: "The Emergence of Nickel."

76. Hervé Pujol, ed., *Tristes mines: Impacts environnementaux et sanitaires de l'industrie extractive* (Bordeaux, France: Les Études hospitalières, 2014).

77. Florence Hachez-Leroy, *L'Aluminium français: L'invention d'un marché (1911–1983)* (Paris: CNRS Éditions, 1999); Hachez-Leroy, "Du métal précieux au matériau invisible: la double vie de l'aluminium," in *Les Chemins de la nouveauté. Innover, inventer au regard de l'histoire*, edited by Anne-Françoise Garçon and Liliane Hilaire-Pérez (Paris: CTHS, 2003), 431–442; Dominique Barjot and Marco Bertilorenzi, eds., *Aluminium: Du métal de luxe au métal de masse (xixe–xxie siècle)* (Paris: Presses de l'université Paris-Sorbonne, 2014); Claire Leymonerie, "L'aluminium, matériau des arts décoratifs à l'Exposition internationale de Paris en 1937," *Cahiers d'histoire de l'aluminium*, vol. 46–47, no 1. (2011): 8–49.

78. Matthew Evenden, "Aluminum, Commodity Chains and the Environmental History of the Second World War," *Environmental History*, vol. 16, no. 1 (2011): 69–93.

79. Florence Hachez-Leroy, "Polémique autour d'un nouveau matériau: l'aluminium dans la cuisine (xixe–xxe siècle)," in *Histoire des innovations alimentaires (xixe–xxe siècle)*, edited by Alain Drouard and Jean-Pierre Williot (Paris: L'Harmattan, 2007), 149–161.

80. Marco Bertilorenzi and Philippe Mioche, "Between Strategy and Diplomacy: History of Alumina Alternative Technologies (1900s–1980s)," *Cahiers d'histoire de l'aluminium*, vol. 51, no. 2 (2013): 42–63.

81. Le Roux, "Les industries chimiques et l'environnement."

82. Le Roux, "Les industries chimiques et l'environnement."

83. Andrea Saba, "La pollution en Val Lagarina (1928–1938): Une nouvelle voie pour l'histoire de l'environnement?" *Histoire, économie et société*, vol. 16, no. 3 (1997): 463–470; De Luigi, Meyer, and Saba, "Industrie, pollution et politique."

84. Quentin R. Skrabec, *Aluminium in America: A History* (Jefferson, NC: McFarland, 2017).

85. Lucie K. Morisset, *Arvida, cité industrielle* (Montréal: Septentrion, 1998); Anne Dalmasso and Lucie K. Morisset, "Cités aluminières en dialogue: Regards

croisés sur la Maurienne et le Saguenay," *Cahiers d'histoire de l'aluminium*, no. 52 (2014): 18–30.

86. René Lesclous, *Histoire des sites producteurs d'aluminium: Les choix stratégiques de Pechiney (1892–1992)* (Paris: Presses de l'École des mines, 1999), 72.

87. Gérard Vindt, *Les Hommes de l'aluminium: Histoire sociale de Pechiney (1921–1973)* (Paris: Éd. de l'Atelier, 2006); Philippe Mioche, *L'Alumine à Gardanne de 1893 à nos jours* (Grenoble, France: Presses universitaires de Grenoble, 1994).

88. Marco Bertilorenzi and Philippe Mioche, "Les résidus de l'alumine à Portovesne (Italie) et à Gardanne/Cassis (France), des années soixante à nos jours," in *Pollutions industrielles*, edited by Laura Centerini and Xavier Daumalin, 275–299; Marie-Claire Loison and Anne Pezet, "L'entreprise verte et les boues rouges: Les pratiques controversées de la responsabilité sociétale à l'usine d'alumine de Gardanne (1960–1966)," *Entreprises et histoire*, vol. 45, no. 4 (2006): 97–115.

89. Louis Chabert, "L'aluminium en Maurienne," *Revue de géographie alpine*, vol. 6, no. 1 (1973): 31–62; Anne Guerin-Henri, *Les Pollueurs: Luttes sociales et pollutions industrielles* (Paris: Le Seuil, 1980).

90. Jacques Donze, "Risque industriel et environnement montagnard: Le cas de Saint-Jean-de-Maurienne," *Bulletin de l'Association des géographes français*, vol. 84, no. 2 (2007): 217–230.

91. *Le Sauvage*, no. 0 (Summer 1972); *La Gueule ouverte*, no. 1 (November 1972) (its editorial board was established in Saint-Jean-de-Maurienne, hometown of its founder, Pierre Fournier).

92. Daniel Ménégoz, "Innover pour protéger l'environnement: La lutte contre la pollution fluorée par les usines d'électrolyse," in *Cent ans d'innovation dans l'industrie de l'aluminium*, edited by Ivan Grinberg, Pascal Griset, and Muriel Le Roux (Paris: L'Harmattan, 1997), 75–86.

93. Ruch Dominic, "Une étonnante longévité: L'histoire d'une usine suisse d'aluminium à Martigny," *Cahiers d'histoire de l'aluminium*, vol. 42–43, no. 1 (2009): 84–107.

94. Centre national des ressources textuelles et lexicales: http://www.cnrtl.fr /etymologie/ (last accessed 15 November 2018).

95. Marcel Mazoyer and Laurence Roudart, *Histoire des agricultures du monde: Du néolithique à la crise contemporaine* (Paris: Le Seuil, 2002), 506; Deléage, *Histoire de l'écologie*, 270.

96. Jean-Paul Diry, "L'industrie française de l'alimentation du bétail," *Annales de géographie*, vol. 88, no. 490 (1979): 671–704.

97. Claas Kirchhelle, "Toxic Confusion: The Dilemma of Antibiotic Regulation in West German Food Production (1951–1990)," *Endeavour*, vol. 40, no. 2 (2016): 114–127; Kirchhelle, *Pyrrhic Progress: Antibiotics and Western Food Production (1949–2013)* (PhD diss., Oxford University, 2015).

98. Stéphane Castonguay, *Protection des cultures, construction de la nature: Agriculture, foresterie et entomologie au Canada (1884–1959)* (Québec: Septentrion, 2004); Fabrice Nicolino and François Veillerette, *Pesticides: Révélations sur un scandale français* (Paris: Fayard, 2007).

99. Ross and Amter, *The Polluters*, 45–58.

100. Wayland J. Hayes, *Pesticides Studied in Man* (Baltimore, MD: Williams & Wilkins, 1982).

101. Russell, *War and Nature*; Russell, "The Strange Career of DDT: Experts, Federal Capacity, and Environmentalism in World War II," *Technology and Culture*, vol. 40, no. 4 (1999): 770–779.

102. Linda Nash, "The Fruits of Ill-Health: Pesticides and Workers Bodies in Post-World War II California," *Osiris*, vol. 19 (2004): 203–219; Nash, *Inescapable Ecologies*.

103. Pete Daniels, *Toxic Drift: Pesticides and Health in the Post-War II South* (Baton Rouge: Louisiana State University Press, 2005).

104. Quoted in Philippe Roger, *Rêves et cauchemars américains. Les États Unis au miroir de l'opinion publique française (1945–1953)* (Lille, France: Presses du Septentrion, 1996), 105.

105. Nathalie Jas, "Pesticides et santé des travailleurs agricoles en France dans les années 1950–1960," in *Sciences, agriculture, alimentation et société en France au xxe siècle*, edited by Christophe Bonneuil, Gilles Denis, and Jean-Luc Mayaud (Paris: L'Harmattan/Quae, 2008); Jas, "Public Health and Pesticides Regulation in France before and after Silent Spring," in "Risk Society in Historical Perspective," edited by Soraya Boudia and Nathalie Jas, special issue of *History and Technology*, vol. 23, no. 4 (2007): 369–388.

106. Henri Mendras, *La Fin des paysans* (Paris: SEDEIS, 1967); Pierre Alphandery, Pierre Bitoun, and Yves Dupont, *Les Champs du départ: Une France rurale sans paysans?* (Paris: La Découverte, 1989); Rémi Fourche, *Contribution à l'histoire de la protection phytosanitaire dans l'agriculture française (1880–1970)* (PhD diss., Université de Lyon 2, 2004).

107. Christophe Bonneuil and Frédéric Thomas, *Gènes, pouvoirs et profits: Recherche publique et régimes de production des savoirs, de Mendel aux OGM* (Paris: Quae, 2009), 78.

108. Marc Elie, "The Soviet Dust Bowl and the Canadian Erosion Experience in the New Lands of Kazakhstan (1950s–1960s)," *Global Environment*, vol. 8, no. 2 (2015): 259–292.

109. Josephson et al., *An Environmental History of Russia*, 211.

110. Susan Rankin Bohme, *Toxic Injustice: A Transnational History of Exposure and Struggle* (Berkeley: University of California Press, 2015).

111. Vandana Shiva, *The Violence of Green Revolution: Third World Agriculture, Ecology and Politics* (Lexington: University Press of Kentucky, 2016); Shiva, *Éthique et agro-industrie: Main basse sur la vie* (Paris: L'Harmattan, 1996); Madhumita Sala and Sigrid Schmalzer, "Green-Revolution

Epistemologies in China and India: Technocracy and Revolution in the Production of Scientific Knowledge and Peasant Identity," *British Journal for the History of Science—Themes* [online], no. 1 (2016): 145–167.

112. Valérie Chansigaud, *La Nature à l'épreuve de l'homme* (Paris: Delachaux & Niestlé, 2015).

113. Roger Heim, *Destruction et protection de la nature* (Paris: Armand Colin, 1952).

114. Gunter Vogt, "The Origins of Organic Farming," in *Organic Farming: An International History*, edited by William Lockeretz (Wallingford, UK: CABI, 2007), 9–29.

115. Fourche, "Contribution à l'histoire de la protection phytosanitaire."

116. Bonneuil, Pessis, and Topçu, eds., *Une autre histoire des "Trente Glorieuses."*

Chapter 10

1. Denis Auribault, "Note sur l'hygiène et la sécurité des ouvriers dans les filatures et tissages d'amiante," *Bulletin de l'inspection du travail* (1906): 120–132.

2. Stradling, *Smokestacks and Progressives*, 66.

3. Brüggemeier, "The Ruhr Basin (1850–1980)"; Brüggemeier, "A Nature Fit for Industry," 48.

4. Joel A. Tarr and Carl Zimring, "The Struggle for Smoke Control in Saint Louis: Achievement and Emulation," in *Common Fields: An Environmental History of St. Louis*, edited by Andrew Hurley (Saint Louis: Missouri Historical Society Press, 1997), 199–220; Uekoetter, *The Age of Smoke*, 74; Tarr, ed., *Devastation and Renewal*.

5. Quoted in Zimmer, *Brouillards toxiques*, 202.

6. Frioux, *Les Batailles de l'hygiène*, 59–62.

7. Uekoetter, *The Age of Smoke*, chap. 3.

8. Quoted in Stéphane Frioux, "Problème global, action locale: les difficultés de la lutte contre les fumées industrielles à Lyon," in *Débordements industriels*, edited by Thomas Le Roux and Michel Letté, 330.

9. Peter Thorsheim, *Inventing Pollution*, 176–184.

10. Stéphane Frioux, "La pollution de l'air, un mal nécessaire? La gestion du problème durant les 'Trente Glorieuses,'" in *Une autre histoire des "Trente Glorieuses,"* edited by Christophe Bonneuil, Céline Pessis, and Sezin Topçu, 99–115.

11. Charvolin et al., *Un air familier?*

12. Stéphane Frioux, "Les batailles de l'eau et de l'air purs : Transferts internationaux et politiques d'amélioration de l'environnement urbain en France des années 1900 aux années 1960," in *Une protection de l'environnement à la française? (xixe–xxe siècle)*, edited by Charles-François Mathis and Jean-François Mouhot (Seyssel, France: Champ Vallon, 2013), 236–245.

13. Alexander Farrell, "Air Pollution in Spain: A 'Peripheral' Nation Transforms," in *Smoke and Mirrors*, edited by E. Melanie DuPuis, 248–249; Pablo Corral Broto, *Protesta y ciudadanía: Conflictos ambientales durante el franquismo en Zaragoza (1939–1979)* (Zaragoza, Spain: Rolde de Estudios Aragoneses, 2015).

14. Uekoetter, *The Age of Smoke*, 89–96.

15. Anna Bramwell, *Blood and Soil: Richard Walther Darre and Hitler's "Green Party"* (London: Kensal Press, 1985).

16. Johann Chapoutot, "Les nazis et la 'nature': Protection ou prédation?" *Vingtième siècle: Revue d'histoire*, vol. 113, no. 1 (2012): 29–39; Frank Uekoetter, "Green Nazis? Reassessing the Environmental History of Nazi Germany," *German Studies Review*, vol. 30, no. 2 (2007): 267–287.

17. N. F. Izmerov, *Lutte contre la pollution de l'air en URSS* (Geneva: WHO, 1974), 16–17; Josephson *et al.*, *An Environmental History of Russia*, 91.

18. Marie-Hélène Mandrillon, "L'expertise d'État, creuset de l'environnement en URSS," *Vingtième siècle: Revue d'histoire*, vol. 113, no. 1 (2012), 107–116; Laurent Coumel, "A Failed Environmental Turn? Khrushchev's Thaw and Nature Protection in Soviet Russia," *The Soviet and Post-Soviet Review*, vol. 40, no. 2 (2013): 167–189.

19. Michel Dupuy, *Histoire de la pollution atmosphérique en Europe et en RDA au xxe siècle* (Paris: L'Harmattan, 2003), 109–116.

20. Ullmann, *Les Installations classées*.

21. Dirk Spierenburg and Raymond Poidevin, *Histoire de la Haute Autorité de la Communauté européenne du charbon et de l'acier: Une expérience supranationale* (Brussels: Bruylant, 1993); Laura Scichilone, *L'Europa e la sfida ecologica: Storia della politica ambientale europea (1969–1998)* (Bologna: Il Mulino, 2008); Jonas Kaesler, "La Pollution environnementale dans l'espace transnational: La lutte des acteurs sarrois contre l'industrie charbonnière française, de 1945 jusqu'aux années 1970" (PhD diss. in progress, CIRED).

22. Laura Grazi and Laura Scichilone, "Environmental Issues and the Improvement of Living and Working Conditions: Innovative Elements in the Process of European Integration During the 1970s," in *Les Trajectoires de l'innovation technologique et la construction européenne: Des voies de structuration durable?*, edited by Christophe Bouneau, David Burigana, and Antonion Varsori (Brussels: Peter Lang, 2010), 57–76.

23. Odette Hardy-Hémery, *Eternit et l'amiante (1922–2000): Aux sources du profit, une industrie du risque* (Villeneuve-d'Ascq, France: Presses universitaires du Septentrion, 2005).

24. Boullet, *Entreprises et environnement en France*; Jean-Claude Daumas and Philippe Mioche, "Histoire des entreprises et environnement. Une frontière pour la recherche," *Entreprises et histoire*, vol. 35, no. 1 (2004), 69–88.

25. *Pétrole progrès*, 1949–1970.

26. Edward Bernays, *Propaganda: Comment manipuler l'opinion en démocratie* (Paris: La Découverte, 2007 [1928]).

27. Robert N. Proctor, *The Nazi War on Cancer* (Princeton, NJ: Princeton University Press, 1999); Proctor, *Cancer Wars: How Politics Shapes What We Know and Don't Know about Cancer* (New York: Oxford University Press, 2001); Proctor, *Golden Holocaust: La conspiration des industriels du tabac* (Sainte-Marguerite-sur-Mer, France: Éd. des Équateurs, 2014); David Michaels, *Doubt Is Their Product: How Industry's Assault on Science Threatens Your Health* (Oxford: Oxford University Press, 2008).

28. Robert N. Proctor and Londa Schiebinger, *Agnotology: The Making and Unmaking of Ignorance* (Stanford, CA: Stanford University Press, 2008); Soraya Boudia and Nathalie Jas, eds., *Powerless Science? Science and Politics in a Toxic World* (New York: Berghahn Books, 2014); Boudia and Jas, *Toxicants, Health and Regulation since 1945* (London: Pickering & Chatto, 2013).

29. Naomi Oreskes and Erik M. Conway, *Les Marchands de doute* (Paris: Le Pommier, 2012 [2010]); Dan Fagin and Marianne Lavelle, *Toxic Deception: How the Chemical Industry Manipulates Science, Bends the Law and Endangers Your Health* (Secaucus, NJ: Carol, 1996); Dan Fagin, *Toms River: A Story of Science and Salvation* (New York: Bantam Books, 2013).

30. Ndiaye, *Du nylon et des bombes*.

31. Ross and Amter, *The Polluters*, 28–34; Gerald Markovitz and David Rosner, *Deceit and Denial: The Deadly Politics of Industrial Pollution* (Berkeley: University of California Press, 2002), 108–138.

32. Markovitz and Rosner, *Deceit and Denial*, see especially 168–194.

33. Laurent Vogel, "L'amiante dans le monde," *HESA Newsletter*, no. 27, June 2005. Online: https://www.researchgate.net/publication/323103561_Newsletter_HESA_n27_-_2005_avec_un_dossier_special_sur_l'amiante_dans_le_monde (last accessed 16 November 2018).

34. Kenichi Miyamoto, Kenji Morinaga, and Mori Hiroyuki, eds., *Asbestos Disaster: Lessons from Japan's Experience* (Tokyo: Springer, 2011), 3.

35. Bruno Ziglioli, *"Sembrava nevicasse"—La Eternit di Casale Monferrato e la Fibronit di Broni: due comunità di fronte all'amianto* (Milan: Franco Angeli, 2016).

36. Jessica van Horssen, *A Town Called Asbestos: Environmental Contamination, Health, and Resilience in a Resource Community* (Vancouver: University of British Columbia Press, 2016).

37. Jock McCulloch and Geoffrey Tweedale, *Defending the Indefensible: The Global Asbestos Industry and Its Fight for Survival* (Oxford: Oxford University Press, 2008); Roger Lenglet, *L'Affaire de l'amiante* (Paris: La Découverte, 1996).

38. Emmanuel Henry, *Amiante, un scandale improbable: Sociologie d'un problème public* (Rennes, France: Presses universitaires de Rennes, 2007); Arthur J. McIvor, *Working Lives: Work in Britain since 1945* (Basingstoke, UK: Palgrave Macmillan, 2013), 159.

39. Christopher Sellers, *Hazards of the Job: From Industrial Disease to Environmental Health Science* (Chapel Hill: University of North Carolina Press, 1997).

40. Michel Dupuy, *Les Cheminements de l'écologie en Europe: Une histoire de la diffusion de l'écologie au miroir de la forêt (1880–1980)* (Paris: L'Harmattan, 2004).

41. Deléage, *Histoire de l'écologie*, 120.

42. Christophe Masutti, *Les Faiseurs de pluie: Dust Bowl, écologie et gouvernement (États-Unis, 1930–1940)*, Creative Commons, 2012.

43. Deléage, *Histoire de l'écologie*, 10.

44. Chip Jacobs and William J. Kelly, *Smogtown: The Lung-Burning History of Pollution in Los Angeles* (New York: Overlook Press, 2008).

45. Stéphane Frioux, "La pollution de l'air," 105.

46. James Rodger Fleming and Ann Johnson, eds., *Toxic Airs: Body, Place, Planet in Historical Perspective* (Pittsburgh, PA: University of Pittsburgh Press, 2014).

47. Michael Bess, *La France vert clair: Écologie et modernité technologique (1960–2000)* (Seyssel, France: Champ Vallon, 2011), 81–82.

48. Philippe Saint-Marc, *La Pollution* (Paris: Robert Laffont, 1975).

49. Francis Chateaureynaud and Didier Torny, *Les Sombres Précurseurs: Une sociologie pragmatique de l'alerte et du risque* (Paris: EHESS, 1999).

50. Douglas R. Weiner, *A Little Corner of Freedom: Russian Nature Protection from Stalin to Gorbachev* (Berkeley: University of California Press, 1999).

51. Judith Rainhorn, "L'épidémiologie de la bottine' ou l'enquête médicale réinventée: Alice Hamilton et la médecine industrielle dans l'Amérique du premier xxᵉ siècle," *Gesnerus: Swiss Journal of the History of Medicine and Sciences*, vol. 69, no. 2 (2012): 330–354.

52. Thomas Morrison Legge, *Industrial Maladies* (Oxford: Oxford University Press, 1934); Vicky Long, *The Rise and Fall of the Healthy Factory: The Politics of Industrial Health in Britain (1914–1960)* (New York: Palgrave Macmillan, 2011).

53. Henry Fairfield Osborn, *La Planète au pillage* (Paris: Acte Sud, 2008 [1948]); Roger Heim, *Destruction et protection de la nature* (Paris: Armand Colin, 1952).

54. Linda Lear, *Rachel Carson: Witness for Nature* (New York: Houghton Mifflin Harcourt, 1991). The first French translation of Rachel Carson's book included a preface by Roger Heim: Rachel Carson, *Printemps silencieux* (Paris: Plon, 1963 [1962]).

55. Cliff I. Davidson, ed., *Clean Hands: Clair Patterson's Crusade against Environmental Lead Contamination* (New York: Nova Science, 1998).

56. Ernst Schumacher, *Small Is Beautiful: Une société à la mesure de l'homme* (Paris: Le Seuil, 1978 [1973]).

57. Laurent Samuel and Dominique Simonnet, "Technologies douces," in Pierre Samuel, Yves Gauthier, and Ignacy Sachs, *L'Homme et son environnement: De*

la démographie à l'écologie (Paris: Encyclopédie moderne, 1976), 477; Langdon Winner, *La Baleine et le réacteur: À la recherche des limites de la haute technologie* (Paris: Descartes & Cie, 2002 [1986]).

58. *Les Aéroports et l'environnement* (Paris: OCDE, 1975).

59. Annie Vallée, *Économie de l'environnement* (Paris: Le Seuil, 2002); Franck-Dominique Vivien, "La pensée économique française dans l'invention de l'environnement et du développement durable," *Annales des Mines: Responsabilité et environnement*, no. 46 (April 2007): 68–72.

60. Karl William Kapp, *Les Coûts sociaux de l'entreprise privée* (Paris: Institut Veblen/Les Petits Matins, 2015 [1950]).

61. Nicholas Georgescu-Roegen, *Demain la décroissance: Entropie, écologie, économie* (Lausanne, Switzerland: Pierre-Marcel Favre, 1979); René Passet, *L'Économique et le Vivant* (Paris: Payot, 1979).

62. Ignacy Sachs, "Environnement et style de développement," *Annales ESC*, vol. 29, no. 3 (1974): 553–570; Sachs, *Pour une économie politique de l'environnement* (Paris: Flammarion, 1977).

63. Élodie Vieille-Blanchard, "Les Limites à la croissance dans un monde global: Modélisations, prospectives, réfutations" (PhD diss., EHESS, 2011).

64. Rémi Barré, Thierry Lavoux, and Vincent Piveteau, eds., *Un demi-siècle d'environnement entre science, politique et prospective: En l'honneur de Jacques Theys* (Versailles, France: Quae, 2015).

65. Masanobu Fukuoka, *La Révolution d'un seul brin de paille: Une introduction à l'agriculture sauvage* (Paris: Guy Trédaniel, 2005 [1975]).

66. Hiroko Amemiya, "Genèse du Teikei: Organisations et groupes de jeunes mères citadines," in *Du Teikei aux Amap: Le renouveau de la vente directe de produits fermiers locaux* (Rennes, Frances: Presses universitaires de Rennes, 2011), 29–54.

67. R. Debroye, "Les insecticides," *L'Abeille de France*, no. 246 (March–April 1946): 9; cited in Léna Humbert, "Les Oppositions des apiculteurs: trices aux pesticides à l'époque de la 'modernisation' agricole (1945–1963)" (PhD diss., EHESS, 2016).

68. Gabrielle Bouleau, "Pollution des rivières. Mesurer pour démoraliser," in *Une autre histoire des "Trente Glorieuses,"* edited by Christophe Bonneuil, Céline Pessis, and Sezin Topçu, 211–229.

69. Frioux and Malange, "L'eau pure pour tous!"

70. Massard-Guilbaud and Rodger, eds., *Environmental and Social Justice in the City*; Aliénor Bertrand, ed., *Justice écologique, justice sociale: Exemples historiques, analogies contemporaines et théorie politique* (Paris: Victoires Éditions, 2015).

71. Andrew Hurley, *Environmental Inequalities: Class, Race and Industrial Pollution in Gary (1945–1980)* (Chapel Hill: University of North Carolina Press, 1995).

72. Ross and Amter, *The Polluters*, 165–166; Joyce Carol Oates, *Les Chutes* (Paris: Philippe Rey, 2005 [2004]).

73. Christopher C. Sellers, *Crabgrass Crucible: Suburban Nature and the Rise of Environmentalism in Twentieth-Century America* (Chapel Hill: University of North Carolina Press, 2012).

74. David Van Reybrouck, *Congo, une histoire* (Arles, France: Actes Sud, 2014 [2012]), see especially 164–174, 253–254.

75. Hecht, "L'empire nucléaire"; Hecht, "L'Afrique et le monde nucléaire: Maladies industrielles et réseaux transnationaux dans l'uranium africain," in *Santé et travail à la mine*, edited by Judith Rainhorn, 173–205.

76. Dominique Taurisson-Mouret, "Exploitation minière dans les colonies et les ex-colonies ou le substrat de l'ineffectivité du droit," in *Justice écologique*, edited by Aliénor Bertrand, 53–90.

77. LeCain, *Mass Destruction*; Smith, *Mining America*; Stiller, *Wounding the West*.

78. Phia Steyn, "Industry, Pollution and the Apartheid State in South Africa," *History Teaching Review Yearbook*, vol. 22 (2008): 67–75.

79. Joan Martínez Alier, *L'Écologisme des pauvres: Une étude des conflits environnementaux dans le monde* (Paris: Les Petits Matins, 2014 [2002]).

80. Maurice Lime, *Pays conquis* (Paris: Éditions sociales internationales, 1936), 43; Lucien Bourgeois, *Faubourgs: Douze récits prolétariens* (Paris: ESI, 1931), 74; Georges Navel, *Passages: Récit* (Paris: Le Sycomore, 1982), 187–188; Albert Soulillou, *Élie ou le Ford-France-580* (Paris: Gallimard, 1933), 151–152. We are grateful to Xavier Vigna for pointing us toward this source: Xavier Vigna, *L'Espoir et l'effroi: Luttes d'écritures et luttes de classes en France au xxe siècle* (Paris: La Découverte, 2016).

81. Bécot, "*Syndicalisme et environnement en France*; Stephania Barca, "Laboring the Earth: Transnational Reflections on the Environmental History of Work," *Environmental History*, vol. 19, no. 1 (2014): 3–27; Barca, "Sur l'écologie de la classe ouvrière: Un aperçu historique et transnational," *Écologie et politique*, no. 50 (2015): 23–40; Renaud Bécot, "Aux racines de l'action environnementale du mouvement syndical québécois (1945–1972)," *Bulletin d'histoire politique*, vol. 23, no. 2 (2015): 48–65; Katrin MacPhee, "Canadian Working-Class Environmentalism (1965–1985)," *Labour/Le Travail*, no. 74 (2014): 123–149.

82. Robert Gordon, "Shell No! OCAW and the Labor-Environmental Alliance," *Environmental History*, vol. 3, no. 4 (1998): 460–487.

83. Robert Gottlieb, *Forcing the Spring: The Transformation of the American Environmental Movement* (Washington, DC: Island Press, 2005), 375.

84. Adam Tompkins, *Ghostworkers and Greens: The Cooperative Campaigns of Farmworkers and Environmentalists for Pesticide Reform* (Ithaca, NY: Cornell University Press, 2016).

85. Laurie Adkin. *The Politics of Sustainable Development: Citizens, Unions and the Corporations* (Tonawanda, NY: Black Rose Books, 1998); Markovitz and Rosner, *Deceit and Denial*, 157–167.

86. Laure Pitti, "Du rôle des mouvements sociaux dans la prévention et la réparation des risques professionnels: Le cas de Penarroya (1971–1988)," in *Cultures du risque au travail et pratiques de prévention au xxe siècle: La France au regard des pays voisins*, edited by Catherine Omnes and Laure Pitti (Rennes, France: Presses universitaires de Rennes, 2009), 217–232; Laure Pitti and Pascal Marichalar, "Réinventer la médecine ouvrière? Retour sur des mouvements médicaux alternatifs dans la France post-1968," *Actes de la recherche en sciences sociales*, no. 196–197 (2013): 114–131; Laure Pitti, "Experts 'bruts' et médecins critiques: Ou comment la mise en débats des savoirs médicaux a modifié la définition du saturnisme durant les années 1970," *Politix*, vol. 23, no. 91 (2010): 103–132; Renaud Bécot, "Syndicalisme et environnement en France," 258–259.

87. Keibo and Masato, *Rowing the Eternal Sea*; Jobin, *Maladies industrielles et renouveau syndical au Japon*; Timothy S. George, *Minamata: Pollution and the Struggle for Democracy in Postwar Japan* (Cambridge, MA: Harvard University Press, 2001).

88. Mandrillon, "L'expertise d'État, creuset de l'environnement en URSS," 107.

89. Anna Trespeuch-Berthelot, "La réception des ouvrages d''alerte environnementale' dans les médias français (1948–1975)," and Alexis Vrignon, "Journalistes et militants: Les périodiques écologistes dans les années 1970," in "De la nature à l'écologie," special issue of *Le Temps des médias*, no. 25 (2015): 104–120 and 121–134.

90. Frank Uekoetter, "A Twisted Road to Earth Day: Air Pollution as an Issue of Social Movements after World War II," in *Natural Protest: Essays on the History of American Environmentalism*, edited by Michael Egan and Jeff Crane (New York: Routledge, 2009), 163–183.

91. Céline Pessis, ed., *Survivre et vivre: Critique de la science, naissance de l'écologie* (Montreuil, France: L'Échappée, 2014).

92. Florian Charvolin, *L'Invention de l'environnement en France: Chronique anthropologique d'une institutionnalisation* (Paris: La Découverte, 2003).

93. Centre national des ressources textuelles et lexicales: http://www.cnrtl.fr /etymologie/.

94. Kirkpatrick Sale, *The Green Revolution: The American Environmental Movement (1962–1992)* (New York: Hill & Wang 1993); Yves Frémion, *Histoire de la révolution écologiste* (Paris: Hoëbeke, 2007); Alexis Vrignon, *La Naissance de l'écologie politique en France: Une nébuleuse au cœur des années 68* (Rennes, France: Presses universitaires de Rennes, 2017).

95. Fabrice Flipo, *Nature et politique: Contribution à une anthropologie de la modernité et de la globalisation* (Paris: Éd. Amsterdam, 2014).

96. The term "green" to designate ecologist political structures was created by German environmentalists Grünen during their first participation in a national election in 1980.

97. René Dumont, *L'Utopie ou la mort!* (Paris: Le Seuil, 1974 [1973]), 5, 14, 85.

98. Uekoetter, *The Age of Smoke*, 246–249.

99. Robert W. Collin, *The Environmental Protection Agency: Cleaning Up America's Act* (Westport, CT: Greenwood Press, 2006).

100. Jacques Theys, "Vingt ans de politique française de l'environnement: les années 70–90. Un essai d'évaluation," in *Les Politiques d'environnement: Évaluation de la première génération (1971–1995)*, edited by Bernard Barraqué and Jacques Theys (Paris: Éd. Recherches, 1998), 27 and 29; Gabriel Ullmann, *Les Installations classées: Deux siècles de législation et de nomenclature*, vol. 2, *La Loi du 19 juillet 1976: La régression accélérée du droit de l'environnement* (Paris: Cogiterra, 2016).

101. Simon Avenell, "From Fearsome Pollution to Fukushima: Environmental Activism and the Nuclear Blind Spot in Contemporary Japan," *Environmental History*, vol. 17, no. 2 (2012): 244–276.

102. Guétondé Touré, *La Politique de l'environnement dans les capitales africaines: Le cas de la ville d'Abidjan en Côte d'Ivoire* (Paris: Publibook, 2006), 226.

103. Shapiro, *Mao's War against Nature*, 192.

104. Wolfram Kaiser and Jan-Henrik Meyer, eds., *International Organizations and Environmental Protection: Conservation and Globalization in the Twentieth Century* (New York: Berghahn Books, 2016); *Organisation mondiale de la santé, La Pollution de l'air*, monograph no. 46 (Geneva: WHO, 1963).

105. Soraya Boudia, "Environnement et construction du global dans le tournant des années 1960–1970: Les infrastructures globales d'observation et d'étude de l'environnement," in *La Mondialisation des risques: Une histoire politique et transnationale des risques sanitaires et environnementaux*, edited by Soraya Boudia and Emmanuel Henry (Rennes, France: Presses universitaires de Rennes, 2015), 61–76.

106. Gérard Bellan and Jean-Marie Pérès, *La Pollution des mers* (Paris: Presses universitaires de France, 1974), 5.

107. Philippe Le Prestre, *Protection de l'environnement et relations internationales: Les défis de l'écopolitique mondiale* (Paris: Dalloz/Armand Colin, 2005).

108. McNeill, *Du nouveau sous le soleil*, 466.

Epilogue

1. Pablo Servigné and Raphael Stevens, *Comment tout peut s'effondrer: Petit manuel de collapsologie à l'usage des générations présentes* (Paris: Le Seuil, 2015).

2. Jean-Noël Salomon, *Danger pollutions!* (Pessac: Presses universitaires de Bordeaux, 2003).

3. Alain Touraine, *La Société post-industrielle: Naissance d'une société* (Paris: Denoël, 1969); Daniel Bell, *Vers la société post-industrielle* (Paris: Robert Laffont, 1976 [1973]).

4. Daniel Cohen, *Trois leçons sur la société post-industrielle* (Paris: Le Seuil, 2016); Jeremy Rifkin, *La Troisième Révolution industrielle: Comment le pouvoir latéral va transformer l'énergie, l'économie et le monde* (Paris, Les Liens qui libèrent, 2012 [2011]).

5. Ulrich Beck, *Risk Society: Towards a New Modernity* (London: Sage Publications, 1992), 21.

6. Yannick Barthe, Michel Callon, and Pierre Lascoumes, *Agir dans un monde incertain: Essai sur la démocratie technique* (Paris: Le Seuil, 2001).

7. Fressoz, *L'Apocalypse joyeuse.*

8. René Passet, "Les fondements bioéconomiques d'un développement durable," *Économie appliquée*, vol. 65, no. 2 (2012): 195–206; Corinne Gendron, *Le Développement durable comme compromise: La modernisation écologique de l'économie à l'ère de la mondialisation* (Québec: Presses de l'université du Québec, 2006); Olivier Godard, *Environnement et développement durable: Une approche méta-économique* (Brussels: De Boeck, 2015).

9. Bernard Méheust, *La Politique de l'oxymore: Comment ceux qui nous gouvernent nous masquent la réalité du monde* (Paris: La Découverte, 2009).

10. Benoît Eugène, "Le 'développement durable,' une pollution mentale au service de l'industrie," *Agone*, no. 34 (2005): 119–134.

11. Jérôme Boissonade, ed., *La Ville durable controversée. Les dynamiques urbaines dans le mouvement critique* (Paris: Petra, 2015), 255–285.

12. Jean-Baptiste Comby, *La Question climatique: Genèse et dépolitisation d'un problème public* (Marseille: Agone, 2015); Spencer R. Weart, *The Discovery of Global Warming* (Cambridge, MA: Harvard University Press, 2003).

13. *Restoring the Quality of Our Environment: Report of the Environmental Pollution Panel President's Science Advisory Committee* (Washington, DC: The White House, November 1965), online: http://climateandcapitalism.com/wp-content/uploads/sites/2/2014/06/Presidents-Advisory-Report-on-warming-1965.pdf.

14. Wallace S. Broecker, "Climatic Change: Are We on the Brink of a Pronounced Global Warming?" *Science*, vol. 189, no. 4201 (1975): 460–463.

15. "Un nouveau nom pour le 'réchauffement climatique?'" M-blogs, *Le Monde.fr*: http://ecologie.blog.lemonde.fr/2010/08/08/faut-il-changer-lexpression-rechauffement-climatique/.

16. Stefan Aykut and Amy Dahan, *Gouverner le climat? Vingt ans de négociations internationales* (Paris: Presses de Sciences Po, 2015); Avkut and Dahan, "La gouvernance du changement climatique. Anatomie d'un schisme de réalité," in *Le Gouvernement des technosciences*, edited by Dominique Pestre, 97–132.

17. Stéphane Foucart, *L'Avenir du climat. Enquête sur les climato-sceptiques* (Paris: Gallimard, 2015); Oreskes and Conway, *Les Marchands de doute*, 282–355; Naomi Klein, *Tout peut changer: Capitalisme et changement climatique* (Arles: Actes Sud, 2015).

18. István Markó, "Pauvre CO$_2$!" 20 April 2015, on the Institut Turgot site; István Markó, ed., *Climat: 15 vérités qui dérangent* (Louvain, Belgium: Texquis, 2013).

19. World Health Organization, *Reducing Global Health Risks Through Mitigation of Short-Lived Climate Pollutants* (Geneva: WHO, 2015).

20. Frédéric Denhez, *Les Nouvelles Pollutions invisibles: Ces poisons qui nous entourent* (Paris: Delachaux & Niestlé, 2011).

21. André Cicolella, *Toxique planète: Le scandale invisible des maladies chroniques* (Paris: Le Seuil, 2013).

22. Theo Colborn, Dianne Dumanoski, and John Peterson Myers, *L'Homme en voie de disparition?* (Mens, France: Terre vivante, 1997 [1996]).

23. Elvire Van Staëvel, *La Pollution sauvage* (Paris: Presses universitaires de France, 2006), 107.

24. Stéphane Horel, *Intoxication: Perturbateurs endocriniens, lobbyistes et eurocrates: une bataille d'influence contre la santé* (Paris: La Découverte, 2015); Marine Jobert and François Veillerette, *Perturbateurs endocriniens: La menace invisible* (Paris: Buchet-Chastel, 2015).

25. Philippe Grandjean, *Cerveaux en danger: Protégeons nos enfants* (Paris: Buchet-Chastel, 2016); Barbara Demeneix, *Losing Our Minds: How Environmental Pollution Impairs Human Intelligence and Mental Health* (Oxford: Oxford University Press, 2014).

26. Federico Robbe, *Seveso 1976: Oltre la diossina* (Castel Bolognese, Italy: Itaca, 2016); Angela Cecilia Pesatori and Pier Alberto Bertazzi, "The Seveso Accident," in *Dioxins and Health: Including Other Persistent Organic Pollutants and Endocrine Disruptors*, edited by Arnold Schecter (Hoboken, NJ: Wiley, 2012), 445–468.

27. World Health Organization, *Les Dioxines et leurs effets sur la santé*, checklist no. 225, October 2016, http://www.who.int/mediacentre/factsheets/fs225/fr/.

28. Philippe Jurgensen, *L'Économie verte: Comment sauver notre planète* (Paris: Odile Jacob, 2009), 90.

29. Christophe Bonneuil and Frédéric Thomas, *Semences: une histoire politique: Amélioration des plantes, agriculture et alimentation en France depuis la Seconde Guerre mondiale* (Paris: Charles Léopold Mayer, 2012).

30. Jacques Testart, *À qui profitent les OGM?* (Paris: CNRS Éditions, 2013); Gilles-Éric Séralini, *Ces OGM qui changent le monde* (Paris: Flammarion, 2010 [2004]).

31. Frédéric Neyrat, *La Part inconstructible de la Terre* (Paris: Le Seuil, 2016), 101; Christophe Bonneuil, "Cultures épistémiques et engagement des chercheurs dans la controverse OGM," *Natures sciences sociétés*, vol. 14, no. 3 (2006): 257–268.

32. Roger Lenglet, *Nanotoxiques: Une enquête* (Arles, France: ActesSud, 2014); *Aujourd'hui le nanomonde: Nanotechnologies, un projet de société totalitaire* (Paris: L'Esprit frappeur, 2006); Association de veille d'information civique sur

les enjeux des nanosciences et des nanotechnologies, *Nanomatériaux et risques pour la santé et l'environnement* (Gap, France: Yves Michel, 2016); Bernadette Bensaude-Vincent, *Les Vertiges de la technoscience: Façonner le monde atome par atome* (Paris: La Découverte, 2009).

33. World Health Organization, *Champs électromagnétiques et santé publique: téléphones portables*, checklist no. 193, October 2014, online: http://www.who .int/mediacentre/factsheets/fs193/fr/; see also the PRIARTEM website: http://www.priartem.fr/accueil.html.

34. Groupe EcoInfo, *Les Impacts écologiques des technologies de l'information et de la communication* (Les Ulis, France: EDP Sciences, 2012).

35. Fabrice Flipo, "Expansion des technologies de l'information et de la communication: vers l'abîme?" *Mouvements*, no. 79 (2014): 115–121.

36. "En Chine, les terres rares tuent des villages," *Le Monde.fr*, 19 July 2012; Guillaume Pitron and Serge Turquier, *La Sale Guerre des terres rares*, documentary, France Télévision, 2012.

37. Philippe Bihouix, *L'Âge des low tech: Vers une civilisation techniquement soutenable* (Paris: Le Seuil, 2014).

38. Fabrice Flipo, Michelle Dobré, and Marion Michot, *La Face cachée du numérique: L'impact environnemental des nouvelles technologies* (Montreuil, France: L'Échappée, 2013); Laurent Lichtenstein and Coline Tison, *Internet, la pollution cachée*, documentary, France 5, 2012; Laurent Lefèvre and Jean-Marc Pierson, "Le big data est-il polluant?" *CNRS Le Journal*, electronic review, 2 April 2015.

39. Arnaud Saint-Martin, "Du *Big Sky* à l'espace pollué: l'effet boomerang des débris spatiaux," *Mouvements*, no. 87 (2016): 36–47.

40. Agence France Presse, 19 April 2017.

41. David Naguib Pellow, *Resisting Global Toxics: Transnational Movement for Environmental Justice*, chapter: "Electronic Waste: The 'Clean Industry' Exports Its Trash," 185–224; Cédric Gossart, "De l'exportation des maux écologiques à l'ère du numérique," *Mouvements*, no. 60 (2009): 23–28.

42. Karin Lundgren, *The Global Impact of e-Waste: Addressing the Challenge* (Geneva: Organisation internationale du travail, 2012); "Les déchets électroniques intoxiquent le Ghana," *Le Monde.fr*, 28 December 2013; Hong-Gang Ni and Eddy Y. Zeng, "Law Enforcement and Global Collaboration Are the Keys to Containing E-Waste Tsunami in China," *Environmental Science and Technology*, vol. 43, no. 11 (2009): 3991–3994.

43. Deborah S. Davis, ed., *The Consumer Revolution in Urban China* (Berkeley: University of California Press, 2000).

44. "Le 'jour de dépassement de la Terre' en infographie," *Le Monde.fr*, 8 August 2016.

45. Mike Davis, *Planet of Slums* (London: Verso Books, 2006), 129; Jeremy Seabrook, *In the Cities of the South: Scenes from a Developing World* (London: Verso, 1996).

46. Davis, *Le Pire des mondes possibles*,135; McNeill, *Du nouveau sous le soleil*, 126–127.

47. Kenneth Pomeranz, "The Transformation of China's Environment (1500–2000)," in *The Environment and World History (1500–2000)*, edited by Kenneth Pomeranz and Edmund T. Burke III (Berkeley: University of California Press, 2009), 118–164.

48. Sébastien Le Belzic, *Chine. Le cauchemar écologique* (Saint-Maur-des-Fossés: Sépia, 2013); Julien Wagner, *Chine-Afrique: le grand pillage* (Paris: Eyrolles, 2014), 59–60; "Study Links Polluted Air in China to 1.6 Million Deaths a Year," *The New York Times*, 13 August 2015.

49. "Chine, un inventaire des 'villages du cancer,'" *Courrier international*, 4 June 2009; "En Chine, 400 'villages du cancer,'" *Le Figaro.fr*, 24 February 2013.

50. Jean-Pierre Favennec and Yves Mathieu, *Atlas mondial des énergies: Ressources, consommation et scénarios d'avenir* (Paris: Armand Colin/IFP, 2014), 42–43.

51. "Le 'nuage brun' d'Asie pourrait menacer le climat de la planète," *Le Monde.fr*, 14 October 2002.

52. Kenneth Pomeranz, "Les eaux de l'Himalaya."

53. Online reports: http://www.worstpolluted.org/.

54. Jacques Charbonnier, *Bhopal, la pire catastrophe industrielle de tous les temps* (Bordeaux: Préventique, 2004); Larry Everest, *Behind the Poison Cloud: Union Carbide's Bhopal Massacre* (Chicago: Banner Press, 1986).

55. Le Belzic, *Chine: Le cauchemar écologique*; "Au Bangladesh, plus de 1 700 morts depuis 1990 dans des ateliers textile," *Le Monde. fr*, 14 May 2013.

56. Hervé Kempf, *Comment les riches détruisent la planète* (Paris: Le Seuil, 2007).

57. Sophie Bernard, Damien Dussaux, Mouez Fodha, and Matthieu Glachant, "Le commerce international des déchets," in *L'Économie mondiale 2013* (Paris: La Découverte, 2012), 104–118.

58. Pellow, *Resisting Global Toxics*, 147–185, 150.

59. Bernard Dussol and Charlotte Nithart, *Le Cargo de la honte: L'effroyable odyssée du "Probo Koala"* (Paris: Stock, 2010); Greenpeace and Amnesty International, *Une vérité toxique: À propos de Trafigura, du "Probo Koala" et du déversement de déchets toxiques en Côte d'Ivoire*, Amnesty International Publications, 2012.

60. Public Eye, *Dirty Diesel: How Swiss Traders Flood Africa with Toxic Fuels* (Lausanne, Switzerland: Public Eye, 2016).

61. Marc Levinson, *The Box: Comment le conteneur a changé le monde* (Paris: Max Milo, 2011), 438.

62. Giljum and Muradian, "Physical Trade Flows of Pollution-Intensive Products."

63. Daniel Faber, *Capitalizing on Environmental Injustice: The Polluter-Industrial Complex in the Age of Globalization* (Lanham, MD: Rowman & Littlefield, 2008).

64. Jean-Pierre Favennec and Yves Mathieu, *Atlas mondial des énergies*, 74.

65. Kuntala Lahiri-Dutt, ed., *The Coal Nation: Histories, Ecologies and Politics of Coal in India* (Surrey, UK: Ashgate, 2014); Gérard Heuzé, *Ouvriers d'un autre monde: L'exemple des travailleurs de la mine dans l'Inde contemporaine* (Paris: MSH, 1989).

66. David Dufresne, Nancy Huston, Naomi Klein, Melina Laboucan-Massimo, and Rudy Wiebe, *Brut: La ruée vers l'or noir* (Montréal: Lux Éditeur, 2015); Maxime Combes, *Sortons de l'âge des fossiles! Manifeste pour la transition* (Paris: Le Seuil, 2015).

67. Danièle Favari (with André Picot and Marc Durand), *Les Vrais Dangers du gaz de schiste* (Paris: Sang de la Terre, 2013); Marine Jobert and François Veillerette, *Le Vrai Scandale des gaz de schiste* (Paris: Les Liens qui libèrent, 2011).

68. WWF website: http://dossier-mer-huile.wwf.fr/dossier.html; Paul Fattal, *Pollutions des côtes par les hydrocarbures* (Rennes, France: Presses universitaires de Rennes, 2008).

69. Abrahm Lustgarten, *Run to Failure: BP and the Making of the "Deepwater Horizon" Disaster* (New York: Norton, 2012).

70. "Nigeria. Les marées noires oubliées du delta du Niger," *Courrier international*, 3 June 2010.

71. Emma Howard, "Middle East Conflict 'Drastically Altered' Air Pollution Levels in Region," *The Guardian*, 21 August 2015; T. M. Hawley, *Against the Fires of Hell: The Environmental Disaster of the Gulf War* (New York: Harcourt Brace, 1992).

72. Baptiste Monsaingeon, "Plastiques: ce continent qui cache nos déchets," *Mouvements*, no. 87 (2016): 48–58; Claude Duval, *Matières plastiques et environnement* (Paris: Dunod, 2009); "Le déversement des plastiques dans les océans pourrait décupler d'ici à dix ans," *Le Monde.fr*, 12 February 2015.

73. Chris Jordan, "Midway: Message from the Gyre," 2009, http://www.chrisjordan.com/gallery/midway/#CF000313%2018x24.

74. Julien Gigault, Boris Pedrono, Benoît Maxit, and Alexandra Ter Halle, "Marine Plastic Litter: The Unanalyzed Nano-Fraction," *Environmental Science: Nano*, vol. 2 (2016): 346–350; "Les océans pollués par des particules invisibles de plastique," *Le Monde.fr*, 22 February 2017; "L'océan, poubelle du globe," *Le Monde.fr*, 23 February 2017.

75. Éric Drezet, "L'énergie des métaux," *EcoInfo*, 2015, http://ecoinfo.cnrs.fr/?p=11396.

76. Andy Bruno, *The Nature of Soviet Power: An Arctic Environmental History* (Cambridge: Cambridge University Press, 2016); "Un cocktail de polluants affecte l'Arctique," *Le Monde.fr*, 19 December 2014.

77. Barbara Landrevie, "La Méditerranée empoisonnée," *Le Monde diplomatique*, May 2015, 8.

78. Andràs Gelencsér *et al.*, "The Red Mud Accident in Ajka (Hungary): Characterization and Potential Health Effects of Fugitive Dust," *Environmental*

Science Technology, vol. 45, no. 4 (2011): 1608–1615; Miroslav Mišík *et al.*, "Red Mud a Byproduct of Aluminum Production Contains Soluble Vanadium That Causes Genotoxic and Cytotoxic Effects in Higher Plants," *The Science of the Total Environment*, vol. 493 (2014): 883–890.

79. Lynch, *Mining in World History*, 280–285; Ivan Grinberg and Maurice Laparra, eds., *Alucam, un destin africain: 50 ans d'aluminium au Cameroun (1957–2007)* (Mirabeau, France: REF.2C Éditions, 2007).

80. Giuseppe Cocco, "La catastrophe du rio Doce, le Tchernobyl brésilien," *Multitudes*, vol. 62, no. 1 (2016): 5–13.

81. François Ramade, *Un monde sans famine? Vers une agriculture durable* (Paris: Dunod, 2014); Claude Bourguignon and Lydia Bourguignon, *Le Sol, la terre et les champs: Pour retrouver une agriculture saine* (Paris: Sang de la Terre, 2015).

82. "Omerta sur les pesticides dans le vignoble bordelais," *Le Monde.fr*, 4 July 2014; Geneviève Teil, Sandrine Barrey, Pierre Floux, and Antoine Hennion, *Le Vin et l'environnement* (Paris: Presses de l'École des mines, 2011).

83. Edwige Charbonnier, Aïcha Ronceux, Anne-Sophie Carpentier, Hélène Soubelet, and Enrique Barriuso, eds., *Pesticides: Des impacts aux changements de pratiques* (Paris: Quae, 2015); Nicolino and Veillerette, *Pesticides*.

84. French Minister of Agriculture communiqué (8 March 2016) on pesticide use: http://agriculture.gouv.fr/produits-phytosanitaires-resultats-pour-lannee -2014-et-lancement-du-plan-ecophyto-2.

85. Paul Degobert, *Automobiles and Pollution* (Paris: Technip, 1995).

86. "La voiture du futur sera propre, communicante, automatique et … partagée," *20minutes.fr*, 10 February 2013.

87. Marina Robin, "La motorisation des ménages continue de s'accroître au prix d'un vieillissement du parc automobile," *La Revue*, Commissariat général au développement durable, December 2010, 99–122.

88. Ugo Bardi and Stefano Caporali, "Precious Metals in Automotive Technology: An Unsolvable Depletion Problem?" *Minerals*, vol. 4, no. 2 (2014): 388–398.

89. Zéhir Kolli, "Dynamique de renouvellement du parc automobile: Projection et impact environnemental" (PhD diss., Université de Paris 1, 2012), 28–30.

90. OCDE, *Le Coût de la pollution de l'air: Impacts sanitaires du transport routier*, s.l., (OCDE, 2014); Bibliographies du Centre de ressources du développement durable, "La qualité de l'air," April 2017, http://www.developpement -durable.gouv.fr/sites/default/files/CRDD%20Biblio%20Qualite%20de%20lair .pdf.

91. François Bost and Gabriel Dupuy, eds., *L'Automobile et son monde* (La Tour-d'Aigues: Éd. de l'Aube, 2000); Marc Chevallier, "Automobile: la fin du rêve?" *Alternatives économiques*, no. 279, April 2009.

92. Frédéric Landy, "Industrie, aménagement du territoire et pollution en Inde: Le cas de l'automobile," in "Les marchés émergents de l'automobile: Une

approche géographique (Inde, Chine et Afrique du Sud)," *2001 Plus* (Paris: Ministère de l'Équipement, des Transports et du Logement, July 2002), 9–22.

93. "Chine: l'alerte rouge de pollution de l'air, dilemme des officiels," *Le Monde .fr*, 10 December 2015.

94. Philippe Ayoun, "Le transport de voyageurs face au défi énergétique et écologique," in *L'Avion: Le rêve, la puissance et le doute*, edited by Gérard Dubey and Alain Gras (Paris: Publications de la Sorbonne, 2009), 279–293; Joyce E. Penner *et al.*, *Aviation and the Global Atmosphere* (Cambridge: Cambridge University Press, 1999).

95. Fabrice Gliszczynski, "Plus d'avions, moins de pollution, l'incroyable pari de l'aéronautique," *La Tribune*, 18 June 2015.

96. Sezin Topçu, *La France nucléaire: L'art de gouverner une technologie contestée* (Paris: Le Seuil, 2013).

97. Kate Brown, *Plutopia: Nuclear Families, Atomic Cities, and the Great Soviet and American Plutonium Disasters* (New York: Oxford University Press, 2013); Jaurès A. Medvedev, *Désastre nucléaire en Oural* (Cherbourg: Isoète, 1988 [1980]).

98. Vladimir K. Savchenko, *The Ecology of Tchernobyl Catastrophe* (Paris: Unesco, 1995); for a discussion of the human losses, see: http://www.dissident -media.org/infonucleaire/estimations.html.

99. Frank Uekoetter, "Fukushima, Europe, and the Authoritarian Nature of Nuclear Technology," *Environmental History*, vol. 17, no. 2 (2012): 277–284; Timothy A. Mousseau and Anders Møller, "Genetic and Ecological Studies of Animals in Chernobyl and Fukushima," *Journal of Heredity*, vol. 105, no. 5 (2014): 704–709; "Fukushima, puits sans fond," *Libération.fr*, 16 February 2017.

100. Jean-Marc Jancovici, "Le nucléaire civil, péché majeur du xxᵉ siècle?" *Le Débat*, vol. 123, no. 1 (2003): 175–192; Jancovici, "Que signifie sortir du nucléaire?" *Le Débat*, vol. 169, no. 2 (2012): 77–86.

101. Annie Thébaud-Mony, "Nucléaire: la catastrophe sanitaire," *LeMonde. fr*, 21 March 2011; Thébaud-Mony, *L'Industrie nucléaire. Sous-traitance et servitude* (Paris: Inserm-EDK, 2000).

102. Topçu, *La France nucléaire*.

103. Laure Noualhat, *Déchets: Le cauchemar du nucléaire* (Paris: Le Seuil, 2009).

104. Annie Thébaud-Mony, *La Science asservie: Les collusions mortifères entre industriels et chercheurs* (Paris: La Découverte, 2014), chap. 12.

105. Franck Boutaric and Pierre Lascoumes, "L'épidémiologie environnementale entre science et politique: Les enjeux de la pollution atmosphérique en France," *Sciences sociales et santé*, vol. 26, no. 4 (2008): 5–38.

106. Yannick Barthe, Madeleine Akrich, and Catherine Rémy, eds., *Sur la piste environnementale: Menaces sanitaires et mobilisations profanes* (Paris: Presses de l'École des mines, 2010).

107. Haut Conseil de la santé publique, *Recommandations pour la gestion du risque amiante dans l'habitat et l'environnement*, 2014, online: http://www.hcsp .fr/explore.cgi/avisrapportsdomaine?clefr=450.

108. Laurent Vogel, "Géopolitique de l'amiante" et "Amiante, crime de masse en temps de paix," *Politique, revue de débats*, no 60, June 2009 (electronic review); "L'amiante sera interdit au Canada d'ici à 2018," *Le Monde.fr*, 16 December 2016.

109. United Nations Environment Programme, "L'heure d'éliminer les HCFC," special issue of *Action ozone* (Paris: UNEP, 2008); "Climat: accord historique pour éliminer les gaz HFC, 14 000 fois plus puissants que le CO_2," *Le Monde.fr*, 15 October 2010.

110. David Kinkela, *DDT and the American Century: Global Health, Environmental Politics, and the Pesticide That Changed the World* (Chapel Hill: University of North Carolina Press, 2011).

111. Mathieu Baudrin, "Historiciser la réflexivité industrielle. Les aérosols et la couche d'ozone (1974–2014)" (PhD diss. in progress, l'École des mines, Paris).

112. Oreskes and Conway, *Les Marchands de doute*; Michaels, *Doubt Is Their Product*; Thomas O. McGarity and Wendy Wagner, *Bending Science: How Special Interests Corrupt Public Health Research* (Cambridge, MA: Harvard University Press, 2010).

113. "Le pneumologue Michel Aubier condamné à six mois de prison avec sursis," *Le Monde.fr*, 5 July 2017.

114. Jean-Noël Jouzel, *Des toxiques invisibles: Sociologie d'une affaire sanitaire oubliée* (Paris: EHESS, 2013).

115. Pierre Lascoumes and Patrick Le Galès, eds., *Gouverner par les instruments* (Paris: Presses de Sciences Po, 2005); "Fauteurs de doute," special issue of *Critique*, vol. 799, no. 12 (2013).

116. Proctor and Schiebinger, *Agnotology*; Foucart, *La Fabrique du mensonge*; Thébaud-Mony, *La Science asservie*.

117. Emmanuel Henry, *Ignorance scientifique et inaction publique: Les politiques de santé au travail* (Paris: Presses de Sciences Po, 2017); Henry, "Militer pour le statu quo. Le Comité permanent amiante ou l'imposition réussie d'un consensus," *Politix*, vol. 70, no. 2 (2005): 29–50.

118. Andrew Hurley, "From Factory Town to Metropolitan Junkyard: Postindustrial Transitions on the Urban Periphery," *Environmental History*, vol. 21, no. 1 (2016): 3–29; Jean-Louis Tornatore, "Beau comme un haut-fourneau: Sur le traitement en monument des restes industriels," *L'Homme*, vol. 170 (2004): 79–116; "Le patrimoine industriel de la chimie," special issue of *Patrimoine industriel*, no. 69 (2016) (published with the financial support of the Solvay group).

119. Amissi Manirabona, "L'affaire Trafigura: vers la répression de graves atteintes environnementales en tant que crimes contre l'humanité," *Revue de droit international et de droit comparé*, vol. 88, no. 4 (2011): 535–576; Valérie

Cabanes, *Un nouveau droit pour la Terre: Pour en finir avec l'écocide* (Paris: Le Seuil, 2016).

120. McNeill, *Du nouveau sous le soleil*, 461; Raphaël Romi, *Droit international et européen de l'environnement* (Paris: Montchrétien, 2012).

121. Martine Rémond-Gouilloud, *Du droit de détruire: Essai sur le droit de l'environnement* (Paris: Presses universitaires de France, 1989).

122. Pierre Lascoumes, *L'éco-pouvoir* (Paris: La Découverte, 1994); Jean-Pierre Galland, "France/Grande-Bretagne: une comparaison entre deux régimes de régulation des risques industriels," *Annales des Mines—Responsabilité et environnement*, vol. 62, no. 2 (2011): 62–66.

123. Keith Hawkins, "Le rôle du droit dans le contrôle de la pollution industrielle: Regard britannique," in *Les politiques d'environnement*, edited by Bernard Barraqué, Jacques Theys, 297–305.

124. Jas, "Gouverner les substances chimiques dangereuses dans les espaces internationaux."

125. Tsayem Demaze Moïse, "Les conventions internationales sur l'environnement: état des ratifications et des engagements des pays développés et des pays en voie de développement," *L'information géographique*, vol. 73, no. 3 (2009): 84–99.

126. Stefan Aykut, "Gouverner le climat, construire l'Europe. L'histoire de la création du marché de carbone (ETS)," *Critique internationale*, vol. 62 (2014): 39–56.

127. Jean-Noël Jouzel et Pierre Lascoumes, "Le règlement REACH: une politique européenne de l'incertain: Un détour de régulation pour la gestion des risques chimiques," *Politique européenne*, vol. 33, no. 1 (2011): 185–214.

128. Eduard B. Vermeer, "Industrial Pollution in China and Remedial Policies," *The China Quarterly*, vol. 156 (1998): 952–985.

129. Shapiro, *Mao's War against Nature*, 209.

130. Jean-François Huchet, *La Crise environnementale en Chine: Évolutions et limites des politiques publiques* (Paris: Presses de Sciences Po, 2016).

131. OCDE, *Le Coût de la pollution de l'air*.

132. "Haro sur les voitures polluantes à Mexico," *Le Monde.fr*, 31 July 2014.

133. Bernard Barraqué and Jacques Theys, eds., *Les Politiques d'environnement*, 27, 74.

134. "Pesticides: l'échec accablant de la 'ferme France,'" *Le Monde.fr*, 9 March 2016.

135. Daniel Faber, "Poisoning American Politics: The Colonization of the State by the Polluter-Industrial Complex," *Socialism and Democracy*, vol. 23, no. 1 (2009): 77–118; "Dieselgate: après Volkswagen et Renault, le tour de FiatChrysler," *Le Point.fr*, 22 March 2017.

136. "Régression: Guerre secrète contre l'environnement," *Courrier international*, 30 October 2003.

137. "Trump veut détricoter les mesures d'Obama sur l'environnement," *Le Monde.fr*, 28 March 2017.

138. Luc Boltanski and Ève Chiapello, *Le Nouvel Esprit du capitalisme* (Paris: Gallimard, 1999).

139. Noam Chomsky, *Le Bien commun: Entretiens avec David Barsamian* (Montréal: Écosociété, 2002); David C. Korten, *Quand les multinationales gouvernent le monde* (Paris: Yves Michel, 2006).

140. "La directive européenne sur le 'secret des affaires' fait polémique," *Le Figaro.fr*, 26 April 2016.

141. Stefan Ambec and Philippe Barla, "Quand la réglementation environnementale profite aux pollueurs: Survol des fondements théoriques de l'hypothèse de Porter," *Revue d'analyse économique*, vol. 83, no. 3 (2007): 399–413.

142. Antonin Pottier, *Comment les économistes réchauffent la planète* (Paris: Le Seuil, 2016).

143. Stephen C. Young, *The Emergence of Ecological Modernisation: Integrating the Environment and the Economy?* (London: Routledge, 2000).

144. http://www.ecomodernism.org/manifesto-english/.

145. Michael Huesemann and Joyce Huesemann, *Techno-Fix: Why Technology Won't Save Us or the Environment* (Gabriola Island, BC: New Society Publishers, 2011).

146. Clive Hamilton, *Requiem pour l'espèce humaine* (Paris: Presses de Sciences Po, 2012); Hamilton, *Les Apprentis sorciers du climat: Raisons et déraisons de la géo-ingénierie* (Paris: Le Seuil, 2013).

147. Frédéric Neyrat, *La Part inconstructible de la Terre: Critique du géo-constructivisme* (Paris: Le Seuil, 2016).

148. Jean-Marc Bournigal et al., *Agriculture-Innovation 2025: 30 projets pour une agriculture compétitive et respectueuse de l'environnement* (Paris: Ministère de l'Agriculture, October 2015).

149. Jurgensen, *L'Economie verte*, 84.

150. Alma Heckenroth *et al.*, "Vers quelles solutions écologiques aux pollutions diffuses?" in *Les Calanques industrielles de Marseille et leurs pollutions*, edited by Xavier Daumalin and Isabelle Laffont-Schwob, 327; Andre F. Clewell and James Aronson, *La Restauration écologique* (Arles, France: Actes Sud, 2010).

151. Jeremy Rifkin, *L'Économie hydrogène: Après la fin du pétrole, la nouvelle révolution économique* (Paris: La Découverte, 2002).

152. Pierre-Étienne Franc (with Pascal Mateo), *Hydrogène: la transition énergétique en marche!* (Paris: Gallimard, 2015).

153. Éric Deville and Alain Prinzhofer, *Hydrogène: La prochaine révolution énergétique* (Paris: Belin, 2015), 154.

154. Marie-Monique Robin, *Notre poison quotidien: La responsabilité de l'industrie chimique dans l'épidémie des maladies chroniques* (Paris: La

Découverte, 2011); Hervé Kempf, *Tout est prêt pour que tout empire: Douze leçons pour éviter la catastrophe* (Paris: Le Seuil, 2017); Fabrice Nicolino, *Ce qui compte vraiment* (Paris: Les Liens qui libèrent, 2017); Jean-Pierre Rogel, *Un paradis de la pollution* (Québec: Presses de l'université du Québec, 1981).

155. Bess, *La France vert clair*, 81–82; Bruno Villalba and Sylvie Ollitrault, "Sous les pavés, la Terre. Mobilisations environnementales en France (1960–2011), entre contestations et expertises," in *Histoire des mouvements sociaux en France, de 1814 à nos jours*, edited by Michel Pigenet and Danielle Tartakowsky (Paris: La Découverte, 2012), 716–723; Sylvie Ollitrault, *Militer pour la planète: Sociologie des écologistes* (Rennes, France: Presses universitaires de Rennes, 2008).

156. Robert Gottlieb, *Forcing the Spring: The Transformation of the American Environmental Movement* (Washington, DC: Island Press, 2005).

157. Frederick Buell, *From Apocalypse to Way of Life: Environmental Crisis in the American Century* (London: Routledge, 2003).

158. Jean-Baptiste Comby, *La Question climatique: Genèse et dépolitisation d'un problème public* (Paris: Raisons d'Agir, 2015).

159. Vaclav Smil, *Making the Modern World: Materials and Dematerialization* (Chichester, UK: Wiley, 2014).

160. Association négaWatt, *Scénario négaWatt (2017–2050): Dossier de synthèse*, s.l., NégaWatt, 2017: https://negawatt.org/IMG/pdf/synthese_scenario-negawatt_2017-2050.pdf.

161. Jason Corburn, *Street Science: Community Knowledge and Environmental Health Justice* (Cambridge, MA: MIT Press, 2005).

Index

Note: Page numbers followed by "n" refer to endnotes.

caustic soda production in, 57,
78–82, 85, 107, 108–110,
123–124, 129
chemical entrepreneurship in, 77
education in, 107
expertise of, 68, 71, 77, 78–79,
116–143
health and public hygiene role of,
64–69, 78–79
industrial capitalism and, 108–109,
111–115, 136
mass consumption and
developments in, 233, 235–236,
238–244, 254–259
political influence of, 69–73, 271
progress/Age of Progress frontiers
for, 106–115, 116–143, 160–165,
168–169
regulations influenced by, 64–73,
77–81, 84–86
scientific mobilization against, 277
sulfuric acid production in, 107,
108–109
synthetic, 106–107, 168–169
technical innovations by, 160–165,
233, 235–236, 238–244,
254–259
water pollution from, 100
Chemical Society of London, 107
Chemical Warfare Service, 193
Chernobyl accident, 317
Chevron, 214
Chicago Citizens' Association for the
Prevention of Smoke, 135
Chile
agrochemical exports from, 111–112
mining industry in, 102, 144, 249
China
agrarian empires in, 13
agrochemical industry in, 302
air pollutions-related deaths in, 2
asbestos in, 268, 319
automotive industry in, 315–316
coal use in, 50, 208–209, 308, 311,
323
contemporary pollutions
management in, 302, 304–308

digital technology and electronic
waste in, 304, 305–306
electricity in, 226
energy consumption in, 230, 308
(*see also coal use* and *electricity
in* subentries)
environmental inequalities or
injustices in, 306, 307–309
fertilizers, pesticides, and herbicides
in, 258
gunpowder production in, 22
industrialization of, 307–308
metallurgy and mining in, 41, 46,
50, 208–209, 313
papermaking industry, 22
population demographics in, 13, 24
preindustrial pollutions in, 20, 21, 22
regulations in, 291, 323
textile industry in, 21
urban areas in, 24, 93, 235
war industry and wartime effects in,
22, 185, 199
Chlorine, 57–58, 67, 167, 194, 195,
196, 247
Chlorofluorocarbon pollutions, 270,
271, 319
Chloropicrin, 197
Cholera, 121, 169
Chomsky, Noam, 325
Christianity, 29–30
Christison, Robert, 85
Cicolella, André, 300
Cinnabar, 45, 55
Circulus theory, 166
Cities. *See* Industrial cities; Urban areas
Citroën, 192
Clark, Thomas, 129
Clean Air Act (1956), 264
Clean Air Act (1963), 263
Cleanliness
access to clean water, 20
clean technologies promotion,
277–278
electricity and fantasy of, 225
health and, 29, 236 (*see also*
Disinfection)
waste collection and, 20